PROTEIN BIOCHEMISTRY, SYNTHESIS, STRUCTURE AND CELLULAR FUNCTIONS

TUBULIN: STRUCTURE, FUNCTIONS AND ROLES IN DISEASE

PROTEIN BIOCHEMISTRY, SYNTHESIS, STRUCTURE AND CELLULAR FUNCTIONS

Additional books in this series can be found on Nova's website under the Series tab.

Additional E-books in this series can be found on Nova's website under the E-book tab.

PROTEIN BIOCHEMISTRY, SYNTHESIS, STRUCTURE AND CELLULAR FUNCTIONS

TUBULIN: STRUCTURE, FUNCTIONS AND ROLES IN DISEASE

WEI YAMAUCHI
AND
ANTONIO SOKIĆ
EDITORS

Nova Science Publishers, Inc.
New York

For permission to use material from this book please contact us:
Telephone 631-231-7269; Fax 631-231-8175
Web Site: http://www.novapublishers.com

Library of Congress Cataloging-in-Publication Data

Tubulin : structure, functions, and roles in disease / editors, Wei Yamauchi and Antonio Sokic.
p. ; cm.
Includes bibliographical references and index.
ISBN 978-1-62100-141-6 (hardcover)
I. Yamauchi, Wei. II. Sokic, Antonio.
[DNLM: 1. Tubulin. QU 55.3]
LC classification not assigned
572'.67--dc23

2011031642

Published by Nova Science Publishers, Inc. ✝ New York

Contents

Preface		**vii**
Chapter I	Microtubule Plus-end Tracking Proteins in Neuronal Development and Nervous System Disorders *C.L. Sayas and J. Avila*	**1**
Chapter II	Microtubule Organization. The Housekeeping Role of Heat-shock Proteins and Immunophilins *Héctor R. Quintá, Natalia M. Galigniana, Cristina Daneri-Becerra, Mariana Lagadari and Mario D. Galigniana*	**41**
Chapter III	Trypanosome Tubulin Molecules as Targets for Drug and Vaccine Development for Protozoan Infections *Barbara A. Katzenback, Debbie A. Plouffe and Miodrag Belosevic*	**69**
Chapter IV	Tubulin: Structure, Function and Role in Disease *Ikuru Yazawa*	**85**
Chapter V	Targeted Disruption of the Tubulin Cytoskeletal Network Activates Expression of the Profibrotic Genes PAI-1 and CTGF: Implications for Vascular Disease *Rohan Samarakoon and Paul J. Higgins*	**101**
Chapter VI	Tubulin Inhibitors as Vascular Disrupting Agents (VDAS) *Jérémie Fournier-Dit-Chabert and Sylvie Ducki*	**113**
Chapter VII	Evolution and Coevolution of Tubulin's Carboxy-Terminal Tails and Mitochondria *Dan L. Sackett*	**143**
Chapter VIII	Myelin Basic Protein Interactions with Actin and Tubulin *in vitro*: Binding, Assembly, and Regulation *Joan M. Boggs*	**167**
Index		**185**

Preface

Tubulin is one of several members of a small family of globular proteins. The most common members of the tubulin family are α-tubulin and β-tubulin, the proteins that make up microtubules. In this book the authors present topical research in the study of the structure, functions and roles in disease of tubulin. Topics discussed in this compilation include microtubule plus-end tracking proteins in neuronal development and nervous system disorders; trypanosome tubulin molecules as targets for drug and vaccine development for protozoan infections; tubulin inhibitors as vascular disrupting agents and evolution and coevolution of tubulin's carboxy-terminal tails and mitochondria.

Chapter I - Regulation of microtubule organization and dynamics is crucial during neuronal differentiation. A plethora of proteins participate in controlling the dynamic behavior of microtubules. Plus-end tracking proteins (+TIPs) are a specialized class of microtubule-associated proteins that specifically accumulate at growing ends. +TIPs comprise a diverse group of structurally unrelated proteins that mediate microtubule interactions with different cellular structures and signaling factors. Several +TIPs, including CLASP1 and CLASP2, CLIP-115 and CLIP-170, EB family members, APC, p150Glued (the large subunit of the dynactin complex), NAV1 and LIS1, have been shown to play important roles during different stages of neuronal development such as migration, axon extension and guidance, dendritic spine formation, and maintenance of proper communication between neurons. Consequently, malfunction of +TIPs can lead to some nervous system disorders. LIS1 is essential for proper migration of cortical and cerebellar neurons. Mieller-Dieker Syndrome (MDS) is a lissencenphaly, characterized by the presence of abnormal cortical layers, corpus callosum defects and ventricle enlargement, is caused by mutations in the LIS1 gene. Mutations in genes encoding proteins that interact with LIS1, such as doublecortin (DCX), lead to another well-characterized neuronal migration disorder. Mutations in p150Glued lead to the disruption of dynein/dynactin and impairment of axonal transport in motor neurons and neuronal cell death, thus causing progressive neurodegeneration. Not surprisingly, mutations in p150Glued were identified in different motor neuron diseases, such as Amyotrophic Lateral Sclerosis (ALS), characterized by progressive paralysis of muscles at adult age provoked by death of motor neurons. In addition, CLIP-115 deficiency has been associated with the neurological symptoms found in Williams Syndrome (WS), a disease characterized by facial abnormalities, cardiovascular problems and mental retardation. Here, we summarize recent studies concerning the involvement of +TIPs in neuronal development and disease.

Chapter II - Cytoskeleton is the basic scaffold of the cell in which other subcellular components are spatially arranged, such that they are able to communicate efficiently between the internal and external environments of the cell. Cytoskeletal structure is continually remodeled to accommodate normal cell growth and to respond to pathophysiological cues. As a consequence, several cytoskeleton-interacting proteins become involved in a variety of cellular processes such as cell growth and division, cell movement, vesicle transportation, cellular organelle location and function, localization and distribution of membrane receptors, and cell-cell communication. Molecular chaperones and immunophilins are counted among the most important proteins that interact closely with the cytoskeleton network, in particular with microtubules and microtubule-associated factors. In several situations, chaperones and immunophilins work together as a functionally active heterocomplex, although both types of proteins also show independent actions. In circumstances where homeostasis is affected by environmental stresses or due to genetic alterations, chaperone proteins help to stabilize the system. Molecular chaperones facilitate the assembly, disassembly and/or folding/refolding of cytoskeletal proteins, so they prevent aberrant protein aggregation. Nonetheless, the role of chaperones is not limited to solve abnormal situations. They also have an active participation during the normal differentiation process of the cell and are key factors for many structural and functional rearrangements during this course of action. A similar observation is also valid for immunophilins. Although the folding system for microtubules was discovered nearly two decades ago, our understanding of the complex quality control pathway of these filaments is still poorly understood and there are many unanswered questions that remain to be elucidated. Microtubule modifications leading to altered localization nuclear factors may result in loss- or gain-of-function of such factors, which affects the cell cycle and cell development. Therefore, microtubules are attractive therapeutic targets, particularly to prevent pathological situations such as rapidly dividing tumors or to favor the process of cell differentiation in other cases. In this chapter we will address some mechanistic aspects of the key regulatory functions of heat-shock proteins and immunophilins in the microtubule network of neurons and cancer cells.

Chapter III - Parasites from the kinetoplastid group (*Trypanosoma* spp., *Leishmania* spp.) are of considerable importance because they are responsible for serious disease and mortality in humans and economically important livestock. The identification of specific parasite molecules responsible for the induction of protective antibodies in the host are of great interest since they represent potential candidates for treatment of, or vaccination against, infections of the host. Despite the high degree of sequence conservation of tubulin molecules among different host and parasite species and the sequestered intracellular location of tubulin, a number of studies involving trypanosomatids have identified parasite tubulin molecules as key targets for drug chemotherapy or for use in vaccination studies to selectively target the parasite while limiting toxicity to the host. These observations were not limited to mammalian trypanosomatids tubulin, as similar observations for trypanosomes of ectothermic vertebrates such as fish have been reported. The combination of these studies suggests that immunization of multiple hosts with parasite tubulin from one trypanosome species may provide protection against homologous and heterologous challenge. In this review, we focus on the use of parasite tubulin as a chemotherapeutic target in mammalian systems, and as a vaccine target in vertebrate hosts, the binding specificity of anti-tubulin antibodies generated during host vaccination and their potential mechanism of action to prevent parasite growth and survival.

Chapter IV - Microtubules are one of the components that make up the filamentous cytoskeleton in the nervous system. A microtubule is formed from subunits of tubulin, which is a heterodimer formed from α-tubulin and β-tubulin. The formation of protein filaments in microtubules allows for regulated filament assembly and disassembly, resulting in the reshaping of the cytoskeleton. Microtubules are dynamic structures that are maintained by a rapid and continuous exchange of tubulins between the soluble and free form and the filamentous form. Neurons regulate the length and stability of microtubules, as well as their number and geometry. Microtubule-associated proteins (MAPs) bind along the sides of microtubules and participate in the regulation of microtubule assembly and function. Drugs that have an effect on microtubule polymerization can be useful for examining the role of microtubules in disease processes. Motor proteins bind to microtubules and transport organelles and particles in neuronal processes. Deficits in microtubule and MAP function contribute to the pathogenesis of multiple diseases, including Alzheimer's disease and hereditary peripheral neuropathy. Multiple system atrophy (MSA), a nonhereditary neurodegenerative disease affecting oligodendrocytes and neurons in the human central nervous system, is characterized by abnormal accumulations of filamentous α-synuclein inclusions. Recent studies in MSA mouse models have identified that microtubule β-III tubulin binds to α-synuclein. In addition, the binding of α-synuclein to β-III tubulin promoted the accumulation of insoluble α-synuclein, leading to neuronal dysfunction and degeneration. Furthermore, α-synuclein accumulation was suppressed by inhibiting microtubule polymerization. Thus, the binding of α-synuclein to β-III tubulin is a key process in neurodegeneration in MSA. Important insights into neurodegeneration in MSA and therapeutic targets have emerged from the MSA mouse model.

Chapter V - Internal constraints imposed by the microtubule, microfilament and intermediate filament cytoskeletal networks, junctional complexes and integrin-extracellular matrix (ECM) interactions are major determinants of cell structure/function and subject to modulation by a variety of external forces. Cell cycle transit, entry into DNA synthetic phase and activation of cell death programs are each influenced by cellular structural dynamics. This is particularly evident in the vascular system which is constantly subjected to a continuum of hemodynamic stimuli (e.g., shear strain, flow disturbances, mechanical or pulsile stretch) that alter microfilament and microtubule organization and associated signaling pathways. These same physical forces that influence cell shape impact gene expression programs that, in turn, modulate cell proliferation, migration and ECM synthesis/deposition resulting in the development of tissue-specific pathologies or wound-healing anomalies (e.g., vessel fibrotic disorders, atherosclerosis).

Chapter VI - Vascular Disrupting Agents (VDAs) are mostly composed of tubulin inhibitors (TIs). They differ from anti-angiogenesis agents as they destroy within minutes to hours newly formed blood vessels. VDAs effectively starve the tumour to death. Deprived from vital nutrients and oxygen, the tumour goes into necrosis. A majority of cancers are solid tumours that require neovasculature to survive. VDAs are therefore emerging as a promising counterpart to conventional cytotoxic chemotherapy.

This chapter presents a new class of promising compounds, VDAs, describing their effect in contrast with anti-angiogenic drugs. Inhibitors of tubulin polymerisation and microtubule-depolymerising agents will be presented. Some of these VDAs are currently under clinical study to treat a variety of solid tumours or exploited in ophthalmology as therapies against Age-Related Macular Degeneration (AMD). We will focus on the largest group of VDAs that are tubulin-binding agents, with the combretastatins (Zybrestat, AVE-8062, Oxi-4503) as the lead compounds. New emerging VDAs will be presented, and progresses through their clinical or preclinical stage will be reviewed. The advantages and the downsides of this new chemotherapy will be explored in terms of efficiency and side-effects.

Chapter VII - Microtubules and mitochondria are two of the defining features of a eukaryotic cell. Tubulin, the subunit protein of microtubules (MT) is found in all eukaryotic species and in essentially all eukaryotic cells, while mitochondria are found in most eukaryotic species though there are a significant number that lack mitochondria sensu stricto. Both tubulin and mitochondria have been part of the eukaryotic spectrum of characteristics for the vast majority of the ~1-2 billion years that eukaryotes have existed, though the details at early time, including when the eukaryotic line began, remain points of vigorous discussion.

Chapter VIII - The sole function of MBP has long been thought to be adhesion of the cytosolic surfaces of the oligodendrocyte membranes of central nervous system (CNS) myelin. However, MBP is known to bind to a number of proteins, including actin and tubulin, and can cause their polymerization and bundling *in vitro*. It can also tether actin filaments and bundles to a membrane surface. These interactions can be regulated by physiological post-translational modifications of MBP (primarily Ser/Thr phosphorylation and Arg deimination to citrulline), by Ca^{2+}-calmodulin binding to MBP, and by physiological changes in membrane surface potential. *In vivo*, both actin and tubulin are isolated together with MBP in a low density, detergent-insoluble, glycosphingolipid-enriched fraction of myelin, which also contains caveolin and kinases, and thus may be a membrane signaling domain. Some MBP is colocalized with actin *in vivo* at the membrane edges of cultured oligodendrocytes and with cytoskeletal veins of microfilaments and microtubules. In OLs from the shiverer mutant mouse, which lacks MBP, actin microfilaments do not form bundles and are not colocalized with microtubules. Moreover, microtubular structures are abnormal in size, and cell processes are usually smaller than normal with a larger cell body. In addition, extracellular signaling via glycosphingolipids to the cytoskeleton does not occur. These observations and changes in shiverer OLs suggest that interactions of MBP with actin microfilaments and microtubules occur also *in vivo* in OLs, and that these interactions are important for cell function, membrane process formation, and myelination.

In: Tubulin: Structure, Functions and Roles in Disease
Editors: Wei Yamauchi and Antonio Sokić
ISBN: 978-1-62100-141-6

Chapter I

Microtubule Plus-end Tracking Proteins in Neuronal Development and Nervous System Disorders

C. L. Sayas and J. Avila
Centro de Biología Molecular "Severo Ochoa", CSIC-UAM,
Universidad Autónoma de Madrid, Spain

Abstract

Regulation of microtubule organization and dynamics is crucial during neuronal differentiation. A plethora of proteins participate in controlling the dynamic behavior of microtubules. Plus-end tracking proteins (+TIPs) are a specialized class of microtubule-associated proteins that specifically accumulate at growing ends. +TIPs comprise a diverse group of structurally unrelated proteins that mediate microtubule interactions with different cellular structures and signaling factors. Several +TIPs, including CLASP1 and CLASP2, CLIP-115 and CLIP-170, EB family members, APC, p150Glued (the large subunit of the dynactin complex), NAV1 and LIS1, have been shown to play important roles during different stages of neuronal development such as migration, axon extension and guidance, dendritic spine formation, and maintenance of proper communication between neurons. Consequently, malfunction of +TIPs can lead to some nervous system disorders. LIS1 is essential for proper migration of cortical and cerebellar neurons. Mieller-Dieker Syndrome (MDS) is a lissencenphaly, characterized by the presence of abnormal cortical layers, corpus callosum defects and ventricle enlargement, is caused by mutations in the LIS1 gene. Mutations in genes encoding proteins that interact with LIS1, such as doublecortin (DCX), lead to another well-characterized neuronal migration disorder. Mutations in p150Glued lead to the disruption of dynein/dynactin and impairment of axonal transport in motor neurons and neuronal cell death, thus causing progressive neurodegeneration. Not surprisingly, mutations in p150Glued were identified in different motor neuron diseases, such as Amyotrophic Lateral Sclerosis (ALS), characterized by progressive paralysis of muscles at adult age provoked by death of motor neurons. In addition, CLIP-115 deficiency has been associated with the neurological symptoms found in Williams Syndrome (WS), a disease characterized by facial abnormalities, cardiovascular problems and mental retardation. Here, we summarize recent studies concerning the involvement of +TIPs in neuronal development and disease.

1. Introduction

Neurons are highly polarized cells with a single long and thin axon that sends information over long distances and multiple shorter dendrites that receive and integrate incoming information. The correct formation and proper function of the nervous system relies on the ability of neurons to polarize and undergo major morphological changes during different processes. Some of these processes are: motility (e.g. neuronal migration), differentiation (e.g. neuronal polarization, neurites/axon/dendrites outgrowth), target site selection (e.g., pathfinding and growth cone turning) and establishment of synaptic contacts. In order for neurons to change shape and move, their cytoskeleton must undergo rearrangements in response to extracellular cues.

Neuronal cytoskeleton is comprised by microtubules (MTs), actin microfilaments (MFs) and neurofilaments (which appear late during development). MTs are structural components of neurites, axons and dendrites that provide tracks for organelle transport. MT assembly and stabilization in a neurite constitute signals that specify axonal identity (Witte et al., 2008). However, MTs are not mere structural scaffolds; they are highly dynamic structures that undergo cycles of growth and disassembly in a process known as *dynamic instability* (Mitchison and Kirschner, 1984). Dynamic instability allows MTs to reorganize rapidly and to generate pushing and pulling forces in response to different signals. Furthermore, dynamic MTs sense the cellular space until a subset of them reaches specific factors at the cell cortex where they become captured and stabilized (Bayley and Martin, 1991; Burbank and Mitchison, 2006; Cassimeris et al., 1987).

MT assembly and dynamics are regulated by a plethora of proteins, such as assembly-promoting factors, stabilizing factors (such as structural or classical MT-associated proteins (MAPs), destabilizing factors, severing proteins, motors and proteins that specifically bind to growing ends (MT plus-end tracking proteins or +TIPs) (Avila et al., 1994; Galjart and Perez, 2003; Hunter and Wordeman, 2000; Tirnauer and Bierer, 2000; Walczak, 2000).

+TIPs have recently attracted much attention since they act as local regulators of MT dynamics at their plus-ends thereby controlling directionality of growth and interactions with organelles and components of the cell cortex (Akhmanova and Hoogenraad, 2005; Galjart, 2010 ; Gouveia and Akhmanova, 2010). Recent reports have shown that +TIPs play important roles in diverse processes involved in neuronal development such as migration, neurite extension, axon outgrowth and pathfinding as well as dendritic spine formation and plasticity (Jaworski et al., 2008). Mutations in some +TIPs have been found to be involved in the onset of different nervous system disorders (Jaworski et al., 2008). In this chapter, we summarize recent findings on +TIPs functions in nervous system development and disorders.

2. MTs in Neuronal Development

MTs are hollow cylinders composed of 13 protofilaments made of α- and β-tubulin heterodimers, polymerized in a head-to-tail fashion (Wade, 2007). This confers an intrinsic polarity to MTs, with one end designated as the "plus-end" and the other as the "minus-end". MT polarity is crucial for motor-driven transport since different motor proteins move along MTs with opposite orientation (e.g. kinesins move towards the plus-ends and dynein towards the minus-ends). *In vitro*, MT plus-ends undergo cycles of growth and shrinkage while MT minus-ends are relatively unstable and shrink without stabilization (Mitchison and Kirschner, 1984). In cells, MT minus-ends are usually embedded in MT-organizing centers known as centrosomes. MT dynamics is regulated by the intrinsic GTPase activity of β-tubulin, by posttranslational modifications of tubulins (acetylation, tyrosination, detyrosination, polyglutamylation, phosphorylation and polyglycilation) and by MT-associated proteins (MAPs) (Amos and Schlieper, 2005; Avila, 1990; Fukushima et al., 2009).

The organization of neuronal MTs differs in axons and dendrites (Figure 1). While axonal MTs display uniform orientation with their plus-ends directed towards the axon tip, dendritic MTs present mixed orientation with their plus-ends facing either the dendritic tip or the cell body (Baas et al., 1988). Moreover, different sets of MAPs are present in axons (e.g. tau) or dendrites (e.g. MAP2) (Matus, 1990).

During development, spherical neuroblasts undergo membrane sprouting that in turn leads to neurite formation (neuritogenesis). Neurites extend further and branch (neurite extension and branching). One single neurite will give rise to the axon (axon specification and extension) that will extend and navigate in response to extracellular signals. Attractive and repulsive signals trigger signaling cascades that promote MT reorganization leading to axon elongation, turning and branching or growth cone collapse and retraction (Conde and Caceres, 2009).

Neuronal MTs are nucleated at the centrosome (Yu et al., 1993) and then are released by the action of katanin, a MT-severing protein (Ahmad et al., 1999; Arevalo and Chao, 2005; Karabay et al., 2004), and transported by motor proteins as short polymers into axons (Hirokawa and Noda, 2008; Vale, 2003). Short MTs move faster than long MTs along axonal shafts (Baas et al., 2005). MT polymerization occurs throughout every neuronal compartment (Stepanova et al., 2003). Non-polymerized tubulin is also transported along the axon (Kimura et al., 2005; Terada et al., 2000).

2.1. Role of MTs in Neuronal Polarity: Neurite Formation and Axon Specification

Cultured primary hippocampal neurons have been widely used as a model to study neuronal polarity, axon specification and outgrowth (Dotti and Banker, 1987). When plated, spherical neuroblasts attach to the substrate and form a broad lamella (stage 1). Then, membrane sprouts at different spots, leading to the initiation of the extension of cytoplasmatic dynamic protrusions, known as neurites (stage 2). Each neurite is tipped by an actin-rich growth cone. Nanomolar concentrations of taxol or nocodazole, drugs that interfere with MT dynamics in opposing manners, reversibly inhibit spontaneous neurite initiation in cultured hippocampal neurons, indicating that MT dynamics is important for neurite formation (Letourneau and

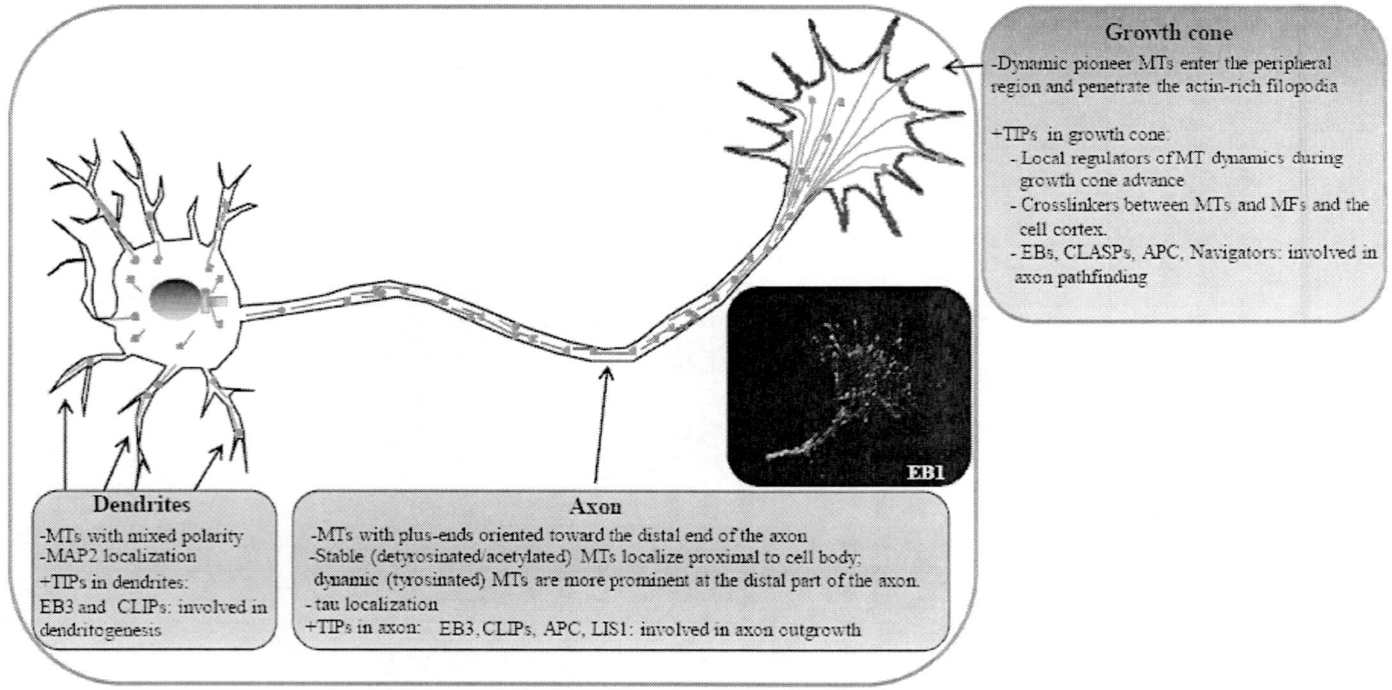

Figure 1. MTs are involved in neuronal differentiation. MTs orientation differs in dendrites and axon, thus contributing to the establishment and maintenance of neuronal polarity. • indicate MT plus-ends. Stable MTs localize to the proximal axon shaft whereas dynamic MTs are present in the more distal part of the axon and the growth cone to allow proper axon outgrowth and navigation. Different sets of classical MAPs and +TIPs participate in the regulation of MT dynamics during different processes involved in neuronal differentiation.

Axonal growth cones are enriched in dynamic MTs that penetrate into the central (C) and peripheral domain (P) (Geraldo and Gordon-Weeks, 2009; Lowery and Van Vactor, 2009). MTs are bundled at the axon shaft and they splay upon entering the growth cone. Upon reception of guidance cues, MTs either remain splayed orbend, loop or bundle (Gordon-Weeks, 2004). Hence, MTs respond to extracellular signals by modifying their configuration in growth cones (Kalil and Dent, 2005), which reflects the relationship between the rate of MT polymerization and growth cone advance (Tanaka and Kirschner, 1991). Advance of growth cones with splayed MTs is balanced with the rate of MT growth. However, looped MTs are usually present in paused growth cones. MT looping could be originated by continuous polymerization in pausing growth cones. These MTs would bend backwards upon reaching the leading edge or when they fail to attach to the growth cone cell cortex. MT bundles present in growth cones arise from coalescence of splayed MTs or by extension of preexisting bundles or loops of MTs (Tanaka and Kirschner, 1991). In growth cones with bundled MTs, growth cone advance exceeds the rate of polymer translocation. MTs constitute thus a major target of signaling cascades involved in neuronal polarization and axon outgrowth. Hence, MT rearrangement in axonal growth cones is essential for the proper advance and elongation of the axon in the right direction. In line with this, high doses of MT-depolymerizing agents induce growth cone collapse and axon retraction whereas low doses that abolish MT dynamics, block axonal extension (Gallo, 1998; McNeil et al., 1999).

Some individual dynamic MTs penetrate actin-rich filopodia at the P region of the growth cone (Figure 1) (Schaefer et al., 2002). These are the so-called pioneer MTs that, far from being mere scaffold elements, act as sensors for specific signals. A subpopulation of pioneer MTs interacts with specific factors at the cell cortex and propagate further the signaling cascades involved in axon outgrowth and pathfinding.

The initial stages of neuronal polarization, neuritogenesis and axon extension implicate the interplay between both cytoskeletal systems, MTs and actin MFs. Actin filaments might act as a "brake" to prevent neurite initiation whereas MTs induce neurite/axon outgrowth. MTs provide the "pushing force" either directly, via polymerization, and/or indirectly, via serving as tracks for motor-based transport of membrane, organelles and macromolecular complexes.

3. MT Plus-end Tracking Proteins (+TIPs)

MT dynamics is regulated by several types of proteins. Among them are MT plus-end tracking proteins (+TIPs), which specifically bind to the plus-ends of growing MTs (Galjart, 2010; Gouveia and Akhmanova, 2010). Since dynamic instability occurs at MT plus-ends +TIPs act as local regulators of MT dynamics. +TIPs influence different dynamic instability parameters: some +TIPs promote MT pauses whereas others function as rescue factors (Galjart, 2010; Gouveia and Akhmanova, 2010). +TIPs also provide a link between MT ends and actin, cell cortex or organelles (Jaworski et al., 2008).

+TIPs accumulate at growing ends in a comet-like pattern, in contrast with classical MT-associated proteins (MAPs) that bind along the MT lattice. Low-affinity binding sites for +TIPs are more numerous at the very tip of MT distal ends and they decay at a constant rate, providing a platform for +TIPs binding with a comet-shape. Comet length has been shown to directly correlate with MT growth speed whereas comet number correlates with the amount of growing MTs (Bieling et al., 2008).

+TIPs are ubiquitous and highly conserved among eukaryotes. +TIPs comprise a large group of functionally and structurally unrelated proteins that range from motor proteins to MT polymerases/depolymerases, adaptor or regulatory proteins (Galjart, 2010; Gouveia and Akhmanova, 2010). All of them accumulate at MT plus-ends, forming multiprotein complexes.

3.1. Structural Classification of +TIPs

Given the high diversity of structural and functional motifs present in different +TIPs, a classification based on the presence of conserved binding domains, repeat sequences and linear motifs has been recently proposed (see Table 1).

3.1.1. EB Family

End binding proteins (EBs) are highly conserved proteins throughout evolution that in mammals are encoded by the MAPRE genes, EB1, EB2 (RP1) and EB3. While EB1 and EB3 are ubiquitous proteins, EB2 is expressed at different levels in various cell lines (Wittmann, 2008) and appears to be the most divergent family member in expression and interacting partners (Morrison, 2007). EB1 was found in a two-hybrid screen as a protein that interacts with APC, another +TIP with a well-known role in Wnt signaling (Morrison et al., 1998). EB3 is enriched in brain, where it interacts with APC2, a brain-specific isoform of APC (Nakagawa et al., 2000). EBs are usually present in cells as globular dimers with two highly conserved domains connected by a linker sequence. The N-terminal part of EB proteins consists on a calponin-homolgy domain (CH) that is necessary and sufficient for binding MTs and tracking growing ends (Coquelle et al., 2009). At the C-terminal part, EB1 contains a coiled-coil motif, responsible for dimerization (Su and Qi, 2001) and an EB1-like motif (EBH), which is involved in the interaction with numerous EB1-binding partners (Coquelle et al., 2009). EBs contain and acidic tail at the C-terminal domain, a 20-30 residues stretch that is implicated in EBs interaction with other proteins as well as EBs self-inhibition (Hayashi et al., 2005; Slep et al., 2005). EB1 and α-tubulin present a highly conserved sequence motif at their C-terminal region, EEY/F (Miller et al., 2006; Weisbrich et al., 2007) (see Table 1).

EB1 is the core protein at MT plus-ends. Every +TIP requires EB1 interaction for MT plus-end tracking (Galjart, 2010). Protein partners interact with EB1 through two major binding modes: either through the interaction of CAP-Gly domains (see below) to the C-terminal EEY/F motif present in EB1 or through binding of a linear SxIP motif with the hydrophobic cavity of EB1. CLIPs and $p150^{Glued}$ (dynactin large subunit) are +TIPs that bind to EB1 via their CAP-Gly domains (Honnappa et al., 2006). An extensive group of structurally diverse proteins are +TIPs that interact with EB1 through SxIP sequences (CLASPs, APC, MCAK, STIM1 and MCAF1) (Honnappa et al., 2009). EBs influence on each specific parameter of dynamic instability remains controversial and is different *in vitro* or *in vivo*. EB1 has been shown to induce MT polymerization as well as to both promote and prevent catastrophe, or promote rescue, depending on the system used. In cells, mammalian EB1 and EB3 have been shown to promote persistent MT growth by suppressing catastrophes

Ressler, 1984). Multiple neurites extend during neuritogenesis and at a certain moment, one of them becomes and axon (stage 3), leading to the establishment of neuronal polarity (Dotti et al., 1988). Both actin and tubulin cytoskeletons play key roles during axon specification. Reorganization of the actin cytoskeleton is crucial at the growth cone level. In this sense, local destabilization of actin MFs at the growth cone of a single neurite leads to the formation of an axon (Bradke and Dotti, 1999). Moreover, in cultured hippocampal neurons, the neurite that will be the future axon is the one that possesses the larger growth cone, with an expanded lamellipodial veil and increased actin dynamics (Bradke and Dotti, 1999).

Localization of a functional centrosome seems to be a crucial determinant for axon specification. During neuritogenesis, the centrosome relocalizes to the base of a neurite that will in turn become the axon (de Anda et al., 2005). Suppression of centrosome function abolishes neuronal polarization and, conversely, neurons with multiple centrosomes develop multiple axons. However, once polarity is achieved, axon extension can occur in neurons lacking centrosomes (Stiess et al., 2010).

Recently, MT stability has been shown to be critical for determining axon identity. Individual axonal MTs comprise two domains with different stabilities in correlation with distinct levels of posttranslationally modified α-tubulin subunits (Baas and Black, 1990; Brown et al., 1993). One domain is enriched in detyrosinated (Glu-α-tubulin) and acetylated α-tubulin, with longer half-lives ($t1/2>2h$), and resistant to the MT-depolymerizing agent nocodazole. The other domain is enriched in tyrosinated α- tubulin, depolymerizes quickly in response to nocodazole and presents a short half-life ($t1/2<5min$). The transition between these two domains is abrupt with the tyrosinated α-tubulin-rich domain at the plus-end of the MT. Tyrosinated α-tubulin is a marker of newly incorporated α-tubulin subunits into the plus-ends of MTs. In growing axons, stabilized (detyrosinated and acetylated) MTs are mainly present in the proximal axon shaft whereas highly dynamic MTs (tyrosinated) are enriched in the distal part of the axon and in growth cones (Figure 1) (Brown et al., 1993). Before the establishment of neuronal polarity, one neurite (the potential axon) contains more stable (acetylated) MTs than the other neurites (Witte et al., 2008), indicating that MT stabilization in one neurite precedes axon formation. Axons possess higher levels of stable (acetylated) MTs than shorter neurites (Witte et al., 2008). Furthermore, MT stabilization by low doses of taxol is sufficient to induce axon formation (Witte et al., 2008). Stable MTs seem to play an important function for these modifications. MT polymerization, needed for axon specification and extension, can occur at the plus-ends of stable MTs that might act as seeds. Moreover, specific motor proteins, responsible for transport of organelles and different macromolecular complexes, bind preferentially to stable MTs (Nakata and Hirokawa, 2003; Reed et al., 2006).

2.2. Dynamic MTs at Growth Cones

Growth cones are specific structures that act as locomotory and sensory "organs" and are located at the tip or growing axons and neurites. Receptors for extracellular cues localize to the membrane of growth cones. Signals trigger cascades that lead to local reorganization of the cytoskeleton, which in turn leads to growth cone advance, turning, collapse or retraction (Tanaka et al., 1995; Tanaka and Kirschner, 1991).

(Komarova et al., 2009). *In vitro*, EB1 increases both catastrophes and rescues (Bieling et al., 2007; Komarova et al., 2009; Vitre et al., 2008). It is likely that EBs prevent catastrophes in cells by counteracting the actions of other regulators of MT dynamics (Bieling et al., 2007; Komarova et al., 2009; Vitre et al., 2008).

3.1.2. CAP-Gly Proteins

The cytoskeleton-associated protein-glycine-rich (CAP-Gly) domain is a specialized protein module comprised by~80 residues, highly conserved in eukaryotes (Steinmetz and Akhmanova, 2008). Different +TIPs contain one or more CAP-Gly domains, such as cytoplasmic linker proteins (CLIPs) and the large subunit of the dynactin complex ($p150^{Glued}$). CAP-Gly domains are mainly involved in protein-protein interactions, specifically with EB proteins, tubulin monomers, dimers and MTs (Steinmetz and Akhmanova, 2008). The C-terminal EEY/F sequence motifs present in α-tubulin and EB proteins tails are recognized by the conserved GKNDG motif of CAP-Gly domains (Honnappa et al., 2006; Weisbrich et al., 2007) (see Table 1).

In most eukaryotic cells, when α-tubulin is synthesized, it contains a tyrosine residue at its C-terminus. This tyrosine can be removed and added back by the action of different enzimes. Tubulin tyrosine ligase (TTL) is the enzyme responsible for this readdition (Erck et al., 2000). In TTL-deficient cells, +TIPs containing CAP-Gly domains are not recruited to MT plus-ends (Peris et al., 2006). This indicates that tubulin tyrosination regulates the interaction of CAP-Gly proteins with MT distal ends.

3.1.2.1. CLIPs

The two main members of the CLIP family are CLIP-170 and CLIP-115. CLIP-170 was the first protein described to have MT plus-end tracking activity (Perez et al., 1999). Structurally, CLIP-170 contains two CAP-Gly domains at the N-terminal region and two metal (zinc) binding domains at the C-terminal region (Lansbergen et al., 2004). CLIP-170 forms dimers with an interaction in a head-to-tail fashion: the zinc knucles at the C-terminus interact with the CAP-Gly domains at the N-terminus (Lansbergen et al., 2004). Both CLIPs display redundant roles as MT rescue factors and this activity is located at their N-termini (Komarova et al., 2002). Phosphorylation of CLIP-170 induces conformational changes leading to the interaction between the N-and the C-terminal domains and to CLIP-170 autoinhibition (Lee et al., 2010). CLIP-115 is a brain-specific CLIP, with an N-terminal region with two CAP-Gly domains very similar to the ones present in CLIP-170, but it lacks the zinc fingers at the C-terminal region (Table 1) (De Zeeuw et al., 1997).

CLIP-170 interacts with MTs and EBs via the CAP-Gly domains (Honnappa et al., 2006). CLIP-170 was originally proposed to link MTs to endosomes in mammalian cells (Pierre et al., 1992). CLIP-170 binds to the large subunit of dynactin, $p150^{Glued}$, through its zinc knucles (Coquelle et al., 2002). In this way, since dynactin interacts with dynein, CLIP-170 recruits the MT minus-end-directed motor dynein to MT plus-ends (Lomakin et al., 2009) and therefore can promote loading of cargo for minus-end-directed transport (Vaughan et al., 2002). In Xenopus melanophores, upon melanosome aggregation, melanosomes are directly loaded by CLIP-170 on MT plus-ends (Lomakin et al., 2009).

Table 1. +TIPs: classification and roles in nervous system development and disease. Schematic table showing a classification of +TIPs based on the presence of different conserved domains (depicted in the box below) and a summary of the known roles of different +TIPs in nervous system development and disorders

+TIPs: classification and roles in nervous system development and disease

Classific.	+TIPs	Structure	Roles in nervous system	Related human nervous system disorders
EB proteins	EB1 EB3	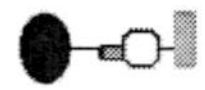	-Neuronal polarity and neuritogenesis (EB1 and EB3). -Dendrite extension, dendritic spine formation and maintenance (EB3). -NMDA-induced LTD (EB3). -Targeting of voltage-gated potassium channes Kv1 to axons (EB1).	
CAP-Gly proteins	CLIP-170		-Axon formation and extension, and dendritogenesis.	
	CLIP-115		-Brain specific +TIP, enriched in dendrites -Both CLIPs affect MT growth rates and distance in neurons.	-Williams Syndrome.
	p150Glued		-Dynein-mediated retrograde transport in neurons.	-Motor neuron disorders (e.g. ALS).
SxIP proteins	APC		-Neuronal polarity, neurite and axon outgrowth (NGFpathway). -Radial glia polarity and neuronal migration. -Localization of neurotransmitter receptors.	-*Austism spectrum disorder.* -Upregulation in Alzheimer disease (in astrocytes).
	CLASP1,2		-Expression enriched in brain (CLASP2). -Role during axon pathfinding (Slit pathway).	
	Navigators		- Neuronal differentiation and axon guidance.	
	ACF7,STIM1, MCAK, Tip150, p140Cap, etc	Diverse structures		
Other +TIPs	LIS1	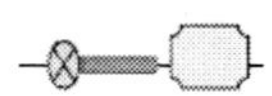	-Neuronal migration. -Axon outgrowth	-Type I lissencephaly -Involved in schizophrenia

Domains

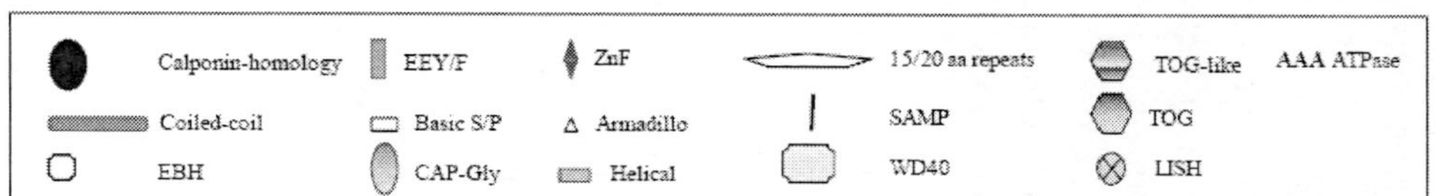

3.1.2.2. p150^{Glued}

p150^{Glued}, the large subunit of the dynactin complex, contains one CAP-Gly domain at its N-terminal region, which is part of the MT-binding domain that also contains a basic and serine-rich domain (Table 1) (Steinmetz and Akhmanova, 2008). p150^{Glued} protein also has two coiled-coil regions required for dimerization and interaction with the dynein intermediate chain and the dynactin shoulder (Zhapparova et al., 2009).

p150^{Glued} interacts with EB proteins (EB1 and EB3) through its CAP-Gly domain. (Honnappa et al., 2006). The C-terminus of EB1 has been crystallized together with the CAP-Gly domain of p150^{Glued} showing that this domain binds to the flexible tail of EB1 via its EEY/F motif (Honnappa et al., 2006). An EEY/F motif present in CLIP-170 is responsible for its interaction with the CAP-Gly domain of p150^{Glued} (Lansbergen et al., 2004). Although p150^{Glued} contains serine-rich sequences implicated in direct binding to MTs (Culver-Hanlon et al., 2006), the protein is recruited to MT plus-ends via its interaction with EB1 and CLIP-170 (Hayashi et al., 2005; Vaughan et al., 2002).

p150^{Glued} is part of the dynactin complex, which is required for targeting dynein to cargo and for dynein motor processivity (Kim et al., 2007; King and Schroer, 2000). The basic and serine-rich domain of p150^{Glued} increases dynein processivity whereas the CAP-Gly domain can inhibit the motility of dynein, thereby acting as a brake (Steinmetz and Akhmanova, 2008).

3.1.3. SxIP Proteins

A diverse group of structurally unrelated +TIP proteins that share the sequence motif SxIP have been shown to interact through it with EB1 (Honnappa et al., 2009). This motif is surrounded by unstructured sequence regions rich in basic and serine residues (Akhmanova and Steinmetz, 2008). Binding of +TIPs to EB1 via the SxIP motif is regulated by phosphorylation (Honnappa et al., 2009).

Among SxIP proteins are APC (adenomatous polyposis coli), CLASPs (CLIP-associating proteins), Navigators, MCAK (mitotic centromere associated kinesin), TIP150, MACF1 (MT-actin crosslinking factor 1, also named ACF7), STIM1 (stromal interaction molecule 1), p140Cap, melanophilin, RhoGEF2, CDK5RAP2 and DDA3 (Honnappa et al., 2009). In this chapter, we focus on APC, CLASPs and Navigators.

3.1.3.1. APC

APC has been extensively studied as a tumor suppressor that takes part in a multiprotein complex in the Wnt signaling pathway. APC, however, localizes at several subcellular regions where it participates in diverse processes (apoptosis, adhesion, chromosome segregation, neuronal differentiation, etc) (Aoki and Taketo, 2007). *In vitro*, APC has been shown to localize to MTs and to promote MT polymerization and stabilization (Munemitsu et al., 1994; Nakamura et al., 2001).

The N-terminus of APC contains a coiled-coiled domain whereas the middle region of the protein contains three to four 15 aa repeats that can interact with β-catenin. A set of seven Armadillo repeats is also present at the N-terminus of APC. Then, there are seven 20-amino acid repeats that also bind to β-catenin and are regulated by phosphorylation. Three stretches that contain SAMP sequences, involved in interaction with Axin, are interspersed between these sites. The C-terminal region of APC contains motifs that mediate interactions with diverse proteins such as EB1. The last 15 amino acids of the protein constitute a binding site for PDZ domains (Table 1) (Nathke, 2004).

APC binds to EB1 via its SxIP motif that localizes to its C-terminal region and which is anchored to the hydrophobic region of EB1 (Honnappa et al., 2009). Mutations in this domain disrupt the interaction with EB1 (Honnappa et al., 2005). APC seems to hitchhike MT plus-ends either via its binding to EB1 or to kinesin II (Berrueta et al., 1998; Jimbo et al., 2002; Mimori-Kiyosue et al., 2000; Morrison et al., 1998). However, a report indicates that APC can interact directly with MTs (Kita et al., 2006). It is likely that APC uses different mechanisms for tracking MT distal ends.

The APC-EB1 interaction is important in different subcellular localizations and processes. EB1 can disrupt the APC-actin interaction and in this way EB1 could control APC-regulation of the actin cytoskeleton (Moseley et al., 2007). In migrating cells, the APC-EB1 interaction is likely to play key roles in MT capping and capture at the cell cortex (Morrison, 2009). mDia has been shown to form part of a complex with APC and EB1 to regulate MT dynamics (Wen et al., 2004). APC and EB1 also colocalize at centrosomes (Berrueta et al., 1998; Etienne-Manneville and Hall, 2003). Moreover, the interaction between EB1 and APC plays a role in tumor formation of intestinal epithelial cells (Mimori-Kiyosue and Tsukita, 2001; Slep et al., 2005).

3.1.3.2. CLASPs

CLASPs are a family of highly conserved proteins from yeasts to humans. Mammalian isoforms CLASP1 and CLASP2 have been shown to interact with CLIPs (Akhmanova et al., 2001). CLASPs regulate MT dynamics and stabilization and link MT distal ends with the cell cortex (Lansbergen et al., 2006; Mimori-Kiyosue et al., 2005). CLASPs not only track MT plus-ends, but also bind to the Golgi apparatus, centrosomes, kinetochores, the mitotic spindle, the midbody and the cell cortex. CLASPs structure contains Dis1/TOG and TOG-like domains and a Ser-Arg-Pro-rich region in the middle of the protein (Mimori-Kiyosue et al., 2005) (Table 1). This Ser-Arg-Pro-rich region of CLASP2 contains two tandemly repeated SxIP motifs, which are required for EB1-binding and targeting to MT ends (Honnappa et al., 2009). CLASPs C-terminal domain is involved in their interaction with numerous proteins such as CLIPs, the cortical protein LL5β, the kinetochore protein CENP-E and the Golgi protein GCC185 (Efimov et al., 2007; Lansbergen et al., 2006; Maffini et al., 2009). CLASPs thus hitchhike on EB proteins to accumulate at MT growing ends and attach to various cellular structures through their C-termini.

Interaction of CLASP2 with MTs is regulated by glycogen synthase kinase-3β (GSK-3β)-mediated phosphorylation at multiple serine residues, located around the SxIP motifs (Akhmanova et al., 2001; Wittmann and Waterman-Storer, 2005). Downregulation or inhibition of GSK-3β (either chemically or downstream of PI3-K activation) lead to an increased binding of CLASP2 to MTs (Akhmanova et al., 2001; Kumar et al., 2009; Wittmann and Waterman-Storer, 2005).

CLASPs act as MT pausing factors in diverse organisms (Al-Bassam et al., 2010) and are involved in several processes such as cell division, persistent migration, linking of MTs with the cell cortex and MT stabilization (Drabek et al., 2006; Lansbergen et al., 2006; Maiato et al., 2003; Mimori-Kiyosue et al., 2005). During mitosis, CLASPs promote MT polymerization and stability and are implicated in chromosome congression, kinetochore-MT attachment, and maintenance of spindle bipolarity (Hannak and Heald, 2006; Inoue et al., 2004; Lemos et al., 2000; Maiato et al., 2003; Maiato et al., 2002).

CLASPs attach MT plus-ends to the cell cortex via their direct interaction with LL5β and ELKS (Lansbergen et al., 2006). The subset of MTs attached to LL5β-ELKS-enriched cortical sites presents an increased stability (Lansbergen et al., 2006). The spectraplakin MACF1 (ACF7), which is also a +TIP, acts in the same MT stabilization pathway and regulates CLASPs localization at the cell cortex (Drabek et al., 2006). Downregulation of both CLASPs leads to a reduction in MT density and stability while CLASPs overexpression increases MT stabilization in non-neuronal cells (Mimori-Kiyosue et al., 2005).

3.1.3.3. Navigators

Mammalian Navigators (NAV1, NAV2 and NAV3) are homologues of Unc-53, a C. elegans protein involved in axon guidance and cell migration, and outgrowth of muscles (Maes et al., 2002). NAV 2 is ubiquitous whereas NAV3 is mainly present in heart and skeletal muscle and NAV1 is brain-specific (Coy et al., 2002; Maes et al., 2002; Peeters et al., 2004). Navigators contain a MT binding domain at the N-terminal region; NAV2 and NAV3 also have a calponin-homology domain at the N-terminal part (Peeters et al., 2004). At the C-terminus the three isoforms contain an AAA ATPase domain (Frickey and Lupas, 2004) (see Table 1). All Navigators have been found to localize to MT plus-ends *in vivo* (Martinez-Lopez et al., 2005; van Haren et al., 2009). Since Navigators contain SxIP motifs surrounded by regions rich in serine and basic residues, their localization at MT distal ends is probably dependent on EB1 binding (Honnappa et al., 2009). Navigators play roles in different processes such as transport, membrane fusion, protein degradation, peroxisome biogenesis, signaling or gene expression regulation (Stringham and Schmidt, 2009)

3.1.4. Other +TIPs: LIS1

Lissencephaly1 (LIS1) is an atypical +TIP that seems to localize to MT plus-ends via its interaction with CLIP-170, dynactin and dynein (Galjart and Perez, 2003; Sapir et al., 1997). CLIP-170 interacts with LIS1 and p150^{Glued} through the same region and therefore, CLIP-170 cannot be simultaneously bound to both proteins (Coquelle et al., 2002; Tai et al., 2002). LIS1 interacts with both dynein (through its heavy and intermediate chains) and dynactin (through dynamitin) (Tai et al., 2002) and could mediate CLIP-170/dynein interactions. Hence, LIS1 might release the dynein-dynactin complex from MT tips (Lansbergen et al., 2004). LIS1 comprises an N-terminal LIS homology domain (LisH) responsible for dimerization and a C-terminal WD40 repeat containing a β-propeller domain (Table 1) (Kim et al., 2004; Tarricone et al., 2004). LIS1 has been shown to reduce MT catastrophe *in vitro* (Sapir et al., 1997). LIS1 plays important roles during MT motor-mediated transport, actin remodeling, neural stem cell division and neuronal migration (Reiner et al., 2002).

4. +TIPs in Neurons

4.1. Migration

Neuronal migration is a key process during nervous system development in which neural progenitors migrate from zones of active neurogenesis to their final destination (Marin and Rubenstein, 2003). Migrating cells extend a long leading process with a swelling full of dynein, followed by asynchronous translocation of the centrosome and the nucleus (Marin and Rubenstein, 2003). MT cytoskeleton reorganization is essential during cell migration (Reiner and Sapir, 2009). Different +TIPs have been implicated in neuronal migration. In particular, LIS1 and dynein have been shown to be crucial for both nucleus and centrosome translocation and, in this way, for proper neuronal migration (Tsai et al., 2007). LIS1 and dynein might act together by pulling the MT plus-ends on the swelling and thereby translocating both centrosome and MTs. As a consequence, dynein bound to pulling MTs reaches the nuclear envelope and pushes the nucleus forward in a process that also requires myosin II contraction (Tsai et al., 2007). A clear function of the interaction LIS1/dynein seems to be the generation of tension forces between nucleus and centrosome (Tanaka et al., 2004) and between MT plus-ends and the cell cortex (Dujardin et al., 2003). LIS1 interaction with CLIP-170 is also crucial for neuronal migration (Kholmanskikh et al., 2006). However, CLIP-170-deficient mice do not have gross ostensible brain abnormalities (Akhmanova and Hoogenraad, 2005), suggesting that CLIP-170 is not essential for neuronal migration. Besides its action on MTs through its interaction with dynein or CLIP-170, LIS1 also regulates actin cytoskeleton dynamics through its interaction with IQGAP1, an effector of Rho GTPases, and with Cdc42 GTPase (Kholmanskikh et al., 2006). IQGAP1 interacts with other +TIPs such as APC, CLIP-170 and CLASP2 (Fukata et al., 2002; Tirnauer, 2004; Watanabe et al., 2009). Therefore, a multiprotein complex might contribute to engage MT plus-ends and actin cytoskeleton to the cell cortex during neuronal migration.

Both NAV1 and NAV2 have been shown to play important roles during neuronal migration. NAV1 is specifically expressed in nervous system in embryonic and postnatal stages (Martinez-Lopez et al., 2005). NAV1 expression is restricted to some specific migratory neurons and to postmitotic postmigratory neurons (Martinez-Lopez et al., 2005). Downregulation of NAV1 induces neurons to lose the directionality in their leading processes and distribute randomly, suggesting an impairment of the mechanisms that generate directional migration (Martinez-Lopez et al., 2005). However, motility itself, which is the other component of neuronal migration, appears to be unaffected.

NAV2 hypomorphic mutant mice lacking the full-length transcript present ataxia and defects in cerebellar development (McNeill et al., 2010). Cerebellum development involves a coordinated program of extension of neuronal processes and migration resulting in a foliated structure crucial in motor function. Granule cells in the germinal zone of the external granule layer (EGL) appear to proliferate normally, however, due to the reduction in cerebellar circumference, they fail to migrate properly toward the interior of the cerebellum. In NAV2 hypomorphs, fewer granule cells migrate out of cerebellar EGL explants. This suggests that neuronal migration is disrupted in NAV2 mutants (McNeill et al., 2010). This work supports an essential role for full-length NAV2 in EGL neuronal migration during cerebellar development.

Different signaling pathways triggered by extracellular factors such as neurotrophins, Wnt or extracellular matrix converge at the kinase GSK-3β and may be involved in the regulation of MT dynamics during cell migration (Arevalo and Chao, 2005; Zhou and Snider, 2006). In particular CLASP2 and APC, two +TIPs that are targets of GSK-3 (Ikeda et al., 2000; Wittmann and Waterman-Storer, 2005), might be important for directed migration. CLASP2 role in neuronal migration remains unexplored; however, CLASP2 promotes directional persistent motility during migration of non-neuronal cells (Drabek et al., 2006). In addition, during neocortex development, APC contributes to centrosome forward movement and neuronal migration downstream of GSK-3β and LKB (Asada and Sanada, 2010). LKB1 phosphorylates and inactivates GSK-3β at the leading process tip of migrating neurons in the developing neocortex. Upon GSK-3β inhibition, APC localizes to MT plus-ends, stabilizing them close to the leading tip of migrating neurons.

Loss of APC in mice induces defective generation and migration of cortical neurons, severely disrupted cortical layer formation, and aberrant axonal tract development (Ivaniutsin et al., 2009). These mice exhibit also a reduction in the proliferative pool, impairment in interkinetic nuclear migration and in cell polarity, and increase in apoptosis (Ivaniutsin et al., 2009). Moreover, APC has been shown to be an essential regulator of radial glial polarity and construction of the cerebral cortex (Yokota et al., 2009). During cortex development, radial glia serve as neuronal progenitors and act as scaffolds for neuronal migration. APC contributes to the polarized radial glial scaffold during brain development. In brains from APC deficient mice, radial glial cells undergo MT destabilization and loss of polarity and responsiveness to the extracellular polarity maintenance cues, such as neuregulin-1 (Yokota et al., 2009). As a result of the defects in radial glial cells, there is also a defective generation and migration of cortical neurons, severely disrupted cortical layer formation, and aberrant axonal tract development. Therefore, APC contributes to the construction of cerebral cortex in mammals by regulating neuronal migration, both directly and indirectly.

4.2. Neuronal Polarization, Neurite and Axon Formation and Extension

As mentioned above, MTs are key players during neuritogenesis, axon initiation and extension. Diverse +TIPs have been shown to be involved in these processes. +TIPs are present at MT plus-ends in developing neurons (Stepanova et al., 2003). GFP-tagged EB3 is widely used as a marker of MT growth in living neurons. MT dynamics has been measured in neurons by tracking EB3-GFP displacements, showing that MTs grow slower than in glia and COS-1 cells (Stepanova et al., 2003). Average MT growth speed and length of EB3-GFP movements are comparable in cell bodies, dendrites, axons, and growth cones (Stepanova et al., 2003). EB3-GFP movements in different orientations in dendrites and axons reflect the highly specific MT organization in neurons. Axonal MTs grow at a constant average speed and they undergo catastrophes at random positions but in a programmed fashion (Stepanova et al., 2010). EB1 and EB3 mark growing MTs throughout extending neurites in differentiating neuroblastoma cells (N2A and N1E-115) (Morrison et al., 2002; Stepanova et al., 2010). EB1 levels increase during axon but not dendrite formation in differentiated N2A neuroblastoma cells (Morrison et al., 2002). EB1 regulates MT growth rate, growth distance, and duration in N1E-115 neuroblastoma cells (Stepanova et al., 2010). Downregulation of

EB1 blocks neurite extension in these cells. EB1 stimulates MT polymerization and structural transitions at MT ends and, in this way, regulates catastrophes and turnover of +TIP binding sites (Stepanova et al., 2010).

EB3 is localized at the plus-ends of MTs that enter filopodia in advancing growth cones where it interacts with the F-actin-binding protein drebrin (Geraldo et al., 2008). This interaction is crucial for the proper formation of growth cones and neurite extension. EB3 might act therefore as a linker between MTs and F-actin. Moreover, EB3 has been shown to be concentrated and stabilized in the axon initial segment (AIS). EB3 and EB1 interact directly with ankyrin G (ankG), a scaffold protein present in the AIS (Leterrier et al., 2010). EBs coordinate a molecular and functional interplay between ankG and the AIS. Therefore, EBs play central roles in the maintenance of neuronal polarity through their actions on the AIS.

CLIPs also participate in MT dynamics during neuronal polarization and neurite and axon extension. In neurons, average MT growth velocity and distance are affected by CLIP-115 and CLIP-170, but not the duration of a MT growth event. CLIPs have been shown to influence the axonal MT/tubulin ratio (Stepanova et al., 2010). CLIPs are enriched in the axonal growth cone and penetrate into its actin-rich domain. CLIPs are necessary and sufficient to induce the formation of an axon (Neukirchen and Bradke, 2010). CLIPs stabilize MTs at the leading edge of the growth cone (Neukirchen and Bradke, 2010). As mentioned above, CLIP-170 interacts with IQGAP1, an effector of Rac1 and Cdc42 GTPases, involved in actin reorganization (Fukata et al., 2002). Moreover, CLIPs positively restrain actin arc formation, necessary for axon extension (Neukirchen and Bradke, 2010). Hence, CLIPsmight serve as a bridge between actin and tubulin cytoskeleton during axon outgrowth. Actually, in neurons without CLIP activity, axon elongation is impaired whereas actin destabilization or myosin II inhibition restores axon outgrowth (Neukirchen and Bradke, 2010). CLIPs are involved in neuronal polarization by controlling MT stabilization and growth cone dynamics.

CLASP homologues have been shown to be implicated in axonal pathfinding in different species. In Xenopus and Drosophila neurons, CLASP is enriched at a subset of MTs in the growth cone that probe the actin-rich peripheral zone (pioneer MTs) (Lee et al., 2004). In this region, guidance signals exert their initial influence on cytoskeletal organization. Orbit/Mast, the Drosophila ortologue of CLASPs, is involved in the regulation of growth cone repulsion by Slit and its receptor Robo, acting downstream of Abl (Lee et al., 2004). Mammalian CLASP1 and CLASP2 interact with actin filaments (Tsvetkov et al., 2007). The spectraplakin MACF1 (ACF7), that connects actin MFs and MTs, regulates CLASP2 cortical localization in non-neuronal cells indicating that it acts upstream of CLASP2 (Drabek et al., 2006). Moreover, CLASP2 interacts directly with the actin-binding protein IQGAP1, in a phosphorylation-dependent manner (Watanabe et al., 2009). CLASP2 phosphorylation by GSK-3β results in the dissociation of CLASP2 from IQGAP1, EB1 and MTs. Phosphorylation of CLASP2 by GSK-3β might control the regional linkage of MTs to actin filaments through IQGAP1. In summary, CLASP2 may facilitate recognition of actin filaments by the MT plus-ends at the initial stages of actin-MT interaction and enable MT capture at cortical sites at the cell edge. As mentioned above, in non-neuronal cells, CLASP2 has been shown to associate with membrane-associated proteins such as LL5β and ELKS (Lansbergen et al., 2006), as well as with EB1 and CLIP-170 (Akhmanova et al., 2001; Mimori-Kiyosue et al., 2005). CLASPs might therefore function as crosslinkers between MTs and the cell cortex during axon outgrowth and navigation.

Navigators were designated after their homology to the C. elegans Unc-53 protein, involved in neuronal differentiation and axon guidance in the worm (Stringham and Schmidt, 2009). Ectopically expressed GFP-tagged Navigators induce the formation of neurite-like extensions in non-neuronal cells (van Haren et al., 2009). In the case of NAV1, this function relies on its ATPase activity and it is not due to a classical type of MT stabilization and bundling (Martinez-Lopez et al., 2005; van Haren et al., 2009). When overexpressed in neurons, EGFPmNAV1 is enriched in growth cones, in particular in filopodia, pointing to a possible role during growth cone navigation by local regulation of MT dynamics (Martinez-Lopez et al., 2005). NAV2 was shown to be retinoic acid-inducible in the developing nervous system (Clagett-Dame et al., 2006). NAV2 is a true orthologue of Unc-53 given that it can rescue the mechanosensory neuron outgrowth defects of *unc-53* mutants, when ectopically expressed in those cells (Muley et al., 2008). NAV3 mRNA is expressed throughout the nervous system, but is especially prominent at the synapses at neuromuscular junctions (Kishi et al., 2005). Expression of NAV3 is reduced in neuroblastoma cells while it is upregulated in nerve cells after brain injury, indicating that NAV3 is involved in neuron growth and regeneration as well as neural tumorigenesis (Coy et al., 2002).

APC is highly expressed in developing rat brain during neuronal maturation (Bhat et al., 1994). APC localizes to sites of neurite growth (Votin et al., 2005). First, it localizes to the tips of every neurite but later on it gets restricted to the tip of the axon to be and finally to the axon distal end (Shi et al., 2004). Actually, in polarizing hippocampal neurons, APC selectively accumulates at the tip of the nascent axon where it colocalizes with mPar3 (Shi et al., 2004). APC is therefore involved in both early neurite outgrowth and neuronal polarization and axon extension. β-catenin has a structural role in inhibiting APC function in neurite growth (Votin et al., 2005). APC is upregulated during NGF-mediated differentiation of PC-12 cells (Dobashi et al., 1996). However, ectopic expression of APC does not lead to PC-12 differentiation. Moreover, it was shown that APC localization at MT plus-ends, downstream of GSK-3β inactivation, is required for proper NGF-induced axon outgrowth (Zhou et al., 2004). Local activation of PI3-K and the subsequent inactivation of GSK-3β occur in the axonal growth cone, where APC is localized (Zhou et al., 2004). Wnt3a is an extracellular guidance cue that regulates MT directionality, resulting in MT looping, growth cone pausing, and remodeling. This effect of Wnt3a on MTs of dorsal root ganglion (DRG) neurons is mediated by the unexpected loss of APC from MT plus-ends (Purro et al., 2008). APC downregulation mimics Wnt3a function in DRG neurons (Purro et al., 2008). Moreover, local inactivation of GSK3β is required for APC localization to MT ends in extension tips in astrocytes (Etienne-Manneville and Hall, 2003). Hence, APC is a converging target for different signaling pathways that lead to GSK-3β inhibition during neuronal differentiation (NGF, Wnt3a, etc). APC is also involved in axon pathfinding since asymmetric inactivation of APC by microscale laser inactivation results in growth cone turning attributable to axon extension on the side containing APC (Koester et al., 2007). Moreover, APC can also bind directly to actin (Moseley et al., 2007) and indirectly via its interaction with IQGAP1, ASEF (a Cdc42-GEF), mDia (activator of actin polymerization) and proteins with a PDZ domain such as Dlg (Kawasaki et al., 2000; Matsumine et al., 1996; Watanabe et al., 2004; Wen et al., 2004). APC might in this way be involved in mediating the interaction between MT distal ends and the cell cortex. Future studies will determine how binding of β-catenin to APC clusters in growth cones in response to diverse signals regulates MT changes necessary for neurite and axon outgrowth, and potentially neuronal targeting and synaptogenesis.

LIS1 and cytoplasmatic dynein also participate in growth cone remodeling since their inhibition impairs the entrance of MTs into the peripheral zone of axonal growth cones (Grabham et al., 2007). In chick sensory neurons treated with laminin, dynein, dynactin, and LIS1 relocalize rapidly to the leading edge of the growth cone, where they associate with MTs. Antibodies against LIS1 and dynein produce an almost complete block of laminin-induced growth cone remodeling and the underlying reorganization of MTs and penetration into the peripheral zone (Grabham et al., 2007). MTs are still able to grow when dynein and LIS1 are inhibited; however, MTs show a reduced ability to resist retrograde actin flow. In hippocampal neurons dynein, dynactin, and LIS1 are enriched in axonal growth cones at stage 3, and both growth cone organization and axon elongation are altered by downregulation of LIS1 (Grabham et al., 2007). Dynein and LIS1 play a role in MT advance during growth cone remodeling associated with axonogenesis.

4.3. Dendritogenesis

Dendrites are neuronal extensions, shorter and thicker than axons, whose function is to receive synaptic inputs. MTs and actin MFs are crucial for proper dendritic morphology. Some +TIPs have recently been found to be implicated in dendrite formation and extension in different species.

EB3 has been shown to interact with PSD-95 in dendrites (Sweet et al., 2010). PSD-95 acts as a brake for dendrite branching. Overexpression of either PSD-95 or mutant EB3 leads to a decreased lifetime of EB3 comets in dendrites. Ectopic expression of PSD-95 results in less organized MTs at dendritic branch points and decreased dendritogenesis (Sweet et al., 2010) . EB3 interaction with PSD-95 may provide a key mechanism for the regulation of MT dynamics during dendrite formation and extension.

CLIP-170, through its interaction with IQGAP1, regulates dendrite morphology by coordinating the interplay between MTs and the actin cytoskeleton (Swiech et al., 2010). mTOR kinase interacts with CLIP-170 and is needed for efficient formation of a protein complex containing CLIP-170 and IQGAP1 (Swiech et al., 2010). CLIP-170, IQGAP1 and dynamic MTs are required for proper dendritogenesis downstream of the PI3K-mTOR signaling pathway. Dendritic arbor morphology may be controlled by enhancing mTOR-mediated crosstalk between dynamic MTs and actin through CLIP-170 and IQGAP1.

Drosophila neurons contain uniform-polarity minus-end-out MTs in dendrites, which are often highly branched. In these neurons, growing MTs navigate dendrite branch points by turning the same way toward the cell body, and track along stable MTs toward their plus-ends (Mattie et al., 2010). +TIPS EB1 and APC are required for uniform dendrite MT polarity in Drosophila neurons (Mattie et al., 2010). Furthermore, APC and APC2 localization and interaction at branch points suggest that these proteins work together at dendrite branches (Mattie et al., 2010). LIS1 has also been shown to be involved in proper formation of dendritic arbors in Drosophila (Liu et al., 2000). Misplaced hippocampal neurons of heterozigous LIS1 mice show diminished dendrite branching (Fleck et al., 2000).

4.4. Synaptic Spine Formation and Synaptic Plasticity

Dendritic spines are specialized regions of the dendrites that constitute the major sites of excitatory synaptic input. Spine morphological changes have been linked to learning and memory processes. Until recently, it was generally believed that MTs were absent from dendritic spines and only the actin cytoskeleton controlled spine morphology and plasticity. However, recent reports have shown that dynamic MTs, tipped with EB3 are present in spines and that EB3 can modulate spine morphology (Gu et al., 2008; Hu et al., 2008; Jaworski et al., 2009). p140Cap, a regulator of Src tyrosine kinase, is an EB3-interacting partner, predominantly localized to spines and enriched in the postsynaptic density (Jaworski et al., 2009). Inhibition of MT dynamics or knockdown of either EB3 or p140Cap, significantly reduces spine formation (Jaworski et al., 2009). p140Cap interacts with the Src substrate and F-actin-binding protein cortactin. Therefore, EB3 modulates actin dynamics within dendritic spines, linking dynamic MTs to changes in spine morphology and synaptic plasticity.

EB3 has also been shown to be involved in the regulation of MT dynamics during NMDA-induced long-term depression (LTD) (Kapitein et al., 2011). NMDA-dependent LTD rapidly attenuates MT dynamics, removing EB3 from MT plus-ends in dendrites. In a second, more prolonged response to NMDA, EB3 accumulates along MAP2-positive microtubule bundles in the dendritic shaft.

4.5. Channels and Receptor Localization in Neurons

+TIPs are not only important regulators of MT dynamics during neuronal differentiation but also in terminally differentiated neurons in which they might act by regulating the positioning of ion channels, neurotransmitter receptors and intracellular organelles at synaptic contact sites.

EB1 is required for targeting of voltage-gated potassium channels Kv1 to axons, where they regulate membrane excitability (Gu et al., 2006). Kv1 axonal targeting depends on the association of the cytoplasmic Kvβ2 subunit with both EB1 and KIF3/kinesin II (Gu et al., 2006). Downregulation of EB1 impairs the targeting of Kv1 vesicles to axons (Gu et al., 2006). EB1, when co-moving with Kv1.2 and Kvbeta2, travels significantly faster than its plus-end tracking (Gu and Gu, 2010). Kvβ2 is phosphorylated by Cdk, and this phosphorylation negatively regulates its interaction with EB1 (Vacher et al., 2011). Endogenous Cdks, EB1 and phosphorylated Kvβ2 colocalize in the axons of cultured hippocampal neurons, with enrichment in the axon initial segment (AIS) (Vacher et al., 2011). Cdk inhibition induces accumulation of EB1, Kvβ2 and Kv1 channel subunits within the AIS. Thus, reversible interaction of EB1 with Kvβ1 might constitute a new regulatory mechanism for the targeting of Kv1 complexes to the axonal membrane.

APC plays key roles in localizing different neurotransmitter receptors to neuronal postsynaptic sites. APC functions in localizing alpha3-nicotinic acetylcholine receptors (nAChRs) to postsynaptic sites (Temburni et al., 2004). APC is selectively enriched at cholinergic synapses. APC forms clusters, juxtaposed to synaptic vesicle clusters, and colocalizes with alpha3-nAChRs but not with the neighboring synaptic glycine receptors or perisynaptic alpha7-nAChRs (Temburni et al., 2004). APC downregulation leads to nAChRs redistribution to the plasma membrane (Potaros et al., 2011). EB1, PSD-93 and β-catenin are also enriched at alpha3-nAChR postsynaptic sites (Rosenberg et al., 2008). EB1 shows close proximity to and partial overlap with alpha3-nAChR and APC surface clusters. In addition, interaction of APC with EB1 is required for alpha3nAChR surface membrane insertion and stabilization (Rosenberg et al., 2008). Multiprotein complexes at nicotinic synapses are formed by APC recruitment of EB1, the cytoskeletal regulators macrophin and IQGAP1, and 14-3-3 adapter protein at nicotinic synapses. 14-3-3 links the alpha3-subunit to APC. Hence, APC functions as a coordinator of nicotinic postsynaptic assembly in vertebrate neurons. APC complex provides retrograde signals that direct presynaptic active zones to develop. Postsynaptic APC multiprotein complex is required for anchoring neuroligin and neurexin at neuronal synapses in vivo (Rosenberg et al., 2010). APC plays a key role in organizing excitatory cholinergic postsynaptic specializations and works as the first non-receptor protein to function in localizing nAChRs to neuronal synapses in vivo

Furthermore, APC associates with PSD-95 (Yanai et al., 2000). Both overexpression of a dominant-negative construct that disrupts the APC/PSD-95 interaction and knockdown of APC, inhibit the clustering of PSD-95 and a glutamate receptor subunit, and reduce AMPA-induced activity of AMPA receptors but not the clustering of NMDA receptors (Shimomura et al., 2007). APC might therefore be involved in the development of glutamatergic synapses.

5. +TIPs in Nervous System Disorders

During the last years, a combination of human genetics studies and research performed in transgenic mouse models, has shed light into the role of +TIPs in the onset of some neurological and neurodegenerative disorders. Here, we summarize the most relevant findings in this field (see also Table 1 for an overview).

5.1. LIS1 and Neuronal Migration Disorders

Some nervous system disorders with defects in brain structure are associated with alterations in neuronal migration during development. Lissencephaly is a severe human neuronal migration defect characterized by a smooth cerebral surface, mental retardation and seizures (Friocourt et al., 2011). As mentioned above, LIS1 protein is crucial for proper neuronal migration during cortex and cerebellum development (Hatten, 2005; Leventer et al., 2001b; Wynshaw-Boris, 2007). The gene coding for LIS1 is one of the most common genes mutated in patients with lissencephaly (Dobyns et al., 1993). Furthermore, the majority of patients with lissencephaly have heterozygous mutations or deletions of LIS1 (Clark, 2001). Mutations in LIS1 were also identified to be responsible for Miller-Dieker Syndrome (MDS) (Isumi et al., 1997; Mizuguchi et al., 1995; Reiner et al., 1993). Dominant point mutations that preserve LIS1 are also found in isolated lissencephaly sequence (Lo Nigro et al., 1997) that together with MDS, comprise Type I lissencephaly.

Type I lissencephaly is a group of disorders characterized by generalized agyria (absence of gyri) and a thickening of the cerebral cortex and can be associated with other brain and visceral anomalies (Friocourt et al., 2011). Type I lissencephaly includes several forms of cerebral malformations, including microcephaly, agyria–pachygyria, heterotopia and polymicrogyria. The term "lissencephalic brain" is applied to human malformed brains that have a significant reduction in the number of cortical convolutions by homology to the brain of inferior vertebrates. Most patients with these defects present pachygyria, with relatively spared gyri instead of agyria characterized by a total absence of gyri and sulci (Dambska et al., 1983). Patients with type I lissencephaly present severe motor and cognitive impairment and epilepsy (Kuchelmeister et al., 1993). Facial dysmorphism and visceral abnormalities are found in patients with MDS at birth (Herman and Siegel, 2008). Usually, patients with lissencephaly present a very poor prognosis with post-natal feeding problems and repeated aspiration pneumonia which lead to a shortened lifespan (Pavone et al., 1993). Living patients present spastic quadriplegia and epilepsy and profound mental retardation (Dobyns, 2010).

Brains of patients with type I lissencephaly are small in size, indicating the presence of some defects in neuron generation and/or cell proliferation (Friocourt et al., 2011). In line with this, LIS1 has been found to participate in the regulation of mitotic spindle orientation that determines the cleavage plane during division of cortical neuroblasts of the ventricular zone (VZ) (Siller and Doe, 2008). Mice deficient in LIS1 show a thinner cortex with dose-dependent disorganization of cortical layers and also in hippocampus, cerebellum and olfactory bulbs (Cahana et al., 2001; Gambello et al., 2003; Hirotsune et al., 1998). Primary cerebellar neurons from LIS1 +/- mice show slower migration and display a increase in the distance between the centrosome and the nucleus, suggesting that LIS1 is required for nuclear movement by coupling the nucleus with the centrosome (Bix and Clark, 1998; Hirotsune et al., 1998; Tanaka et al., 2004). LIS1 has been shown to play important roles in cortical interneuron migration in both mouse and human (Marcorelles et al., 2010; McManus et al., 2004; Nasrallah et al., 2006; Pancoast et al., 2005). Altered distribution of neurons in human lissencephaly might be the result from both defects in cell migration and division since the timing of the last division affects the final destination of cortical neurons.

Heterozygous mutations in the APC gene increase the migration deficit and the incidence of hydrocephalus in LIS1+/- mice (Hebbar et al., 2008). APC and LIS1 interact during mammalian brain development (Hebbar et al., 2008). The C-terminus of APC interacts with both dynein and LIS1 and their distribution is altered in cells derived from APC mutant mice (Min mice). LIS1 has been proposed to be involved in the retrograde transport of vesicles and organelles from neuronal extensions of migrating neurons to the nucleus through its actions on dynein (Sasaki et al., 2000). The complex formed by LIS1, Ndel1 and dynein may influence anterograde transport directed toward MT plus-ends, to the periphery of the migrating neuron, hence contributing to the extension of the leading process (Yamada et al., 2008). Defective vesicle transport during development may result in impaired formation of the leading process.

Mutations in doublecortin (DCX), another MT-associated protein, have been linked to another well-characterized neuronal migration disorder that results in X-linked lissencephaly in males and subcortical heterotipia or double cortex in females (LoTurco and Bai, 2006). Heterozygous mutations in DCX in females is probably due to random inactivation of one of the X chromosomes resulting in two genetically distinct populations of cells that segregate in a cell-autonomous fashion (Guerrini and Marini, 2006). Thus, affected females present epilepsy and mild-to-moderate intellectual disability that correlate with a mosaic phenotype that is intermediate in severity (Gleeson et al., 2000; Guerrini et al., 2003; Leventer et al.,

2001a). These patients present hypothalamic dysfunction, respiratory distress at birth, neonatal-onset epilepsy, severe hypotonia from birth, chronic diarrhea and deficient temperature regulation. Underdeveloped or ambiguous genitalia are found in this type of lissencephaly. Most of these patients show craniofacial dysmorphism including psychomotor delay and congenital microcephaly, and die before the age of 18 months (Dobyns et al., 1999; Kato et al., 2004; Kitamura et al., 2002; Ogata et al., 2000). Contrary to LIS1, missense mutations of DCX are more common than truncating mutations. DCX itself is not a +TIP but interacts with the +TIP LIS1, and they crosstalk during the development of cerebral cortex (Caspi et al., 2000). Moreover, DCX protein is characterized by the presence of two evolutionary conserved tandem repeats which constitute MT-binding modules (Francis et al., 1999; Gleeson et al., 1999). Each of these repeats binds to tubulin but both are necessary for MT polymerization and stabilization (Cierpicki et al., 2006; Horesh et al., 1999; Moores et al., 2004; Moores et al., 2006). Mutations in DCX linked to lissencephaly are clustered in the two MT-binding domains and most of them have been shown to impair MT polymerization both *in vitro* and *in vivo* (Sapir et al., 2000; Taylor et al., 2000). The N-terminal region of DCX binds only to MT polymers whereas the C-terminal domain binds to both unpolymerized tubulin and MTs. In line with this, missense mutations in the C-terminal MT-binding domain have been suggested to cause a less severe lissencephaly phenotype compared with those located in the N-terminal region (Leger et al., 2008). Deletions in one or more DCX exons have also been linked to lissencephaly (Haverfield et al., 2009).

DCX knockout mice present spindle orientation abnormalities, similar to the ones found in LIS1 mutants, which lead to moderate proliferation defects (Couillard-Despres et al., 2004). DCX might act together with LIS1 and dynein to mediate coupling between nucleus and centrosome (Tanaka et al., 2004). DCX or LIS1 overexpression increases migration rates, and DCX ectopic expression can rescue the coupling defect provoked by diminished levels of LIS1 or by dynein inhibition (Tanaka et al., 2004). DCX promotes neuronal intracellular transport (Deuel et al., 2006), probably by stabilizing MTs and thereby providing stable tracks for kinesin motors, directed toward the MT plus-end. It may also be involved in clathrin-vesicular trafficking (Friocourt et al., 2001).

DCX and LIS1 mutant mice present abnormal network activity and synaptogenesis in the hippocampus and cortex, thus providing useful models to study seizures and altered cognitive function in type I lissencephaly.

In summary, all the proteins involved in classical lissencephalies have been shown to regulate both radial and tangential migration, as well as cell proliferation and differentiation of neurons during development. Moreover, most of them are MT-associated proteins, and, in particular, +TIPs or +TIPs-interacting proteins, indicating that local regulation of MT dynamics plays key roles in the pathophysiology of type I lissencephaly.

LIS1 has also been found to be related to neurological disorders such as schizophrenia. LIS1, together with its partner NUDEL, interacts with DISC, a protein that confers susceptibility to schizophrenia (Mackie et al., 2007). DISC1 is involved in axon extension induced by neurotrophins. Moreover, truncated DISC1 causes defects in neurite elongation and neuronal positioning (Clapcote et al., 2007; Mackie et al., 2007; Taya et al., 2007). Neuronal migration might therefore play a role in certain psychiatric disorders (Kamiya et al., 2005; Ross et al., 2006; Tomita et al.).

5.2. p150^{Glued} Mutations and Motor Neuron Disorders

Several motor neuron disorders are linked to mutations in genes that encode proteins required for axonal transport and intracellular trafficking. Amyotrophic Lateral Sclerosis (ALS) is the most common motor neuron disease, characterized by death of motor neurons and progressive paralysis of muscles in adult age. Recently, several mutations in p150^{Glued} have been found in ALS patients (Munch et al., 2004). A mutation in p150^{Glued} also gives rise to the human neurodegenerative disease distal Spinal and Bulbar Muscular Atrophy (dSBMA) (Puls et al., 2005). A transgenic mouse model of dSBMA shows late-onset with slow progressive muscle weakness and non-shortened lifespan, consistent with the human phenotype (Chevalier-Larsen et al., 2008). Neurons of these mice present deficits in axonal caliber and neuromuscular junction (NMJ) integrity, signs of distal degeneration of motor neurons. In these mice, mutant p150^{Glued} causes neurodegeneration in the absence of significant changes in axonal transport. However, in other mouse models with mutations in p150^{Glued}, disruption of dynein/dynactin inhibits axonal transport in motor neurons causing late-onset progressive degeneration. These mice show elevated levels of dynamitin (a dynactin subunit, which causes dynactin complex disassembly when overexpressed) concomitantly with decreased strength and endurance as well as muscle denervation (Burkhardt et al., 1997; LaMonte et al., 2002). Some of the pathological mutations in p150^{Glued} involve a Gly to Ser substitution in the CAP-Gly motif, responsible for binding of p150^{Glued} to EB1 and MT plus-ends (Askham et al., 2002; Puls et al., 2005; Vaughan et al., 2002). Mutated p150^{Glued} binds less efficiently to MTs as well and presents a slightly disturbed dynactin function (Levy et al., 2006; Puls et al., 2005). Moreover, mutated p150^{Glued} forms aggregates that can induce neuronal cell death (Levy et al., 2006). Cytoplasmic dynein, which localizes to MT plus-ends through its interaction with p150^{Glued}, has also been linked to motor neuron disorders. Mutations in cytoplasmic dynein also lead to axonal transport defects and motor neuron degeneration (Hafezparast et al., 2003).

Other human disorders have been found to be associated to p150^{Glued}. Five different mutations in p150^{Glued} that affect its interaction with MTs have been linked to Perry Syndrome (Wider et al., 2010), clinically characterized by autosomal dominantly inherited parkinsonism, rapidly progressive depression, hypoventilation and weight loss (Wider and Wszolek, 2008). Interestingly, p150^{Glued} interacts with huntingtin-associated protein (HAP) and huntingtin (the protein mutated in Huntington´s disease) in a complex that may participate in dynein–dynactin-associated intracellular transport (Engelender et al., 1997; Li et al., 1998). In line with this, huntingtin controls neurotrophic support and survival of neurons by enhancing BDNF vesicular transport along MTs, indicating that the interaction with p150^{Glued} might be crucial in this process (Gauthier et al., 2004). Moreover, huntingtin is localized at spindle poles during mitosis and huntingtin knockdown in cells disrupts spindle orientation by mislocalizing p150^{Glued}, dynein, and the large nuclear mitotic apparatus NuMA protein (Godin et al., 2010). Hence, defects in the p150^{Glued}/Huntingtin/HAP interaction might be involved in Huntington disease.

+TIP motor proteins that associate to MT plus-ends and move towards the minus-ends are involved in motor-driven retrograde transport in axons. Malfunctioning of these proteins would lead to deficient cargo loading and disturbed retrograde transport. In summary, defects in different factors involved in axonal transport may be the main cause of motor neuron death, the hallmark of motor neuron disorders.

5.3. CLIP-115 Deficiency in Williams Syndrome

Williams Syndrome (WS) is a human syndrome characterized by cardiovascular disease, distinctive facial abnormalities, mild mental retardation, a specific cognitive profile, unique personality characteristics, growth and endocrine abnormalities (Morris, 1993). Patients usually present hypotonia and hyperextensible joints as well as feeding difficulties that often lead to failure to thrive in infancy (Morris, 1993). Among the genes involved in this syndrome is the one that codes for the brain-specific +TIP CLIP-115 (termed *CYLND2* gene) (Hoogenraad et al., 1998). Deletions in this gene seem to contribute to the motor coordination defects in the disease since patients with a smaller genomic deletion that does not affect the *CYLND2* gene, present better cognitive and motor coordination functions than typical WS patients with large deleted regions in the CLIP-115 gene (van Hagen et al., 2007). Studies performed in mice suggest that CLIP-115 is responsible for part of the observed neurological abnormalities in WS patients. CLIP-115 knockout mice present features reminiscent of WS, such as mild growth deficiency, brain abnormalities, hippocampal dysfunction and particular deficits in motor coordination (Hoogenraad et al., 2002). Behavioral and electrophysiological analysis of CLIP-115 deficient mice revealed defects in contextual memory, motor coordination and hippocampal long-term potentiation (LTP) (Hoogenraad et al., 2002). CLIP-115 mediates the interaction between MTs and specific organelles (De Zeeuw et al., 1997; Hoogenraad et al., 2000). CLIP-115-deficient cells exhibit increased levels of CLIP-170 and dynactin at the tips of growing MTs. Redistribution of these proteins may affect dynein motor regulation and this, together with the loss of CLIP-115-specific functions, may underlie neurological alterations in WS.

5.4. APC and Nervous System Disorders

Given the important roles of APC protein during nervous system development, its role in the onset of human nervous system disorders is under active investigation. In line with this, some single nucleotide polymorphisms (SNPs) have been found in the APC gene in association with Autism Spectrum Disorder (ASD) (Zhou et al., 2007). This might be related to the role of APC as part of a multiprotein complex required for anchoring neuroligin and neurexin at neuronal synapses *in vivo* (Rosenberg et al., 2010). APC complex also might provide retrograde signals that direct presynaptic active zones to develop.

Moreover, APC expression seems to be upregulated in astrocytes upon their activation by different types of pathological insults (Leroy et al., 2001). Astrocytoma and glioblastoma present high levels of APC. APC is expressed in neuronal cell bodies and dendrites both in control subjects and in patients with Alzheimer´s disease (AD). However, high APC levels are found in a proportion of fibrillary and glial fibrillary acidic protein-positive astrocytes in AD, mainly in astrocytes associated with β-amyloid (Aβ) deposits in senile plaques in AD. Aβ, and/or the inflammatory process associated with Aβ deposits, leads to the preferential increase of APC expression in astrocytes in AD (Leroy et al., 2001) .

Conclusion

MTs play key functions in developing and mature neurons. MTs not only provide structural tracks for motor-mediated directional transport but also contribute to neuronal polarization and differentiation in response to extracellular cues. +TIPs are proteins that specifically bind to the plus-ends of growing MTs. Different +TIPs contribute to the local regulation of MT dynamics in different processes such as neurite and axon elongation, synaptic spine formation or the correct localization of receptors at synapses. Increasing evidence indicate that correct functioning of some of these proteins seem to be crucial for the proper development and maintenance of nervous system. Some neurological and neurodegenerative human disorders have been associated with mutations in various +TIPs. Human genetics studies as well as mouse models, deficient in different +TIPs, will contribute to further elucidate the roles of +TIPs in nervous system.

References

Ahmad, F.J., W. Yu, F.J. McNally, and P.W. Baas. 1999. An essential role for katanin in severing microtubules in the neuron. *J Cell Biol*. 145:305-15.

Akhmanova, A., and C.C. Hoogenraad. 2005. Microtubule plus-end-tracking proteins: mechanisms and functions. *Curr Opin Cell Biol*. 17:47-54.

Akhmanova, A., C.C. Hoogenraad, K. Drabek, T. Stepanova, B. Dortland, T. Verkerk, W. Vermeulen, B.M. Burgering, C.I. De Zeeuw, F. Grosveld, and N. Galjart. 2001. Clasps are CLIP-115 and -170 associating proteins involved in the regional regulation of microtubule dynamics in motile fibroblasts. *Cell*. 104:923-35.

Akhmanova, A., and M.O. Steinmetz. 2008. Tracking the ends: a dynamic protein network controls the fate of microtubule tips. *Nat Rev Mol Cell Biol*. 9:309-22.

Al-Bassam, J., H. Kim, G. Brouhard, A. van Oijen, S.C. Harrison, and F. Chang. 2010. CLASP promotes microtubule rescue by recruiting tubulin dimers to the microtubule. *Dev Cell*. 19:245-58.

Amos, L.A., and D. Schlieper. 2005. Microtubules and maps. *Adv Protein Chem*. 71:257-98.

Aoki, K., and M.M. Taketo. 2007. Adenomatous polyposis coli (APC): a multi-functional tumor suppressor gene. *J Cell Sci*. 120:3327-35.

Arevalo, J.C., and M.V. Chao. 2005. Axonal growth: where neurotrophins meet Wnts. *Curr Opin Cell Biol*. 17:112-5.

Asada, N., and K. Sanada. 2010. LKB1-mediated spatial control of GSK3beta and adenomatous polyposis coli contributes to centrosomal forward movement and neuronal migration in the developing neocortex. *J Neurosci*. 30:8852-65.

Askham, J.M., K.T. Vaughan, H.V. Goodson, and E.E. Morrison. 2002. Evidence that an interaction between EB1 and p150(Glued) is required for the formation and maintenance of a radial microtubule array anchored at the centrosome. *Mol Biol Cell*. 13:3627-45.

Avila, J. 1990. Microtubule dynamics. *Faseb J*. 4:3284-90.

Avila, J., J. Dominguez, and J. Diaz-Nido. 1994. Regulation of microtubule dynamics by microtubule-associated protein expression and phosphorylation during neuronal development. *Int J Dev Biol*. 38:13-25.

Baas, P.W., and M.M. Black. 1990. Individual microtubules in the axon consist of domains that differ in both composition and stability. *J Cell Biol*. 111:495-509.

Baas, P.W., J.S. Deitch, M.M. Black, and G.A. Banker. 1988. Polarity orientation of microtubules in hippocampal neurons: uniformity in the axon and nonuniformity in the dendrite. *Proc Natl Acad Sci U S A*. 85:8335-9.

Baas, P.W., A. Karabay, and L. Qiang. 2005. Microtubules cut and run. *Trends Cell Biol*. 15:518-24.

Bayley, P.M., and S.R. Martin. 1991. Microtubule dynamic instability: some possible physical mechanisms and their implications. *Biochem Soc Trans*. 19:1023-8.

Berrueta, L., S.K. Kraeft, J.S. Tirnauer, S.C. Schuyler, L.B. Chen, D.E. Hill, D. Pellman, and B.E. Bierer. 1998. The adenomatous polyposis coli-binding protein EB1 is associated with cytoplasmic and spindle microtubules. *Proc Natl Acad Sci U S A*. 95:10596-601.

Bhat, R.V., J.M. Baraban, R.C. Johnson, B.A. Eipper, and R.E. Mains. 1994. High levels of expression of the tumor suppressor gene APC during development of the rat central nervous system. *J Neurosci*. 14:3059-71.

Bieling, P., S. Kandels-Lewis, I.A. Telley, J. van Dijk, C. Janke, and T. Surrey. 2008. CLIP-170 tracks growing microtubule ends by dynamically recognizing composite EB1/tubulin-binding sites. *J Cell Biol*. 183:1223-33.

Bieling, P., L. Laan, H. Schek, E.L. Munteanu, L. Sandblad, M. Dogterom, D. Brunner, and T. Surrey. 2007. Reconstitution of a microtubule plus-end tracking system in vitro. *Nature*. 450:1100-5.

Bix, G.J., and G.D. Clark. 1998. Platelet-activating factor receptor stimulation disrupts neuronal migration In vitro. *J Neurosci*. 18:307-18.

Bradke, F., and C.G. Dotti. 1999. The role of local actin instability in axon formation. *Science*. 283:1931-4.

Brown, A., Y. Li, T. Slaughter, and M.M. Black. 1993. Composite microtubules of the axon: quantitative analysis of tyrosinated and acetylated tubulin along individual axonal microtubules. *J Cell Sci*. 104 (Pt 2):339-52.

Burbank, K.S., and T.J. Mitchison. 2006. Microtubule dynamic instability. *Curr Biol*. 16:R516-7.

Burkhardt, J.K., C.J. Echeverri, T. Nilsson, and R.B. Vallee. 1997. Overexpression of the dynamitin (p50) subunit of the dynactin complex disrupts dynein-dependent maintenance of membrane organelle distribution. *J Cell Biol*. 139:469-84.

Cahana, A., T. Escamez, R.S. Nowakowski, N.L. Hayes, M. Giacobini, A. von Holst, O. Shmueli, T. Sapir, S.K. McConnell, W. Wurst, S. Martinez, and O. Reiner. 2001. Targeted mutagenesis of Lis1 disrupts cortical development and LIS1 homodimerization. *Proc Natl Acad Sci U S A*. 98:6429-34.

Caspi, M., R. Atlas, A. Kantor, T. Sapir, and O. Reiner. 2000. Interaction between LIS1 and doublecortin, two lissencephaly gene products. *Hum Mol Genet*. 9:2205-13.

Cassimeris, L.U., R.A. Walker, N.K. Pryer, and E.D. Salmon. 1987. Dynamic instability of microtubules. *Bioessays*. 7:149-54.

Cierpicki, T., M.H. Kim, D.R. Cooper, U. Derewenda, J.H. Bushweller, and Z.S. Derewenda. 2006. The DC-module of doublecortin: dynamics, domain boundaries, and functional implications. *Proteins*. 64:874-82.

Clagett-Dame, M., E.M. McNeill, and P.D. Muley. 2006. Role of all-trans retinoic acid in neurite outgrowth and axonal elongation. *J Neurobiol*. 66:739-56.

Clapcote, S.J., T.V. Lipina, J.K. Millar, S. Mackie, S. Christie, F. Ogawa, J.P. Lerch, K. Trimble, M. Uchiyama, Y. Sakuraba, H. Kaneda, T. Shiroishi, M.D. Houslay, R.M. Henkelman, J.G. Sled, Y. Gondo, D.J. Porteous, and J.C. Roder. 2007. Behavioral phenotypes of Disc1 missense mutations in mice. *Neuron*. 54:387-402.

Clark, G.D. 2001. Cerebral gyral dysplasias: molecular genetics and cell biology. *Curr Opin Neurol*. 14:157-62.

Conde, C., and A. Caceres. 2009. Microtubule assembly, organization and dynamics in axons and dendrites. *Nat Rev Neurosci*. 10:319-32.

Coquelle, F.M., M. Caspi, F.P. Cordelieres, J.P. Dompierre, D.L. Dujardin, C. Koifman, P. Martin, C.C. Hoogenraad, A. Akhmanova, N. Galjart, J.R. De Mey, and O. Reiner. 2002. LIS1, CLIP-170's key to the dynein/dynactin pathway. *Mol Cell Biol*. 22:3089-102.

Coquelle, F.M., B. Vitre, and I. Arnal. 2009. Structural basis of EB1 effects on microtubule dynamics. *Biochem Soc Trans*. 37:997-1001.

Couillard-Despres, S., G. Uyanik, S. Ploetz, C. Karl, H. Koch, J. Winkler, and L. Aigner. 2004. Mitotic impairment by doublecortin is diminished by doublecortin mutations found in patients. *Neurogenetics*. 5:83-93.

Coy, J.F., S. Wiemann, I. Bechmann, D. Bachner, R. Nitsch, O. Kretz, H. Christiansen, and A. Poustka. 2002. Pore membrane and/or filament interacting like protein 1 (POMFIL1) is predominantly expressed in the nervous system and encodes different protein isoforms. *Gene*. 290:73-94.

Culver-Hanlon, T.L., S.A. Lex, A.D. Stephens, N.J. Quintyne, and S.J. King. 2006. A microtubule-binding domain in dynactin increases dynein processivity by skating along microtubules. *Nat Cell Biol*. 8:264-70.

Chevalier-Larsen, E.S., K.E. Wallace, C.R. Pennise, and E.L. Holzbaur. 2008. Lysosomal proliferation and distal degeneration in motor neurons expressing the G59S mutation in the p150Glued subunit of dynactin. *Hum Mol Genet*. 17:1946-55.

Dambska, M., K. Wisniewski, and J.H. Sher. 1983. Lissencephaly: two distinct clinico-pathological types. *Brain Dev*. 5:302-10.

de Anda, F.C., G. Pollarolo, J.S. Da Silva, P.G. Camoletto, F. Feiguin, and C.G. Dotti. 2005. Centrosome localization determines neuronal polarity. *Nature*. 436:704-8.

De Zeeuw, C.I., C.C. Hoogenraad, E. Goedknegt, E. Hertzberg, A. Neubauer, F. Grosveld, and N. Galjart. 1997. CLIP-115, a novel brain-specific cytoplasmic linker protein, mediates the localization of dendritic lamellar bodies. *Neuron*. 19:1187-99.

Deuel, T.A., J.S. Liu, J.C. Corbo, S.Y. Yoo, L.B. Rorke-Adams, and C.A. Walsh. 2006. Genetic interactions between doublecortin and doublecortin-like kinase in neuronal migration and axon outgrowth. *Neuron*. 49:41-53.

Dobashi, Y., R.N. Bhattacharjee, K. Toyoshima, and T. Akiyama. 1996. Upregulation of the APC gene product during neuronal differentiation of rat pheochromocytoma PC12 cells. *Biochem Biophys Res Commun*. 224:479-83.

Dobyns, W.B. 2010. The clinical patterns and molecular genetics of lissencephaly and subcortical band heterotopia. *Epilepsia*. 51 Suppl 1:5-9.

Dobyns, W.B., E. Berry-Kravis, N.J. Havernick, K.R. Holden, and D. Viskochil. 1999. X-linked lissencephaly with absent corpus callosum and ambiguous genitalia. *Am J Med Genet*. 86:331-7.

Dobyns, W.B., O. Reiner, R. Carrozzo, and D.H. Ledbetter. 1993. Lissencephaly. A human brain malformation associated with deletion of the LIS1 gene located at chromosome 17p13. *Jama*. 270:2838-42.

Dotti, C.G., and G.A. Banker. 1987. Experimentally induced alteration in the polarity of developing neurons. *Nature*. 330:254-6.

Dotti, C.G., C.A. Sullivan, and G.A. Banker. 1988. The establishment of polarity by hippocampal neurons in culture. *J Neurosci*. 8:1454-68.

Drabek, K., M. van Ham, T. Stepanova, K. Draegestein, R. van Horssen, C.L. Sayas, A. Akhmanova, T. Ten Hagen, R. Smits, R. Fodde, F. Grosveld, and N. Galjart. 2006. Role of CLASP2 in microtubule stabilization and the regulation of persistent motility. *Curr Biol*. 16:2259-64.

Dujardin, D.L., L.E. Barnhart, S.A. Stehman, E.R. Gomes, G.G. Gundersen, and R.B. Vallee. 2003. A role for cytoplasmic dynein and LIS1 in directed cell movement. *J Cell Biol*. 163:1205-11.

Efimov, A., A. Kharitonov, N. Efimova, J. Loncarek, P.M. Miller, N. Andreyeva, P. Gleeson, N. Galjart, A.R. Maia, I.X. McLeod, J.R. Yates, 3rd, H. Maiato, A. Khodjakov, A. Akhmanova, and I. Kaverina. 2007. Asymmetric CLASP-dependent nucleation of noncentrosomal microtubules at the trans-Golgi network. *Dev Cell*. 12:917-30.

Engelender, S., A.H. Sharp, V. Colomer, M.K. Tokito, A. Lanahan, P. Worley, E.L. Holzbaur, and C.A. Ross. 1997. Huntingtin-associated protein 1 (HAP1) interacts with the p150Glued subunit of dynactin. *Hum Mol Genet*. 6:2205-12.

Erck, C., R. Frank, and J. Wehland. 2000. Tubulin-tyrosine ligase, a long-lasting enigma. *Neurochem Res*. 25:5-10.

Etienne-Manneville, S., and A. Hall. 2003. Cdc42 regulates GSK-3beta and adenomatous polyposis coli to control cell polarity. *Nature*. 421:753-6.

Fleck, M.W., S. Hirotsune, M.J. Gambello, E. Phillips-Tansey, G. Suares, R.F. Mervis, A. Wynshaw-Boris, and C.J. McBain. 2000. Hippocampal abnormalities and enhanced excitability in a murine model of human lissencephaly. *J Neurosci*. 20:2439-50.

Francis, F., A. Koulakoff, D. Boucher, P. Chafey, B. Schaar, M.C. Vinet, G. Friocourt, N. McDonnell, O. Reiner, A. Kahn, S.K. McConnell, Y. Berwald-Netter, P. Denoulet, and J. Chelly. 1999. Doublecortin is a developmentally regulated, microtubule-associated protein expressed in migrating and differentiating neurons. *Neuron*. 23:247-56.

Frickey, T., and A.N. Lupas. 2004. Phylogenetic analysis of AAA proteins. *J Struct Biol*. 146:2-10.

Friocourt, G., P. Chafey, P. Billuart, A. Koulakoff, M.C. Vinet, B.T. Schaar, S.K. McConnell, F. Francis, and J. Chelly. 2001. Doublecortin interacts with mu subunits of clathrin adaptor complexes in the developing nervous system. *Mol Cell Neurosci*. 18:307-19.

Friocourt, G., P. Marcorelles, P. Saugier-Veber, M.L. Quille, S. Marret, and A. Laquerriere. 2011. Role of cytoskeletal abnormalities in the neuropathology and pathophysiology of type I lissencephaly. *Acta Neuropathol*. 121:149-70.

Fukata, M., T. Watanabe, J. Noritake, M. Nakagawa, M. Yamaga, S. Kuroda, Y. Matsuura, A. Iwamatsu, F. Perez, and K. Kaibuchi. 2002. Rac1 and Cdc42 capture microtubules through IQGAP1 and CLIP-170. *Cell*. 109:873-85.

Fukushima, N., D. Furuta, Y. Hidaka, R. Moriyama, and T. Tsujiuchi. 2009. Post-translational modifications of tubulin in the nervous system. *J Neurochem*. 109:683-93.

Galjart, N. 2010. Plus-end-tracking proteins and their interactions at microtubule ends. *Curr Biol*. 20:R528-37.

Galjart, N., and F. Perez. 2003. A plus-end raft to control microtubule dynamics and function. *Curr Opin Cell Biol*. 15:48-53.

Gallo, G. 1998. Involvement of microtubules in the regulation of neuronal growth cone morphologic remodeling. *J Neurobiol*. 35:121-40.

Gambello, M.J., D.L. Darling, J. Yingling, T. Tanaka, J.G. Gleeson, and A. Wynshaw-Boris. 2003. Multiple dose-dependent effects of Lis1 on cerebral cortical development. *J Neurosci*. 23:1719-29.

Gauthier, L.R., B.C. Charrin, M. Borrell-Pages, J.P. Dompierre, H. Rangone, F.P. Cordelieres, J. De Mey, M.E. MacDonald, V. Lessmann, S. Humbert, and F. Saudou. 2004. Huntingtin controls neurotrophic support and survival of neurons by enhancing BDNF vesicular transport along microtubules. *Cell*. 118:127-38.

Geraldo, S., and P.R. Gordon-Weeks. 2009. Cytoskeletal dynamics in growth-cone steering. *J Cell Sci*. 122:3595-604.

Geraldo, S., U.K. Khanzada, M. Parsons, J.K. Chilton, and P.R. Gordon-Weeks. 2008. Targeting of the F-actin-binding protein drebrin by the microtubule plus-tip protein EB3 is required for neuritogenesis. *Nat Cell Biol*. 10:1181-9.

Gleeson, J.G., P.T. Lin, L.A. Flanagan, and C.A. Walsh. 1999. Doublecortin is a microtubule-associated protein and is expressed widely by migrating neurons. *Neuron*. 23:257-71.

Gleeson, J.G., R.F. Luo, P.E. Grant, R. Guerrini, P.R. Huttenlocher, M.J. Berg, S. Ricci, R. Cusmai, J.W. Wheless, S. Berkovic, I. Scheffer, W.B. Dobyns, and C.A. Walsh. 2000. Genetic and neuroradiological heterogeneity of double cortex syndrome. *Ann Neurol*. 47:265-9.

Godin, J.D., K. Colombo, M. Molina-Calavita, G. Keryer, D. Zala, B.C. Charrin, P. Dietrich, M.L. Volvert, F. Guillemot, I. Dragatsis, Y. Bellaiche, F. Saudou, L. Nguyen, and S. Humbert. 2010. Huntingtin is required for mitotic spindle orientation and mammalian neurogenesis. *Neuron*. 67:392-406.

Gordon-Weeks, P.R. 2004. Microtubules and growth cone function. *J Neurobiol*. 58:70-83.

Gouveia, S.M., and A. Akhmanova. 2010. Cell and molecular biology of microtubule plus end tracking proteins: end binding proteins and their partners. *Int Rev Cell Mol Biol*. 285:1-74.

Grabham, P.W., G.E. Seale, M. Bennecib, D.J. Goldberg, and R.B. Vallee. 2007. Cytoplasmic dynein and LIS1 are required for microtubule advance during growth cone remodeling and fast axonal outgrowth. *J Neurosci*. 27:5823-34.

Gu, C., W. Zhou, M.A. Puthenveedu, M. Xu, Y.N. Jan, and L.Y. Jan. 2006. The microtubule plus-end tracking protein EB1 is required for Kv1 voltage-gated K+ channel axonal targeting. *Neuron*. 52:803-16.

Gu, J., B.L. Firestein, and J.Q. Zheng. 2008. Microtubules in dendritic spine development. *J Neurosci*. 28:12120-4.

Gu, Y., and C. Gu. 2010. Dynamics of Kv1 channel transport in axons. *PLoS One*. 5:e11931.

Guerrini, R., and C. Marini. 2006. Genetic malformations of cortical development. *Exp Brain Res*. 173:322-33.

Guerrini, R., F. Moro, E. Andermann, E. Hughes, D. D'Agostino, R. Carrozzo, A. Bernasconi, F. Flinter, L. Parmeggiani, A. Volzone, E. Parrini, D. Mei, J.M. Jarosz, R.G. Morris, P. Pratt, G. Tortorella, F. Dubeau, F. Andermann, W.B. Dobyns, and S. Das. 2003.

Nonsyndromic mental retardation and cryptogenic epilepsy in women with doublecortin gene mutations. *Ann Neurol*. 54:30-7.

Hafezparast, M., R. Klocke, C. Ruhrberg, A. Marquardt, A. Ahmad-Annuar, S. Bowen, G. Lalli, A.S. Witherden, H. Hummerich, S. Nicholson, P.J. Morgan, R. Oozageer, J.V. Priestley, S. Averill, V.R. King, S. Ball, J. Peters, T. Toda, A. Yamamoto, Y. Hiraoka, M. Augustin, D. Korthaus, S. Wattler, P. Wabnitz, C. Dickneite, S. Lampel, F. Boehme, G. Peraus, A. Popp, M. Rudelius, J. Schlegel, H. Fuchs, M. Hrabe de Angelis, G. Schiavo, D.T. Shima, A.P. Russ, G. Stumm, J.E. Martin, and E.M. Fisher. 2003. Mutations in dynein link motor neuron degeneration to defects in retrograde transport. *Science*. 300:808-12.

Hannak, E., and R. Heald. 2006. Xorbit/CLASP links dynamic microtubules to chromosomes in the Xenopus meiotic spindle. *J Cell Biol*. 172:19-25.

Hatten, M.E. 2005. LIS-less neurons don't even make it to the starting gate. *J Cell Biol*. 170:867-71.

Haverfield, E.V., A.J. Whited, K.S. Petras, W.B. Dobyns, and S. Das. 2009. Intragenic deletions and duplications of the LIS1 and DCX genes: a major disease-causing mechanism in lissencephaly and subcortical band heterotopia. *Eur J Hum Genet*. 17:911-8.

Hayashi, I., A. Wilde, T.K. Mal, and M. Ikura. 2005. Structural basis for the activation of microtubule assembly by the EB1 and p150Glued complex. *Mol Cell*. 19:449-60.

Hebbar, S., A.M. Guillotte, M.T. Mesngon, Q. Zhou, A. Wynshaw-Boris, and D.S. Smith. 2008. Genetic enhancement of the Lis1+/- phenotype by a heterozygous mutation in the adenomatous polyposis coli gene. *Dev Neurosci*. 30:157-70.

Herman, T.E., and M.J. Siegel. 2008. Miller--Dieker syndrome, type 1 lissencephaly. *J Perinatol*. 28:313-5.

Hirokawa, N., and Y. Noda. 2008. Intracellular transport and kinesin superfamily proteins, KIFs: structure, function, and dynamics. *Physiol Rev*. 88:1089-118.

Hirotsune, S., M.W. Fleck, M.J. Gambello, G.J. Bix, A. Chen, G.D. Clark, D.H. Ledbetter, C.J. McBain, and A. Wynshaw-Boris. 1998. Graded reduction of Pafah1b1 (Lis1) activity results in neuronal migration defects and early embryonic lethality. *Nat Genet*. 19:333-9.

Honnappa, S., S.M. Gouveia, A. Weisbrich, F.F. Damberger, N.S. Bhavesh, H. Jawhari, I. Grigoriev, F.J. van Rijssel, R.M. Buey, A. Lawera, I. Jelesarov, F.K. Winkler, K. Wuthrich, A. Akhmanova, and M.O. Steinmetz. 2009. An EB1-binding motif acts as a microtubule tip localization signal. *Cell*. 138:366-76.

Honnappa, S., C.M. John, D. Kostrewa, F.K. Winkler, and M.O. Steinmetz. 2005. Structural insights into the EB1-APC interaction. *Embo J*. 24:261-9.

Honnappa, S., O. Okhrimenko, R. Jaussi, H. Jawhari, I. Jelesarov, F.K. Winkler, and M.O. Steinmetz. 2006. Key interaction modes of dynamic +TIP networks. *Mol Cell*. 23:663-71.

Hoogenraad, C.C., A. Akhmanova, F. Grosveld, C.I. De Zeeuw, and N. Galjart. 2000. Functional analysis of CLIP-115 and its binding to microtubules. *J Cell Sci*. 113 (Pt 12):2285-97.

Hoogenraad, C.C., B.H. Eussen, A. Langeveld, R. van Haperen, S. Winterberg, C.H. Wouters, F. Grosveld, C.I. De Zeeuw, and N. Galjart. 1998. The murine CYLN2 gene: genomic organization, chromosome localization, and comparison to the human gene that is located within the 7q11.23 Williams syndrome critical region. *Genomics*. 53:348-58.

Hoogenraad, C.C., B. Koekkoek, A. Akhmanova, H. Krugers, B. Dortland, M. Miedema, A. van Alphen, W.M. Kistler, M. Jaegle, M. Koutsourakis, N. Van Camp, M. Verhoye, A. van der Linden, I. Kaverina, F. Grosveld, C.I. De Zeeuw, and N. Galjart. 2002. Targeted mutation of Cyln2 in the Williams syndrome critical region links CLIP-115 haploinsufficiency to neurodevelopmental abnormalities in mice. *Nat Genet.* 32:116-27.

Horesh, D., T. Sapir, F. Francis, S.G. Wolf, M. Caspi, M. Elbaum, J. Chelly, and O. Reiner. 1999. Doublecortin, a stabilizer of microtubules. *Hum Mol Genet.* 8:1599-610.

Hu, X., C. Viesselmann, S. Nam, E. Merriam, and E.W. Dent. 2008. Activity-dependent dynamic microtubule invasion of dendritic spines. *J Neurosci.* 28:13094-105.

Hunter, A.W., and L. Wordeman. 2000. How motor proteins influence microtubule polymerization dynamics. *J Cell Sci.* 113 Pt 24:4379-89.

Ikeda, S., M. Kishida, Y. Matsuura, H. Usui, and A. Kikuchi. 2000. GSK-3beta-dependent phosphorylation of adenomatous polyposis coli gene product can be modulated by beta-catenin and protein phosphatase 2A complexed with Axin. *Oncogene.* 19:537-45.

Inoue, Y.H., M.S. Savoian, T. Suzuki, E. Mathe, M.T. Yamamoto, and D.M. Glover. 2004. Mutations in orbit/mast reveal that the central spindle is comprised of two microtubule populations, those that initiate cleavage and those that propagate furrow ingression. *J Cell Biol.* 166:49-60.

Isumi, H., S. Takashima, A. Kakita, M. Yamada, K. Ikeda, and M. Mizuguchi. 1997. Expression of the LIS-1 gene product in brain anomalies with a migration disorder. *Pediatr Neurol.* 16:42-4.

Ivaniutsin, U., Y. Chen, J.O. Mason, D.J. Price, and T. Pratt. 2009. Adenomatous polyposis coli is required for early events in the normal growth and differentiation of the developing cerebral cortex. *Neural Dev.* 4:3.

Jaworski, J., C.C. Hoogenraad, and A. Akhmanova. 2008. Microtubule plus-end tracking proteins in differentiated mammalian cells. *Int J Biochem Cell Biol.* 40:619-37.

Jaworski, J., L.C. Kapitein, S.M. Gouveia, B.R. Dortland, P.S. Wulf, I. Grigoriev, P. Camera, S.A. Spangler, P. Di Stefano, J. Demmers, H. Krugers, P. Defilippi, A. Akhmanova, and C.C. Hoogenraad. 2009. Dynamic microtubules regulate dendritic spine morphology and synaptic plasticity. *Neuron.* 61:85-100.

Jimbo, T., Y. Kawasaki, R. Koyama, R. Sato, S. Takada, K. Haraguchi, and T. Akiyama. 2002. Identification of a link between the tumour suppressor APC and the kinesin superfamily. *Nat Cell Biol.* 4:323-7.

Kalil, K., and E.W. Dent. 2005. Touch and go: guidance cues signal to the growth cone cytoskeleton. *Curr Opin Neurobiol.* 15:521-6.

Kamiya, A., K. Kubo, T. Tomoda, M. Takaki, R. Youn, Y. Ozeki, N. Sawamura, U. Park, C. Kudo, M. Okawa, C.A. Ross, M.E. Hatten, K. Nakajima, and A. Sawa. 2005. A schizophrenia-associated mutation of DISC1 perturbs cerebral cortex development. *Nat Cell Biol.* 7:1167-78.

Kapitein, L.C., K.W. Yau, S.M. Gouveia, W.A. van der Zwan, P.S. Wulf, N. Keijzer, J. Demmers, J. Jaworski, A. Akhmanova, and C.C. Hoogenraad. 2011. NMDA Receptor Activation Suppresses Microtubule Growth and Spine Entry. *J Neurosci.* 31:8194-209.

Karabay, A., W. Yu, J.M. Solowska, D.H. Baird, and P.W. Baas. 2004. Axonal growth is sensitive to the levels of katanin, a protein that severs microtubules. *J Neurosci.* 24:5778-88.

Kato, M., S. Das, K. Petras, K. Kitamura, K. Morohashi, D.N. Abuelo, M. Barr, D. Bonneau, A.F. Brady, N.J. Carpenter, K.L. Cipero, F. Frisone, T. Fukuda, R. Guerrini, E. Iida, M. Itoh, A.F. Lewanda, Y. Nanba, A. Oka, V.K. Proud, P. Saugier-Veber, S.L. Schelley, A. Selicorni, R. Shaner, M. Silengo, F. Stewart, N. Sugiyama, J. Toyama, A. Toutain, A.L. Vargas, M. Yanazawa, E.H. Zackai, and W.B. Dobyns. 2004. Mutations of ARX are associated with striking pleiotropy and consistent genotype-phenotype correlation. *Hum Mutat*. 23:147-59.

Kawasaki, Y., T. Senda, T. Ishidate, R. Koyama, T. Morishita, Y. Iwayama, O. Higuchi, and T. Akiyama. 2000. Asef, a link between the tumor suppressor APC and G-protein signaling. *Science*. 289:1194-7.

Kholmanskikh, S.S., H.B. Koeller, A. Wynshaw-Boris, T. Gomez, P.C. Letourneau, and M.E. Ross. 2006. Calcium-dependent interaction of Lis1 with IQGAP1 and Cdc42 promotes neuronal motility. *Nat Neurosci*. 9:50-7.

Kim, H., S.C. Ling, G.C. Rogers, C. Kural, P.R. Selvin, S.L. Rogers, and V.I. Gelfand. 2007. Microtubule binding by dynactin is required for microtubule organization but not cargo transport. *J Cell Biol*. 176:641-51.

Kim, M.H., D.R. Cooper, A. Oleksy, Y. Devedjiev, U. Derewenda, O. Reiner, J. Otlewski, and Z.S. Derewenda. 2004. The structure of the N-terminal domain of the product of the lissencephaly gene Lis1 and its functional implications. *Structure*. 12:987-98.

Kimura, T., H. Watanabe, A. Iwamatsu, and K. Kaibuchi. 2005. Tubulin and CRMP-2 complex is transported via Kinesin-1. *J Neurochem*. 93:1371-82.

King, S.J., and T.A. Schroer. 2000. Dynactin increases the processivity of the cytoplasmic dynein motor. *Nat Cell Biol*. 2:20-4.

Kishi, M., T.T. Kummer, S.J. Eglen, and J.R. Sanes. 2005. LL5beta: a regulator of postsynaptic differentiation identified in a screen for synaptically enriched transcripts at the neuromuscular junction. *J Cell Biol*. 169:355-66.

Kita, K., T. Wittmann, I.S. Nathke, and C.M. Waterman-Storer. 2006. Adenomatous polyposis coli on microtubule plus ends in cell extensions can promote microtubule net growth with or without EB1. *Mol Biol Cell*. 17:2331-45.

Kitamura, K., M. Yanazawa, N. Sugiyama, H. Miura, A. Iizuka-Kogo, M. Kusaka, K. Omichi, R. Suzuki, Y. Kato-Fukui, K. Kamiirisa, M. Matsuo, S. Kamijo, M. Kasahara, H. Yoshioka, T. Ogata, T. Fukuda, I. Kondo, M. Kato, W.B. Dobyns, M. Yokoyama, and K. Morohashi. 2002. Mutation of ARX causes abnormal development of forebrain and testes in mice and X-linked lissencephaly with abnormal genitalia in humans. *Nat Genet*. 32:359-69.

Koester, M.P., O. Muller, and G.E. Pollerberg. 2007. Adenomatous polyposis coli is differentially distributed in growth cones and modulates their steering. *J Neurosci*. 27:12590-600.

Komarova, Y., C.O. De Groot, I. Grigoriev, S.M. Gouveia, E.L. Munteanu, J.M. Schober, S. Honnappa, R.M. Buey, C.C. Hoogenraad, M. Dogterom, G.G. Borisy, M.O. Steinmetz, and A. Akhmanova. 2009. Mammalian end binding proteins control persistent microtubule growth. *J Cell Biol*. 184:691-706.

Komarova, Y.A., A.S. Akhmanova, S. Kojima, N. Galjart, and G.G. Borisy. 2002. Cytoplasmic linker proteins promote microtubule rescue in vivo. *J Cell Biol*. 159:589-99.

Kuchelmeister, K., M. Bergmann, and F. Gullotta. 1993. Neuropathology of lissencephalies. *Childs Nerv Syst*. 9:394-9.

Kumar, P., K.S. Lyle, S. Gierke, A. Matov, G. Danuser, and T. Wittmann. 2009. GSK3beta phosphorylation modulates CLASP-microtubule association and lamella microtubule attachment. *J Cell Biol*. 184:895-908.

LaMonte, B.H., K.E. Wallace, B.A. Holloway, S.S. Shelly, J. Ascano, M. Tokito, T. Van Winkle, D.S. Howland, and E.L. Holzbaur. 2002. Disruption of dynein/dynactin inhibits axonal transport in motor neurons causing late-onset progressive degeneration. *Neuron*. 34:715-27.

Lansbergen, G., I. Grigoriev, Y. Mimori-Kiyosue, T. Ohtsuka, S. Higa, I. Kitajima, J. Demmers, N. Galjart, A.B. Houtsmuller, F. Grosveld, and A. Akhmanova. 2006. CLASPs attach microtubule plus ends to the cell cortex through a complex with LL5beta. *Dev Cell*. 11:21-32.

Lansbergen, G., Y. Komarova, M. Modesti, C. Wyman, C.C. Hoogenraad, H.V. Goodson, R.P. Lemaitre, D.N. Drechsel, E. van Munster, T.W. Gadella, Jr., F. Grosveld, N. Galjart, G.G. Borisy, and A. Akhmanova. 2004. Conformational changes in CLIP-170 regulate its binding to microtubules and dynactin localization. *J Cell Biol*. 166:1003-14.

Lee, H., U. Engel, J. Rusch, S. Scherrer, K. Sheard, and D. Van Vactor. 2004. The microtubule plus end tracking protein Orbit/MAST/CLASP acts downstream of the tyrosine kinase Abl in mediating axon guidance. *Neuron*. 42:913-26.

Lee, H.S., Y.A. Komarova, E.S. Nadezhdina, R. Anjum, J.G. Peloquin, J.M. Schober, O. Danciu, J. van Haren, N. Galjart, S.P. Gygi, A. Akhmanova, and G.G. Borisy. 2010. Phosphorylation controls autoinhibition of cytoplasmic linker protein-170. *Mol Biol Cell*. 21:2661-73.

Leger, P.L., I. Souville, N. Boddaert, C. Elie, J.M. Pinard, P. Plouin, M.L. Moutard, V. des Portes, H. Van Esch, S. Joriot, J.L. Renard, J. Chelly, F. Francis, C. Beldjord, and N. Bahi-Buisson. 2008. The location of DCX mutations predicts malformation severity in X-linked lissencephaly. *Neurogenetics*. 9:277-85.

Lemos, C.L., P. Sampaio, H. Maiato, M. Costa, L.V. Omel'yanchuk, V. Liberal, and C.E. Sunkel. 2000. Mast, a conserved microtubule-associated protein required for bipolar mitotic spindle organization. *Embo J*. 19:3668-82.

Leroy, K., C. Duyckaerts, L. Bovekamp, O. Muller, B.H. Anderton, and J.P. Brion. 2001. Increase of adenomatous polyposis coli immunoreactivity is a marker of reactive astrocytes in Alzheimer's disease and in other pathological conditions. *Acta Neuropathol*. 102:1-10.

Leterrier, C., H. Vacher, M.P. Fache, S.A. d'Ortoli, F. Castets, A. Autillo-Touati, and B. Dargent. 2010. End-binding proteins EB3 and EB1 link microtubules to ankyrin G in the axon initial segment. *Proc Natl Acad Sci U S A*. 108:8826-31.

Letourneau, P.C., and A.H. Ressler. 1984. Inhibition of neurite initiation and growth by taxol. *J Cell Biol*. 98:1355-62.

Leventer, R.J., C. Cardoso, D.H. Ledbetter, and W.B. Dobyns. 2001a. LIS1 missense mutations cause milder lissencephaly phenotypes including a child with normal IQ. *Neurology*. 57:416-22.

Leventer, R.J., C. Cardoso, D.H. Ledbetter, and W.B. Dobyns. 2001b. LIS1: from cortical malformation to essential protein of cellular dynamics. *Trends Neurosci*. 24:489-92.

Levy, J.R., C.J. Sumner, J.P. Caviston, M.K. Tokito, S. Ranganathan, L.A. Ligon, K.E. Wallace, B.H. LaMonte, G.G. Harmison, I. Puls, K.H. Fischbeck, and E.L. Holzbaur.

2006. A motor neuron disease-associated mutation in p150Glued perturbs dynactin function and induces protein aggregation. *J Cell Biol.* 172:733-45.

Li, S.H., C.A. Gutekunst, S.M. Hersch, and X.J. Li. 1998. Interaction of huntingtin-associated protein with dynactin P150Glued. *J Neurosci.* 18:1261-9.

Liu, Z., R. Steward, and L. Luo. 2000. Drosophila Lis1 is required for neuroblast proliferation, dendritic elaboration and axonal transport. *Nat Cell Biol.* 2:776-83.

Lo Nigro, C., C.S. Chong, A.C. Smith, W.B. Dobyns, R. Carrozzo, and D.H. Ledbetter. 1997. Point mutations and an intragenic deletion in LIS1, the lissencephaly causative gene in isolated lissencephaly sequence and Miller-Dieker syndrome. *Hum Mol Genet.* 6:157-64.

Lomakin, A.J., I. Semenova, I. Zaliapin, P. Kraikivski, E. Nadezhdina, B.M. Slepchenko, A. Akhmanova, and V. Rodionov. 2009. CLIP-170-dependent capture of membrane organelles by microtubules initiates minus-end directed transport. *Dev Cell.* 17:323-33.

LoTurco, J.J., and J. Bai. 2006. The multipolar stage and disruptions in neuronal migration. *Trends Neurosci.* 29:407-13.

Lowery, L.A., and D. Van Vactor. 2009. The trip of the tip: understanding the growth cone machinery. *Nat Rev Mol Cell Biol.* 10:332-43.

Mackie, S., J.K. Millar, and D.J. Porteous. 2007. Role of DISC1 in neural development and schizophrenia. *Curr Opin Neurobiol.* 17:95-102.

Maes, T., A. Barcelo, and C. Buesa. 2002. Neuron navigator: a human gene family with homology to unc-53, a cell guidance gene from Caenorhabditis elegans. *Genomics.* 80:21-30.

Maffini, S., A.R. Maia, A.L. Manning, Z. Maliga, A.L. Pereira, M. Junqueira, A. Shevchenko, A. Hyman, J.R. Yates, 3rd, N. Galjart, D.A. Compton, and H. Maiato. 2009. Motor-independent targeting of CLASPs to kinetochores by CENP-E promotes microtubule turnover and poleward flux. *Curr Biol.* 19:1566-72.

Maiato, H., E.A. Fairley, C.L. Rieder, J.R. Swedlow, C.E. Sunkel, and W.C. Earnshaw. 2003. Human CLASP1 is an outer kinetochore component that regulates spindle microtubule dynamics. *Cell.* 113:891-904.

Maiato, H., P. Sampaio, C.L. Lemos, J. Findlay, M. Carmena, W.C. Earnshaw, and C.E. Sunkel. 2002. MAST/Orbit has a role in microtubule-kinetochore attachment and is essential for chromosome alignment and maintenance of spindle bipolarity. *J Cell Biol.* 157:749-60.

Marcorelles, P., A. Laquerriere, C. Adde-Michel, S. Marret, P. Saugier-Veber, C. Beldjord, and G. Friocourt. 2010. Evidence for tangential migration disturbances in human lissencephaly resulting from a defect in LIS1, DCX and ARX genes. *Acta Neuropathol.* 120:503-15.

Marin, O., and J.L. Rubenstein. 2003. Cell migration in the forebrain. *Annu Rev Neurosci.* 26:441-83.

Martinez-Lopez, M.J., S. Alcantara, C. Mascaro, F. Perez-Branguli, P. Ruiz-Lozano, T. Maes, E. Soriano, and C. Buesa. 2005. Mouse neuron navigator 1, a novel microtubule-associated protein involved in neuronal migration. *Mol Cell Neurosci.* 28:599-612.

Matsumine, A., A. Ogai, T. Senda, N. Okumura, K. Satoh, G.H. Baeg, T. Kawahara, S. Kobayashi, M. Okada, K. Toyoshima, and T. Akiyama. 1996. Binding of APC to the human homolog of the Drosophila discs large tumor suppressor protein. *Science.* 272:1020-3.

Mattie, F.J., M.M. Stackpole, M.C. Stone, J.R. Clippard, D.A. Rudnick, Y. Qiu, J. Tao, D.L. Allender, M. Parmar, and M.M. Rolls. 2010. Directed microtubule growth, +TIPs, and kinesin-2 are required for uniform microtubule polarity in dendrites. *Curr Biol.* 20:2169-77.

Matus, A. 1990. Microtubule-associated proteins and the determination of neuronal form. *J Physiol (Paris).* 84:134-7.

McManus, M.F., I.M. Nasrallah, M.M. Pancoast, A. Wynshaw-Boris, and J.A. Golden. 2004. Lis1 is necessary for normal non-radial migration of inhibitory interneurons. *Am J Pathol.* 165:775-84.

McNeil, R.S., J.W. Swann, B.R. Brinkley, and G.D. Clark. 1999. Neuronal cytoskeletal alterations evoked by a platelet-activating factor (PAF) analogue. *Cell Motil Cytoskeleton.* 43:99-113.

McNeill, E.M., K.P. Roos, D. Moechars, and M. Clagett-Dame. 2010. Nav2 is necessary for cranial nerve development and blood pressure regulation. *Neural Dev.* 5:6.

Miller, R.K., S. D'Silva, J.K. Moore, and H.V. Goodson. 2006. The CLIP-170 orthologue Bik1p and positioning the mitotic spindle in yeast. *Curr Top Dev Biol.* 76:49-87.

Mimori-Kiyosue, Y., I. Grigoriev, G. Lansbergen, H. Sasaki, C. Matsui, F. Severin, N. Galjart, F. Grosveld, I. Vorobjev, S. Tsukita, and A. Akhmanova. 2005. CLASP1 and CLASP2 bind to EB1 and regulate microtubule plus-end dynamics at the cell cortex. *J Cell Biol.* 168:141-53.

Mimori-Kiyosue, Y., N. Shiina, and S. Tsukita. 2000. The dynamic behavior of the APC-binding protein EB1 on the distal ends of microtubules. *Curr Biol.* 10:865-8.

Mimori-Kiyosue, Y., and S. Tsukita. 2001. Where is APC going? *J Cell Biol.* 154:1105-9.

Mitchison, T., and M. Kirschner. 1984. Dynamic instability of microtubule growth. *Nature.* 312:237-42.

Mizuguchi, M., S. Takashima, A. Kakita, M. Yamada, and K. Ikeda. 1995. Lissencephaly gene product. Localization in the central nervous system and loss of immunoreactivity in Miller-Dieker syndrome. *Am J Pathol.* 147:1142-51.

Moores, C.A., M. Perderiset, F. Francis, J. Chelly, A. Houdusse, and R.A. Milligan. 2004. Mechanism of microtubule stabilization by doublecortin. *Mol Cell.* 14:833-9.

Moores, C.A., M. Perderiset, C. Kappeler, S. Kain, D. Drummond, S.J. Perkins, J. Chelly, R. Cross, A. Houdusse, and F. Francis. 2006. Distinct roles of doublecortin modulating the microtubule cytoskeleton. *Embo J.* 25:4448-57.

Morris, C.A. 1993. Williams Syndrome:Autosomal Dominant Inheritance. Am J Med Genet 47:478-481, 1993.

Morrison, E.E. 2007. Action and interactions at microtubule ends. *Cell Mol Life Sci.* 64:307-17.

Morrison, E.E. 2009. The APC-EB1 interaction. *Adv Exp Med Biol.* 656:41-50.

Morrison, E.E., P.M. Moncur, and J.M. Askham. 2002. EB1 identifies sites of microtubule polymerisation during neurite development. *Brain Res Mol Brain Res.* 98:145-52.

Morrison, E.E., B.N. Wardleworth, J.M. Askham, A.F. Markham, and D.M. Meredith. 1998. EB1, a protein which interacts with the APC tumour suppressor, is associated with the microtubule cytoskeleton throughout the cell cycle. *Oncogene.* 17:3471-7.

Moseley, J.B., F. Bartolini, K. Okada, Y. Wen, G.G. Gundersen, and B.L. Goode. 2007. Regulated binding of adenomatous polyposis coli protein to actin. *J Biol Chem.* 282:12661-8.

Muley, P.D., E.M. McNeill, M.A. Marzinke, K.M. Knobel, M.M. Barr, and M. Clagett-Dame. 2008. The atRA-responsive gene neuron navigator 2 functions in neurite outgrowth and axonal elongation. *Dev Neurobiol.* 68:1441-53.

Munch, C., R. Sedlmeier, T. Meyer, V. Homberg, A.D. Sperfeld, A. Kurt, J. Prudlo, G. Peraus, C.O. Hanemann, G. Stumm, and A.C. Ludolph. 2004. Point mutations of the p150 subunit of dynactin (DCTN1) gene in ALS. *Neurology.* 63:724-6.

Munemitsu, S., B. Souza, O. Muller, I. Albert, B. Rubinfeld, and P. Polakis. 1994. The APC gene product associates with microtubules in vivo and promotes their assembly in vitro. *Cancer Res.* 54:3676-81.

Nakagawa, H., K. Koyama, Y. Murata, M. Morito, T. Akiyama, and Y. Nakamura. 2000. EB3, a novel member of the EB1 family preferentially expressed in the central nervous system, binds to a CNS-specific APC homologue. *Oncogene.* 19:210-6.

Nakamura, M., X.Z. Zhou, and K.P. Lu. 2001. Critical role for the EB1 and APC interaction in the regulation of microtubule polymerization. *Curr Biol.* 11:1062-7.

Nakata, T., and N. Hirokawa. 2003. Microtubules provide directional cues for polarized axonal transport through interaction with kinesin motor head. *J Cell Biol.* 162:1045-55.

Nasrallah, I.M., M.F. McManus, M.M. Pancoast, A. Wynshaw-Boris, and J.A. Golden. 2006. Analysis of non-radial interneuron migration dynamics and its disruption in Lis1+/- mice. *J Comp Neurol.* 496:847-58.

Nathke, I. 2004. APC at a glance. *J Cell Sci.* 117:4873-5.

Neukirchen, D., and F. Bradke. 2010. Cytoplasmic linker proteins regulate neuronal polarization through microtubule and growth cone dynamics. *J Neurosci.* 31:1528-38.

Ogata, T., N. Matsuo, N. Hiraoka, and J.I. Hata. 2000. X-linked lissencephaly with ambiguous genitalia: delineation of further case. *Am J Med Genet.* 94:174-6.

Pancoast, M., W. Dobyns, and J.A. Golden. 2005. Interneuron deficits in patients with the Miller-Dieker syndrome. *Acta Neuropathol.* 109:400-4.

Pavone, L., R. Rizzo, and W.B. Dobyns. 1993. Clinical manifestations and evaluation of isolated lissencephaly. *Childs Nerv Syst.* 9:387-90.

Peeters, P.J., A. Baker, I. Goris, G. Daneels, P. Verhasselt, W.H. Luyten, J.J. Geysen, S.U. Kass, and D.W. Moechars. 2004. Sensory deficits in mice hypomorphic for a mammalian homologue of unc-53. *Brain Res Dev Brain Res.* 150:89-101.

Perez, F., G.S. Diamantopoulos, R. Stalder, and T.E. Kreis. 1999. CLIP-170 highlights growing microtubule ends in vivo. *Cell.* 96:517-27.

Peris, L., M. Thery, J. Faure, Y. Saoudi, L. Lafanechere, J.K. Chilton, P. Gordon-Weeks, N. Galjart, M. Bornens, L. Wordeman, J. Wehland, A. Andrieux, and D. Job. 2006. Tubulin tyrosination is a major factor affecting the recruitment of CAP-Gly proteins at microtubule plus ends. *J Cell Biol.* 174:839-49.

Pierre, P., J. Scheel, J.E. Rickard, and T.E. Kreis. 1992. CLIP-170 links endocytic vesicles to microtubules. *Cell.* 70:887-900.

Potaros, T., S. Phornchirasilp, S.B. McKay, T.F. Gonzalez-Cestari, R.T. Boyd, and D.B. McKay. 2011. Evidence for the involvement of adenomatous polyposis coli (APC) protein in maintaining cellular distributions of alpha3beta4 nicotinic receptors. *Neurosci Lett.* 489:105-9.

Puls, I., S.J. Oh, C.J. Sumner, K.E. Wallace, M.K. Floeter, E.A. Mann, W.R. Kennedy, G. Wendelschafer-Crabb, A. Vortmeyer, R. Powers, K. Finnegan, E.L. Holzbaur, K.H.

Fischbeck, and C.L. Ludlow. 2005. Distal spinal and bulbar muscular atrophy caused by dynactin mutation. *Ann Neurol*. 57:687-94.

Purro, S.A., L. Ciani, M. Hoyos-Flight, E. Stamatakou, E. Siomou, and P.C. Salinas. 2008. Wnt regulates axon behavior through changes in microtubule growth directionality: a new role for adenomatous polyposis coli. *J Neurosci*. 28:8644-54.

Reed, N.A., D. Cai, T.L. Blasius, G.T. Jih, E. Meyhofer, J. Gaertig, and K.J. Verhey. 2006. Microtubule acetylation promotes kinesin-1 binding and transport. *Curr Biol*. 16:2166-72.

Reiner, O., A. Cahana, T. Escamez, and S. Martinez. 2002. LIS1-no more no less. *Mol Psychiatry*. 7:12-6.

Reiner, O., R. Carrozzo, Y. Shen, M. Wehnert, F. Faustinella, W.B. Dobyns, C.T. Caskey, and D.H. Ledbetter. 1993. Isolation of a Miller-Dieker lissencephaly gene containing G protein beta-subunit-like repeats. *Nature*. 364:717-21.

Reiner, O., and T. Sapir. 2009. Polarity regulation in migrating neurons in the cortex. *Mol Neurobiol*. 40:1-14.

Rosenberg, M.M., F. Yang, M. Giovanni, J.L. Mohn, M.K. Temburni, and M.H. Jacob. 2008. Adenomatous polyposis coli plays a key role, in vivo, in coordinating assembly of the neuronal nicotinic postsynaptic complex. *Mol Cell Neurosci*. 38:138-52.

Rosenberg, M.M., F. Yang, J.L. Mohn, E.K. Storer, and M.H. Jacob. 2010. The postsynaptic adenomatous polyposis coli (APC) multiprotein complex is required for localizing neuroligin and neurexin to neuronal nicotinic synapses in vivo. *J Neurosci*. 30:11073-85.

Ross, C.A., R.L. Margolis, S.A. Reading, M. Pletnikov, and J.T. Coyle. 2006. Neurobiology of schizophrenia. *Neuron*. 52:139-53.

Sapir, T., M. Elbaum, and O. Reiner. 1997. Reduction of microtubule catastrophe events by LIS1, platelet-activating factor acetylhydrolase subunit. *Embo J*. 16:6977-84.

Sapir, T., D. Horesh, M. Caspi, R. Atlas, H.A. Burgess, S.G. Wolf, F. Francis, J. Chelly, M. Elbaum, S. Pietrokovski, and O. Reiner. 2000. Doublecortin mutations cluster in evolutionarily conserved functional domains. *Hum Mol Genet*. 9:703-12.

Sasaki, S., A. Shionoya, M. Ishida, M.J. Gambello, J. Yingling, A. Wynshaw-Boris, and S. Hirotsune. 2000. A LIS1/NUDEL/cytoplasmic dynein heavy chain complex in the developing and adult nervous system. *Neuron*. 28:681-96.

Schaefer, A.W., N. Kabir, and P. Forscher. 2002. Filopodia and actin arcs guide the assembly and transport of two populations of microtubules with unique dynamic parameters in neuronal growth cones. *J Cell Biol*. 158:139-52.

Shi, S.H., T. Cheng, L.Y. Jan, and Y.N. Jan. 2004. APC and GSK-3beta are involved in mPar3 targeting to the nascent axon and establishment of neuronal polarity. *Curr Biol*. 14:2025-32.

Shimomura, A., M. Ohkuma, A. Iizuka-Kogo, K. Kohu, R. Nomura, E. Miyachi, T. Akiyama, and T. Senda. 2007. Requirement of the tumour suppressor APC for the clustering of PSD-95 and AMPA receptors in hippocampal neurons. *Eur J Neurosci*. 26:903-12.

Siller, K.H., and C.Q. Doe. 2008. Lis1/dynactin regulates metaphase spindle orientation in Drosophila neuroblasts. *Dev Biol*. 319:1-9.

Slep, K.C., S.L. Rogers, S.L. Elliott, H. Ohkura, P.A. Kolodziej, and R.D. Vale. 2005. Structural determinants for EB1-mediated recruitment of APC and spectraplakins to the microtubule plus end. *J Cell Biol*. 168:587-98.

Steinmetz, M.O., and A. Akhmanova. 2008. Capturing protein tails by CAP-Gly domains. *Trends Biochem Sci*. 33:535-45.

Stepanova, T., J. Slemmer, C.C. Hoogenraad, G. Lansbergen, B. Dortland, C.I. De Zeeuw, F. Grosveld, G. van Cappellen, A. Akhmanova, and N. Galjart. 2003. Visualization of microtubule growth in cultured neurons via the use of EB3-GFP (end-binding protein 3-green fluorescent protein). *J Neurosci*. 23:2655-64.

Stepanova, T., I. Smal, J. van Haren, U. Akinci, Z. Liu, M. Miedema, R. Limpens, M. van Ham, M. van der Reijden, R. Poot, F. Grosveld, M. Mommaas, E. Meijering, and N. Galjart. 2010. History-dependent catastrophes regulate axonal microtubule behavior. *Curr Biol*. 20:1023-8.

Stiess, M., N. Maghelli, L.C. Kapitein, S. Gomis-Ruth, M. Wilsch-Brauninger, C.C. Hoogenraad, I.M. Tolic-Norrelykke, and F. Bradke. 2010. Axon extension occurs independently of centrosomal microtubule nucleation. *Science*. 327:704-7.

Stringham, E.G., and K.L. Schmidt. 2009. Navigating the cell: UNC-53 and the navigators, a family of cytoskeletal regulators with multiple roles in cell migration, outgrowth and trafficking. *Cell Adh Migr*. 3:342-6.

Su, L.K., and Y. Qi. 2001. Characterization of human MAPRE genes and their proteins. *Genomics*. 71:142-9.

Sweet, E.S., M.L. Previtera, J.R. Fernandez, E.I. Charych, C.Y. Tseng, M. Kwon, V. Starovoytov, J.Q. Zheng, and B.L. Firestein. 2010. PSD-95 alters microtubule dynamics via an association with EB3. *J Neurosci*. 31:1038-47.

Swiech, L., M. Blazejczyk, M. Urbanska, P. Pietruszka, B.R. Dortland, A.R. Malik, P.S. Wulf, C.C. Hoogenraad, and J. Jaworski. 2010. CLIP-170 and IQGAP1 cooperatively regulate dendrite morphology. *J Neurosci*. 31:4555-68.

Tai, C.Y., D.L. Dujardin, N.E. Faulkner, and R.B. Vallee. 2002. Role of dynein, dynactin, and CLIP-170 interactions in LIS1 kinetochore function. *J Cell Biol*. 156:959-68.

Tanaka, E., T. Ho, and M.W. Kirschner. 1995. The role of microtubule dynamics in growth cone motility and axonal growth. *J Cell Biol*. 128:139-55.

Tanaka, E.M., and M.W. Kirschner. 1991. Microtubule behavior in the growth cones of living neurons during axon elongation. *J Cell Biol*. 115:345-63.

Tanaka, T., F.F. Serneo, C. Higgins, M.J. Gambello, A. Wynshaw-Boris, and J.G. Gleeson. 2004. Lis1 and doublecortin function with dynein to mediate coupling of the nucleus to the centrosome in neuronal migration. *J Cell Biol*. 165:709-21.

Tarricone, C., F. Perrina, S. Monzani, L. Massimiliano, M.H. Kim, Z.S. Derewenda, S. Knapp, L.H. Tsai, and A. Musacchio. 2004. Coupling PAF signaling to dynein regulation: structure of LIS1 in complex with PAF-acetylhydrolase. *Neuron*. 44:809-21.

Taya, S., T. Shinoda, D. Tsuboi, J. Asaki, K. Nagai, T. Hikita, S. Kuroda, K. Kuroda, M. Shimizu, S. Hirotsune, A. Iwamatsu, and K. Kaibuchi. 2007. DISC1 regulates the transport of the NUDEL/LIS1/14-3-3epsilon complex through kinesin-1. *J Neurosci*. 27:15-26.

Taylor, K.R., A.K. Holzer, J.F. Bazan, C.A. Walsh, and J.G. Gleeson. 2000. Patient mutations in doublecortin define a repeated tubulin-binding domain. *J Biol Chem*. 275:34442-50.

Temburni, M.K., M.M. Rosenberg, N. Pathak, R. McConnell, and M.H. Jacob. 2004. Neuronal nicotinic synapse assembly requires the adenomatous polyposis coli tumor suppressor protein. *J Neurosci*. 24:6776-84.

Terada, S., M. Kinjo, and N. Hirokawa. 2000. Oligomeric tubulin in large transporting complex is transported via kinesin in squid giant axons. *Cell.* 103:141-55.

Tirnauer, J.S. 2004. A new cytoskeletal connection for APC: linked to actin through IQGAP. *Dev Cell.* 7:778-80.

Tirnauer, J.S., and B.E. Bierer. 2000. EB1 proteins regulate microtubule dynamics, cell polarity, and chromosome stability. *J Cell Biol.* 149:761-6.

Tomita, K., K.I. Kubo, K. Ishii, and K. Nakajima. 2010. Disrupted-in-Schizophrenia-1 (Disc1) is necessary for migration of the pyramidal neurons during mouse hippocampal development. *Hum Mol Genet.* 20(14):2834-45

Tsai, J.W., K.H. Bremner, and R.B. Vallee. 2007. Dual subcellular roles for LIS1 and dynein in radial neuronal migration in live brain tissue. *Nat Neurosci.* 10:970-9.

Tsvetkov, A.S., A. Samsonov, A. Akhmanova, N. Galjart, and S.V. Popov. 2007. Microtubule-binding proteins CLASP1 and CLASP2 interact with actin filaments. *Cell Motil Cytoskeleton.* 64:519-30.

Vacher, H., J.W. Yang, O. Cerda, A. Autillo-Touati, B. Dargent, and J.S. Trimmer. 2011. Cdk-mediated phosphorylation of the Kvbeta2 auxiliary subunit regulates Kv1 channel axonal targeting. *J Cell Biol.* 192:813-24.

Vale, R.D. 2003. The molecular motor toolbox for intracellular transport. *Cell.* 112:467-80.

van Hagen, J.M., J.N. van der Geest, R.S. van der Giessen, G.C. Lagers-van Haselen, H.J. Eussen, J.J. Gille, L.C. Govaerts, C.H. Wouters, I.F. de Coo, C.C. Hoogenraad, S.K. Koekkoek, M.A. Frens, N. van Camp, A. van der Linden, M.C. Jansweijer, S.S. Thorgeirsson, and C.I. De Zeeuw. 2007. Contribution of CYLN2 and GTF2IRD1 to neurological and cognitive symptoms in Williams Syndrome. *Neurobiol Dis.* 26:112-24.

van Haren, J., K. Draegestein, N. Keijzer, J.P. Abrahams, F. Grosveld, P.J. Peeters, D. Moechars, and N. Galjart. 2009. Mammalian Navigators are microtubule plus-end tracking proteins that can reorganize the cytoskeleton to induce neurite-like extensions. *Cell Motil Cytoskeleton.* 66:824-38.

Vaughan, P.S., P. Miura, M. Henderson, B. Byrne, and K.T. Vaughan. 2002. A role for regulated binding of p150(Glued) to microtubule plus ends in organelle transport. *J Cell Biol.* 158:305-19.

Vitre, B., F.M. Coquelle, C. Heichette, C. Garnier, D. Chretien, and I. Arnal. 2008. EB1 regulates microtubule dynamics and tubulin sheet closure in vitro. *Nat Cell Biol.* 10:415-21.

Votin, V., W.J. Nelson, and A.I. Barth. 2005. Neurite outgrowth involves adenomatous polyposis coli protein and beta-catenin. *J Cell Sci.* 118:5699-708.

Wade, R.H. 2007. Microtubules: an overview. *Methods Mol Med.* 137:1-16.

Walczak, C.E. 2000. Microtubule dynamics and tubulin interacting proteins. *Curr Opin Cell Biol.* 12:52-6.

Watanabe, T., J. Noritake, M. Kakeno, T. Matsui, T. Harada, S. Wang, N. Itoh, K. Sato, K. Matsuzawa, A. Iwamatsu, N. Galjart, and K. Kaibuchi. 2009. Phosphorylation of CLASP2 by GSK-3beta regulates its interaction with IQGAP1, EB1 and microtubules. *J Cell Sci.* 122:2969-79.

Watanabe, T., S. Wang, J. Noritake, K. Sato, M. Fukata, M. Takefuji, M. Nakagawa, N. Izumi, T. Akiyama, and K. Kaibuchi. 2004. Interaction with IQGAP1 links APC to Rac1, Cdc42, and actin filaments during cell polarization and migration. *Dev Cell.* 7:871-83.

Weisbrich, A., S. Honnappa, R. Jaussi, O. Okhrimenko, D. Frey, I. Jelesarov, A. Akhmanova, and M.O. Steinmetz. 2007. Structure-function relationship of CAP-Gly domains. *Nat Struct Mol Biol.* 14:959-67.

Wen, Y., C.H. Eng, J. Schmoranzer, N. Cabrera-Poch, E.J. Morris, M. Chen, B.J. Wallar, A.S. Alberts, and G.G. Gundersen. 2004. EB1 and APC bind to mDia to stabilize microtubules downstream of Rho and promote cell migration. *Nat Cell Biol.* 6:820-30.

Wider, C., J.C. Dachsel, M.J. Farrer, D.W. Dickson, Y. Tsuboi, and Z.K. Wszolek. 2010. Elucidating the genetics and pathology of Perry syndrome. *J Neurol Sci.* 289:149-54.

Wider, C., and Z.K. Wszolek. 2008. Rapidly progressive familial parkinsonism with central hypoventilation, depression and weight loss (Perry syndrome)--a literature review. *Parkinsonism Relat Disord.* 14:1-7.

Witte, H., D. Neukirchen, and F. Bradke. 2008. Microtubule stabilization specifies initial neuronal polarization. *J Cell Biol.* 180:619-32.

Wittmann, T. 2008. EBs clip CLIPs to growing microtubule ends. *J Cell Biol.* 183:1183-5.

Wittmann, T., and C.M. Waterman-Storer. 2005. Spatial regulation of CLASP affinity for microtubules by Rac1 and GSK3beta in migrating epithelial cells. *J Cell Biol.* 169:929-39.

Wynshaw-Boris, A. 2007. Lissencephaly and LIS1: insights into the molecular mechanisms of neuronal migration and development. *Clin Genet.* 72:296-304.

Yamada, M., S. Toba, Y. Yoshida, K. Haratani, D. Mori, Y. Yano, Y. Mimori-Kiyosue, T. Nakamura, K. Itoh, S. Fushiki, M. Setou, A. Wynshaw-Boris, T. Torisawa, Y.Y. Toyoshima, and S. Hirotsune. 2008. LIS1 and NDEL1 coordinate the plus-end-directed transport of cytoplasmic dynein. *Embo J.* 27:2471-83.

Yanai, H., K. Satoh, A. Matsumine, and T. Akiyama. 2000. The colorectal tumour suppressor APC is present in the NMDA-receptor-PSD-95 complex in the brain. *Genes Cells.* 5:815-22.

Yokota, Y., W.Y. Kim, Y. Chen, X. Wang, A. Stanco, Y. Komuro, W. Snider, and E.S. Anton. 2009. The adenomatous polyposis coli protein is an essential regulator of radial glial polarity and construction of the cerebral cortex. *Neuron.* 61:42-56.

Yu, W., V.E. Centonze, F.J. Ahmad, and P.W. Baas. 1993. Microtubule nucleation and release from the neuronal centrosome. *J Cell Biol.* 122:349-59.

Zhapparova, O.N., S.A. Bryantseva, L.V. Dergunova, N.M. Raevskaya, A.V. Burakov, O.B. Bantysh, N.A. Shanina, and E.S. Nadezhdina. 2009. Dynactin subunit p150Glued isoforms notable for differential interaction with microtubules. *Traffic.* 10:1635-46.

Zhou, F.Q., and W.D. Snider. 2006. Intracellular control of developmental and regenerative axon growth. *Philos Trans R Soc Lond B Biol Sci.* 361:1575-92.

Zhou, F.Q., J. Zhou, S. Dedhar, Y.H. Wu, and W.D. Snider. 2004. NGF-induced axon growth is mediated by localized inactivation of GSK-3beta and functions of the microtubule plus end binding protein APC. *Neuron.* 42:897-912.

Zhou, X.L., M. Giacobini, B.M. Anderlid, H. Anckarsater, D. Omrani, C. Gillberg, M. Nordenskjold, and A. Lindblom. 2007. Association of adenomatous polyposis coli (APC) gene polymorphisms with autism spectrum disorder (ASD). *Am J Med Genet B Neuropsychiatr Genet.* 144B:351-4.

In: Tubulin: Structure, Functions and Roles in Disease
Editors: Wei Yamauchi and Antonio Sokić ISBN: 978-1-62100-141-6

Chapter II

Microtubule Organization. The Housekeeping Role of Heat-shock Proteins and Immunophilins

Héctor R. Quintá, Natalia M. Galigniana, Cristina Daneri-Becerra, Mariana Lagadari and Mario D. Galigniana*

Instituto de Biología y Medicina Experimental-CONICET

Facultad de Ciencias Exactas y Naturales, Universidad de Buenos Aires, Argentina

Abstract

Cytoskeleton is the basic scaffold of the cell in which other subcellular components are spatially arranged, such that they are able to communicate efficiently between the internal and external environments of the cell. Cytoskeletal structure is continually remodeled to accommodate normal cell growth and to respond to pathophysiological cues. As a consequence, several cytoskeleton-interacting proteins become involved in a variety of cellular processes such as cell growth and division, cell movement, vesicle transportation, cellular organelle location and function, localization and distribution of membrane receptors, and cell-cell communication. Molecular chaperones and immunophilins are counted among the most important proteins that interact closely with the cytoskeleton network, in particular with microtubules and microtubule-associated factors. In several situations, chaperones and immunophilins work together as a functionally active heterocomplex, although both types of proteins also show independent actions. In circumstances where homeostasis is affected by environmental stresses or due to genetic alterations, chaperone proteins help to stabilize the system. Molecular chaperones facilitate the assembly, disassembly and/or folding/refolding of cytoskeletal proteins, so they prevent aberrant protein aggregation. Nonetheless, the role of chaperones is not limited to solve abnormal situations. They also have an active participation during the normal differentiation process of the cell and are key factors for

* E-mail address: mgaligniana@conicet.gov.ar; Tel: +54 (11) 4783-2869; Fax: +54 (11) 4786-2564; Address: Instituto de Biología y Medicina Experimental, Facultad de Ciencias Exactas y Naturales, Universidad de Buenos Aires, Vuelta de Obligado 2490, Buenos Aires C1428ADN, Argentina. (Corresponding author).

many structural and functional rearrangements during this course of action. A similar observation is also valid for immunophilins. Although the folding system for microtubules was discovered nearly two decades ago, our understanding of the complex quality control pathway of these filaments is still poorly understood and there are many unanswered questions that remain to be elucidated. Microtubule modifications leading to altered localization nuclear factors may result in loss- or gain-of-function of such factors, which affects the cell cycle and cell development. Therefore, microtubules are attractive therapeutic targets, particularly to prevent pathological situations such as rapidly dividing tumors or to favor the process of cell differentiation in other cases. In this chapter we will address some mechanistic aspects of the key regulatory functions of heat-shock proteins and immunophilins in the microtubule network of neurons and cancer cells.

Introduction

The Cytoskeleton

The cytoskeleton constitutes the structural support of the living matter. It also plays key functional roles in the life time of the cells and during their replication. The cytoskeleton is present in all eukaryotic cells and was once thought to be an exclusive structure of eukaryotes, but homologues of the major scaffold proteins of the eukaryotic cytoskeleton have also been found in prokaryotes (Erickson, 2007; Shih *et al.*, 2006). The cytoskeleton is a dynamic three dimensional filamentous structure that fills the cytoplasm maintaining the cell shape, enables the cell to move, and has cardinal roles in the intracellular transport of vesicles, organelles and soluble proteins, during the cell division, the segregation of chromosomes, maintenance of proper cell shape, cell polarity and assembly of intracellular organelle-like structures. The three types of cytoskeletal elements that have been characterized in eukaryotic cells are tubulins, actins, and intermediate filaments (IFs).

Tubulin forms microtubules of about 25 nm in diameter consisting of straight 13 protofilaments that assemble into hollow tubules through lateral contacts to both sides of the protofilaments. The cylinder yields a helical arrangement in which each turn of the helix spans three tubulin monomers (e.g., α, β, α). This generates a seam in the microtubule wall where, instead of the predominant αα and ββ lateral contacts, subunits are laterally adjacent to β subunits. Microtubules are organized from the minus end (in close vicinity with the centrosome) and extend through association of GTP-bound α-tubulin and β-tubulin heterodimers onto the plus end, which faces the cell periphery. The dimers are released from the minus end after hydrolysis, which destabilizes the intrafilament contacts. However, microtubules can rapidly release dimers from both ends and thus shrink, a stochastic event called *catastrophic collapse*, giving rise to dynamic instability of the filaments. Thus, tubulin filaments are highly dynamic elements within the cells. γ-Tubulin, which is homologous to α and β tubulins, nucleates microtubule assembly within the centrosome. Around 12 to14 copies of γ-tubulin associate in a complex with other proteins called GRIPS (for gamma-ring proteins), this complex being capped at one end and thought to be the minus end terminal. Microtubule polymerization at this end is inhibited, and the GRIP proteins of the cap may be involved in mediating binding to the centrosome. Phosphorylation of a conserved tyrosine residue of γ-tubulin has been shown to regulate microtubule nucleation (Vogel *et al.*, 2001).

Microtubules set up the tracks of the mitotic spindle apparatus that are used to segregate chromosomes during mitosis and meiosis through dedicated motor proteins. Microtubules also serve to provide tracks for the transport of intracellular vesicles and other cargoes that are moved through cells by motor proteins such as dynein and kinesin, which show predominant retrograde and anterograde direction of movement on microtubules.

Actin forms a two-stranded, right-handed helical filament of about 7 nm in diameter with an axial rise of 5.4 nm per monomer that has a plus-/minus-end polarity and is also dynamic (Mogilner *et al.*, 2003). Owing to filament asymmetry, ATP-bound actin adds to the plus end (often called the *barbed end*) much faster than it does to the minus end (also called the *pointed end*). ATPase activity leads to conversion to ADP-actin within the filament, and ADP-actin is released at the minus end. This process is called *treadmilling* and leads to a net polymerization at one end and depolymerization at the other end. Thus, while the center of the filament (and each subunit) remains stationary, the tip of the filament can push objects by limiting them in their diffusion, such that they can diffuse only in the direction of the extended filament. Actin polymerization is required for cell movement via extension of pseudopods, where other additional proteins induce branching of actin filaments generating a network that pushes against the leading edge of the membrane in a brush-like manner (Defeu Soufo *et al.*, 2004). Actin is also required for the movement of some types of vesicles through cells (Wiesner *et al.*, 2003). Actin is also a contractile protein and one of the most abundant proteins in cells. In muscle cells, the association of microfilaments with myosin is responsible for muscle contraction. Microfilaments can also carry out cellular movements including gliding, contraction, and cytokinesis. They are mostly concentrated just beneath the plasma membrane, as they keep cellular shape, form cytoplasmatic protuberances such as pseudopodia and microvilli, and participate in cell-to-cell or cell-to-matrix junctions and in the transduction of signals. Recently, it was demonstrated the existence of a subset of actomyosin fibers that form a perinuclear actin cap, a cytoskeletal structure that tightly wraps around the nucleus (Khatau *et al.*, 2010). Fibers in the actin cap are distinct from the conventional actin fibers at the basal and dorsal surfaces of adherent cells in their organization, dynamics, response to cytoskeletal pharmacological treatments and biochemical stimuli, and response to disease-associated mutations of lamins A/C. The perinuclear actin cap precisely shapes the nucleus in interphase cells and is a mediator of microenvironment mechanosensing and mechanotransduction, as well as a regulator of cell motility, polarization and differentiation (Khatau *et al.*, 2010; Wirtz *et al.*, 2010).

IFs are composed of extended coiled-coil proteins that assemble into rigid sheets. IFs form 8- to 10-nm-thick cytoskeletal elements that provide internal mechanical support for the cell and position different organelles (Capetanaki *et al.*, 2007; Toivola *et al.*, 2005). They form a large superfamily of proteins containing 73 gene products (Szeverenyi *et al.*, 2008) that are expressed in tissue-, cell-, differentiation-, and development-dependent manners, and are related to more than 90 human pathologies (www.interfil.org). IFs provide the tensile strength of a cell and determine its internal structure; for example, they are structural components of the nuclear envelope and of the sarcomeres. They also participate in some cell-cell and cell-matrix junctions. Many intermediate filaments are made of vimentins; others, in cells of the skinheair and nails, are made of keratin. Some filaments, made of laminin, give structural support to the nuclear envelope. IFs are much less conserved in sequence than are tubulins or actin, because different sequences can make up coiled-coil motifs. IFs display a tripartite structure consisting of an α-helical central rod flanked by nonhelical head and tail domains, and assemble to give rise to nuclear (lamins) or to different cytoplasmic IF networks such as keratins, vimentin, neurofilaments, peripherin, desmin, and glial fibrillary acidic protein (GFAP) (Herrmann *et al.*, 2009; Omary, 2009; Toivola *et al.*, 2010). IFs are regulated by both post-translational modifications and through interactions with multiple IF-associated signaling cascade proteins such as 14-3-3 proteins, apoptosis-related proteins, kinases and phosphatases (Green *et al.*, 2005; Pallari *et al.*, 2006), all of them being processes that may affect from IF solubility to sensitivity to be degraded by caspases or proteosomal degradation .

Before the 90's, the cytoskeleton was thought to have evolved only in eukaryotes. However, cytological techniques developed for eukaryotic cells were adapted and used for the study of the cellular organization of bacteria and archaea, revealing that these organisms also contain a number of subcellular counterparts of actin, tubulin, and intermediate filament proteins (Aylett *et al.*, 2010; Cabeen *et al.*, 2010; Michie *et al.*, 2006; Shih *et al.*, 2006). FtsZ was the first of these proteins to be identified. Like tubulin, FtsZ forms filaments in the presence of GTP, although the filaments do not group into tubules. During cell division, FtsZ moves to the division site and recruits other proteins required to generate a new cell wall between the dividing cells. On the other hand, MreB is considered the prokaryotic actin-like protein. As such, it is involved in maintaining cell shape. Non-spherical bacteria also encode other actin-like proteins that form a helical network beneath the cell membrane, which in turn guides the proteins required for the synthesis of a new cell wall. Other actin-like protein is ParM, whose filaments participate in the division of plasmidic DNA into the daughter cells by a mechanism similar to that used by microtubules during eukaryotic mitosis. Finally, crescentin is a bacterial counterpart related to IFs of eukaryotic cells and is involved in maintaining cell shape.

Cytoskeleton proteins of both eukaryotic and prokaryotic cells require the assistance of associated factors for the maintenance of their structures and functions. In this regard, molecular chaperones are essential for this process. This is not surprising if two essential facts are considered: a) there is a structural similarity between actins and the Hsp70 class of molecular chaperones (Aravind *et al.*, 1999; Bork *et al.*, 1992), and b) heat-shock has deleterious effects on the internal organization of the cell beyond the unfolding of individual proteins and, especially in eukaryotes, one of the major damages observed in response to stress conditions are defects of the cytoskeleton and loss of the proper localization of organelles and breakdown of intracellular transport processes.

Molecular Chaperones

The predominant class of heat-shock proteins is the molecular chaperones. This superfamily comprises five major and broadly conserved subfamilies—Hsp100s, Hsp90s, Hsp70s, Hsp60s, and small heat shock proteins (sHsps). Several other heat-inducible molecular chaperones, like Hsp33 (Jakob *et al.*, 1999) are also part of the superfamily, but since they are not ubiquitous, they were not included in a separate subfamily. Because the stability of cellular proteins is relatively low and aggregation competes with their efficient folding, even at physiological temperatures, there is a constant need for chaperone assistance during de novo protein folding and refolding of non-native polypeptide chains (Mayer, 2010; Richter *et al.*, 2010). What discriminates a native protein from its non-native unfolded (partially or globally) counterpart is the exposure of hydrophobic residues, which are recognized by molecular chaperones. Normally, binding to hydrophobic patches, specific peptide sequences, or structural elements of the non-native protein takes place to protect these proteins from aggregation and degradation. Generally, molecular chaperones do not contribute to the structural information for protein folding, but prevent those unwanted intermolecular interactions. Chaperones do this through controlled binding and release of non-native proteins, which is usually accomplished by a change of the affinity of the chaperone for its client substrate. This exchange between two (or more) affinity states is controlled by the binding to the chaperone and/or subsequent hydrolysis of ATP molecules, with the sole exception of small Hsps. The latter chaperones are optimal for an efficient first line of defense. This is why they are also known as *holdases*. On the other hand, the *foldases* come in stress-induced and constitutively expressed versions (e.g., Hsp70 versus Hsc70).

A subset of molecular chaperones is the so called subfamily of chaperonines. Chaperonines are ring-shaped chaperones that encapsulate non-native proteins in an ATP-dependent manner. In bacteria, the most prominent chaperonin is the GroE machinery, which consists of fourteen GroEL subunits arranged in a cylinder of two heptametrical rings, to which the cochaperone GroES, also a heptametrical ring, binds (Horwich *et al.*, 2006). The closely related proteins in mitochondria are called Hsp60 and Hsp10. Non-native protein chains up to 60-kDa are bound in the central cavity of the GroEL cylinder, and the cavity is then closed by the binding of the GroES cofactor in the presence of ATP. As a result, the client protein may fold or it could gain their native structure after release from GroE, depending on the folding properties of the respective protein. Therefore, a GroE-bound protein can be refolded isolated from other non-native polypeptide chains, a wise mechanism that prevents massive aggregation. Such a potent protein-folding machine, which is essential in bacteria, should also have enormous importance for the proper management of stressing situations in eukaryotic cells. However, there is no GroE in the eukaryotic or archaeal cells. Nonetheless, the GroE function has been replaced by a distant relative machinery, the CCT (chaperonin-containing the T-complex polypeptide-1), also named TRiC (tailless complex polypeptide 1-ring complex), whose architecture and function resembles that of GroE. The CCT/TRiC machinery is the eukaryotic Hsp60 cytoplasmic chaperonin essential for growth of worms and yeasts. Mutations in individual subunits affect assembly of tubulin and actin (Dunn *et al.*, 2001) and have been implicated in neurodegenerative diseases (Morimoto, 2008; Muchowski *et al.*, 2005). The main cellular function of CCT/TRiC is to promote the folding of newly synthesized polypeptides, which may be presented by Hsp70 and/or the cochaperone prefoldin (Broadley *et al.*, 2009; Spiess *et al.*, 2004). Perhaps due to this reason, CCT/TRiC is not induced by stress but is instead transcriptionally and functionally linked to protein synthesis (Albanese *et al.*, 2006). CCT/TRiC forms a large complex composed of eight homologous subunits arranged in two octameric rings stacked back-to-back that form a 1-MDa complex that functions like a cage for protein folding to occur unimpaired by aggregation. Whereas all CCT/TRiC subunits have essentially identical ATPase domains, their polypeptide-binding regions have significantly diverged during evolution to create substrate binding specificity. As a result, a diverse set of polypeptide substrates, estimated to be 5-10% of newly synthesized cytoplasmic proteins, including actin and tubulin, flux through CCT/TRiC (Thulasiraman *et al.*, 1999).

The term "heat-shock protein" stems from the original observation that heat-stress greatly enhances the production of a specific class of proteins, subsequently identified as molecular chaperones. Hsp90 is the major soluble cytosolic protein that is expressed under both normal and stressful conditions (Echeverria *et al.*, 2010; Galigniana *et al.*, 2010a; Richter *et al.*, 2010; Trepel *et al.*, 2010). In mammals there are two isoforms, Hsp90α, Hsp90β, encoded by separate genes and usually forming homodimers (Pratt *et al.*, 2004b). Hsp90 displays only weak ATPase and chaperoning activity, i.e., prevention of aggregation of unfolded polypeptides and assistance in their refolding. It is more selective than other chaperones and chaperonines for the recognition of its substrates and shows low affinity for unfolded proteins. Actually, its principal role in the cell is to provide biological activity to properly folded client proteins and functions as a delicate and refined sensor of protein function rather than a gross folding factor. In particular, the cooperation of Hsp90 with other chaperones such as the Hsp70/Hsp40 complex and cochaperones such as p23 and immunophilins makes a vital contribution to the maturation of signal-transduction proteins such as steroid receptors and protein-kinases (Pratt *et al.*, 1997). Thus, highly orchestrated interactions between the immature, low affinity steroid receptor protein and the Hsp90-based heterocomplex is required to shape the mature and biologically active forms of the receptors that transduce steroid signaling. Importantly, the Hsp90-based heterocomplex associated to immunophilin FKB52 is responsible for the retrotransport of soluble factors using microtubules as cellular tracts.

The Hsp70 subfamily of chaperones is the most conserved of the entire family. Human Hsp70 also shows to isoforms —Hsp72 (gene name *HspA1*), which is inducible upon the onset of stressing conditions, and the other, more properly named Hsp73 or Hsc70 (heat-shock cognate-70), is constitutive (gene name *HspA8*). Human Hsc70 has 85% identity with human Hsp70. Unlike canonical heat-shock proteins that prevent the aggregation of unfolding proteins and can refold aggregated proteins, Hsc70 is constitutively expressed and performs functions related to normal cellular processes. Hsc70 was placed in the heat-shock protein family due to homology with other heat-shock proteins, but the biological functions of both isoforms differ. A good example is the case of the murine epithelial sodium channel, whose surface expression and function are decreased by Hsc70, whereas both properties are favored by Hsp70 (Goldfarb *et al.*, 2006). The activity of Hsp70s is regulated by cofactors, the most important being Hsp40. The J-domains of these proteins interact with the ATPase domain of Hsp70 and stimulate the hydrolysis of bound ATP (Kampinga *et al.*, 2010). The release of nucleotide and substrate is further accelerated by nucleotide-exchange factors. It was originally shown that Hsp70 and Hsp40 are required for ubiquitin-dependent degradation of short-lived and abnormal proteins (Bercovich *et al.*, 1997). The overexpression of the Hsp70/Hsp40 complex decreases the level of abnormal proteins and improves viability in cellular models of certain neurodegenerative diseases characterized by the accumulation of aberrant proteins, such as Parkinson's disease, Huntington's disease, and spinal and bulbar muscular atrophy (Adachi *et al.*, 2003; Bailey *et al.*, 2002; Chan *et al.*, 2000; Jana *et al.*, 2000; Klucken *et al.*, 2004). It was originally thought that both Hsp90 and Hsp70 worked to promote refolding of proteins that aggregated in stressed cells, but it has subsequently become clear that it is Hsp70 the chaperone that focuses unfolded proteins on proteasomal degradation, whereas Hsp90 shows a protective role, as it is suggested by the fact that its inhibition with geldanamycin promotes protein degradation.

Hsp90 is one of the most abundant proteins in the eukaryotic cells, comprising ~2-3% of the total soluble proteins in normal conditions, and raises to 6-10% under stress conditions (Pratt *et al.*, 2003). It is evolutionarily conserved among species, and is proven essential for cell survival. Its contribution to various cellular processes, including signal transduction, protein folding and degradation, and morphological evolution, has been extensively studied and reviewed (DeZwaan *et al.*, 2010; Kadota *et al.*, 2010; Luo; Pratt *et al.*, 1997; Sreedhar *et al.*, 2004; Trepel *et al.*, 2010). Hsp90 is primarily a cytosolic protein, but a small portion rapidly accumulates in cell nuclei upon stress. There are two isoforms for this protein identified as Hsp90-α (inducible form/major form) and Hsp90-β (constitutive form/minor form). Its analogues include Grp94 in the endoplasmic reticulum (ER), and Hsp75/TRAP1 in the mitochondrial matrix. The full functional activity of Hsp90 is displayed in concert with other co-chaperones, playing an important role in providing specific biologic activity to proteins that already shows a tertiary structure and stable conformation. For example, Hsp90-binding to steroid receptors or protein-kinases is essential for hormone binding or enzymatic activity, activities that were not missing before Hsp90 binding as a consequence of a denatured state. Hsp90 forms dimers and is phosphorylated, containing two to three phosphate molecules per monomer. It is thought that these phosphorylations enhance its function.

Two types of cochaperone interactions have been described for hsp90. Some proteins that contain TPR (tetratricopeptide repeats) domains have been shown to bind near the C-terminus of Hsp90, and it appears that these proteins compete for the same site or overlapping sites on Hsp90, which has functional implications (Pratt *et al.*, 2004a). Thus, binding of TPR proteins with inhibitory or stimulatory action of the client protein can regulate diverse aspects of the cell physiology (Galigniana *et al.*, 2010b; Gallo *et al.*, 2007). The other binding site for cochaperones is where the small acidic cochaperone p23 binds, and comprises the N-terminal ATP-binding site plus a downstream domain, all in an Hsp90-dimer arrangement.(Chadli *et al.*, 2000). It enters at a late stage of the cycle and enhances the maturation of client proteins stabilizing their interactions with Hsp90. It appears that p23 binding to Hsp90 changes the environment of the ATP bound to Hsp90 and prevents the intrinsic ATPase activity of the chaperone (Karagoz *et al.*, 2011), which in turn affects the biological activity of the Hsp90-bound client factor. In addition to this stabilizing function, p23 also shows independent functions, the most surprising being its cytosolic activity of glutathione-dependent prostaglandin E2 synthase (Tanioka *et al.*, 2000). It has been shown in mice that p23 is not essential for prenatal development and morphogenesis, but it is absolutely necessary for perinatal survival (Grad *et al.*, 2006). Thus, $p23^{-/-}$ mice show an incomplete skin barrier, the lungs display underdeveloped airspaces and substantially reduced expression of surfactant genes. This correlates with the known function of glucocorticoids in promoting lung maturation and the role of p23 in the assembly of a hormone-responsive GR•Hsp90 complex. Thus, p23 contributes a non-redundant, temporally restricted, and tissue-specific function during mouse development.

TPR-Domain Immunophilins

Immunophilins comprise a family of intracellular proteins with peptidyl-prolyl-*cis/trans*-isomerase (PPIase) activity, i.e., *cis↔trans* interconvertion of Xaa-Pro bonds. Immunophilins are classified by their ability to bind immunosuppressant drugs –*cyclophilins* bind cyclosporine A, and *FKBPs* (*FK506-binding proteins*) bind FK506 (Kang *et al.*, 2008; Pratt *et al.*, 1997). The signature domain of the family is the PPIase domain. Only the low molecular weight immunophilins FKBP12 and CyPA are related to the immunosuppressive effect when the drug•IMM complex inhibits the Ser/Thr-phosphatase activity of PP2B/calcineurin. High molecular weight immunophilins have three additional domains –the nucleotide-binding domain, where ATP binds, the calmodulin-binding domain, a poorly characterized domain able to interact with calmodulin, and TPR domains , through which they bind to Hsp90 (Davies *et al.*, 2005). TPR-domain immunophilins are abundant and ubiquitous proteins that were first discovered associated to steroid receptors. To date, the biological function of these proteins is poorly understood. In 2001, it was reported that the dynein/dynactin complex binds to FKBP52 favoring the glucocorticoid retrotransport (Galigniana *et al.*, 2001). Complexes of immunophilins and motor proteins have also been related to other proteins' trafficking, such as mineralocorticoid receptor (Galigniana *et al.*, 2010b), androgen receptor (Thomas *et al.*, 2006), p53 (Galigniana *et al.*, 2004a), the AIF/Rac3 complex (Colo *et al.*, 2008), adeno-associated virus 2 (Zhao *et al.*, 2006), etc., suggesting that factors able to interact with the hsp90•FKBP52 complex could share a similar molecular machinery of movement. Once the transported complex has reached the nuclear envelope via microtubules, it must traverse the nuclear pore. Interestingly, the Hsp90-based heterocomplex (including FKBP52, but not the highly homologous partner FKBP51) is able to interact with the nuclear import machinery, including nucleoporins and importins (Echeverria *et al.*, 2010).

Most TPR-domain immunophilins can interact with motor proteins (Galigniana *et al.*, 2002), but this property or the ability of favoring the retrotransport of proteins is also shared by low molecular weight cyclophilins (Galigniana *et al.*, 2004b; Rycyzyn *et al.*, 2000). Nonetheless, the best characterized are the TPR-domain immunophilins because they were found associated to the steroid receptor•Hsp90 complex, in particular FKBP52 and FKBP51. Both proteins show 60% identity and 75% similarity (Cox *et al.*, 2007), bind hsp90 via TPR domains, contain an active PPIase domain, and adopt similar conformations according to X-ray crystallography images (Sinars *et al.*, 2003; Wu *et al.*, 2004). Nonetheless, their functions differ drastically. FKBP52 is a positive regulator of the activity of the glucocorticoid receptor, progesterone receptor and androgen receptor, but not of estrogen receptor and mineralocorticoid receptor, whereas FKBP51 has generally been regarded to be a negative regulator of steroid receptor activity as its overexpression prevents the regulation by FKBP52 due to its competition for binding to the receptor complex. Importantly, FKB52 links steroid receptor with microtubules and favors their retrotransport, whereas FKBP52 itself is able to interact with cytoskeletal structures (Czar *et al.*, 1994; Chambraud *et al.*, 2007; Perrot-Applanat *et al.*, 1995). Perhaps related to this property and that on protein trafficking, it was shown recently that the macrolide FK506 promotes neuroregeneration via FKBP52, the early subcellular relocalization of FKBP52 being required for the efficient outgrowth of neurites (Quintá *et al.*, 2010). On the other hand, FKBP51 antagonizes the effects of FKB52.

Cytoskeletal Architecture is Shaped by Molecular Chaperones

A small increase in temperature cause rapid and significant protein unfolding along with entanglement of cell architecture and, in a relatively extreme situation, non-specific aggregation of proteins. Therefore, many of the morphological and phenotypic effects of heat stress could be explained as a side effect of the aggregation of proteins with the consequent loss of protein homeostasis. In other words, it is reasonable to assume that the deleterious accumulation of improperly folded proteins is a signal that must trigger homeostatic countermeasures, which indirectly implies that the cell itself could not recognize changes of temperature as such, but via the balance between molecular chaperones and unfolded proteins. There is a plethora of stresses beyond the heat-shock that trigger similar responses, such as metabolic toxics, oxidations, heavy metals, ethanol, irradiation, glucose deprivation, disturbance of calcium homeostasis, etc. One of the major damages observed in response to most of these stress conditions are defects of the cytoskeleton. Mild heat stress leads to the reorganization of actin filaments into stress fibers, while severe heat stress results in the aggregation of vimentin or other filament-forming proteins, leading to the collapse of intermediary, actin, and tubulin networks. Along with the disruption of the cytoskeleton, the loss of the correct localization of organelles and a breakdown of intracellular transport processes are also observed. The cytoskeletal network is not independent for the different types of filaments. When microfilaments are inhibited at the periphery of the cell, an immediate response is the depolymerization of microtubules from the cell edge (Kaverina *et al.*, 2002).

Unlike heat-shock proteins, changes in IF levels in response to stress are dependent on both the tissue and gene involved. For example, the astrocytic GFAP expression remain unaltered after heat exposure (Goldbaum *et al.*, 2009), while vimentin is heat-sensitive since it is elevated in non-neuronal cells and in the microvasculature of the hippocampus (Bechtold *et al.*, 2003), but it is diminished in heat-exposed rat embryos (Fisher *et al.*, 1996) whose de novo synthesis of Hsp70 and Hsp90 is very rapidly induced (1-8 h after exposure) (Fisher *et al.*, 1995). Hsp70 is thought to participate in de novo folding pathways of cytoskeletal proteins along with their cognate chaperonines (Hartl *et al.*, 2009). Notably, the mechanism in eukaryotes involves a physical interaction between Hsp70 and CCT/TRiC near the opening of the chaperonin cavity, indicative of a direct hand-off of substrates (Cuellar *et al.*, 2008). It is unclear whether or not nascent actin and tubulin proteins use the Hsp70 chaperone under certain circumstances.

Although αβ-tubulin heterodimers are very stable, the mean half-life of tubulin does not reflect the fact that tubulin is partitioned into multiple states that could have very different turnover rates (i.e. chaperone–monomer complexes, GTP- or GDP-bound dimers, dynamic or stable microtubules, etc.). Tubulin can be rapidly degraded in response to destabilizing drugs such as colcemid, which increases the concentration of soluble tubulin. From the biological perspective, it appears that chaperonin CCT/TRiC is a major factor responsible for the stability of microtubules. In the absence of CCT/TRiC function, tubulin cannot fold efficiently and non-native tubulin is removed by proteolysis (Grantham *et al.*, 2006). The degree of growth arrest strongly correlates with the extent of CCT/TRiC depletion, indicating that its full activity is required for normal cell growth and division. Depletion of CCT/TRiC does not affect actin polypeptide synthesis, but causes a reduction in levels of native actin and perturbation of actin-based cell motility, and there are no large-scale effects on cytoplasmic protein synthesis or a general heat-shock response during periods of low CCT/TRiC activity. The integrity of the entire cytoskeleton is closely linked to this chaperonin via the folding requirements of actin and tubulin, but the role of this chaperonin in cytoskeletal organization remains still unclear. Recent evidence suggests that some CCT/TRiC monomer subunits can influence cytoskeletal organization and polymerization, so the biological activity of the chaperonin may extend beyond the folding of newly synthesized polypeptides, representing other functions for individual subunits distinct from their role in the oligomer (Brackley *et al.*, 2010).

Hsp90 also protects tubulin against thermal denaturation and keeps it in a state compatible with microtubule polymerization (Weis *et al.*, 2010). While Hsp90 cannot resolve tubulin aggregates, the chaperone binds early unfolding intermediates preventing their further aggregation. Protection is maximal at a stoichiometry of two molecules of Hsp90 per molecule of tubulin and does not require ATP, although it is counteracted by specific Hsp90 inhibitors such as geldanamycin. More generally, besides stress conditions, it is known that microtubules reorganize at the interphase-mitosis (G_2/M) transition leading to the decrease of the cellular microtubule quantity (Zhai *et al.*, 1996). Concomitantly, the free tubulin cytosolic concentration temporarily increases such that, at this point of cell cycle, Hsp90 can participate in maintaining tubulin dimer integrity and making it available to form the mitotic spindle.

Tau proteins also interact with tubulin to promote tubulin assembly into microtubules, which in turn are stabilized by this association by two ways, due to isoform-specific binding and phosphorylation by serine/threonine kinases. There are six tau isoforms in human brain tissue that are identified by the number of binding domains. Tau is thought to be an inherently unfolded protein that undergoes aggregation by hyperphosphorylation, folding and protein cleavage (Guillozet-Bongaarts *et al.*, 2005; Weaver *et al.*, 2000). These post-translational modifications can impact the interaction of tau with microtubules and thus, there may be specific isoforms of tau that are preferred chaperone substrates relative to others. It has recently been reported that chaperone proteins, including Hsp27, Hsp70 and CHIP (carboxy-terminus of Hsc70-interacting protein), can recognize abnormal tau and reduce its concentration by facilitating its degradation and dephosphorylation. Hsp27 preferentially binds to hyperphosphorylated tau as well as paired helical filamentous tau, but not to non-phosphorylated tau (Shimura *et al.*, 2004). The expression of another small Hsp, alphaB-crystallin has been found in glial inclusions of tauopathies (Dabir *et al.*, 2004). Hsp27 also increases tau phosphorylation at Ser^{262}, whereas alphaB-crystallin decreases phosphorylated tau and GSK-3β levels (Bjorkdahl *et al.*, 2008). Interestingly, Hsp27 is cross-linked with tau in neurofilaments from Alzheimer's disease brains, and positive correlations have been found in the soluble protein levels from of these brain tissues between tau and molecular Hsp27, Hsp40, Hsp90, alphaB-crystallin and CHIP (Sahara *et al.*, 2007). Conversely, the levels of heat-shock proteins were inversely correlated with the levels of granulated tau oligomers, an intermediate of tau filaments. In a separate study, increased levels of Hsp70 and Hsp90 were found to promote tau solubility and microtubule binding in various cellular models (Dou *et al.*, 2003). Subsequent studies show that tau binds directly to Hsp70, and Bag1 has a role in this interaction (Elliott *et al.*, 2007). Taken together, these findings suggest that chaperones are necessary to maintain tau in a non-aggregated state, a consequence that may ultimately be deleterious for the brain. A novel mechanism demonstrating that there are two pathways that can lead to opposing outcomes for tau biology with regard to Hsp90; some chaperones preserve tau whereas others promote its degradation (Dickey *et al.*, 2007). Furthermore, mutant, but not wild-type tau, is maintained in tauopathies by Hsp90; and the inhibition of Hsp90 leads to reductions in the pathogenicity of these mutant species (Luo).

In addition to chaperone-assisted folding and assembly, covalent modification is often an integral part of the biogenesis pathway leading to functional proteins. These post-translational modifications affect folding or they can act as reversible switches able to regulate the features of cytoskeletal proteins. In the case of actin, there is little or no evidence in this regard, whereas tubulin is subject to a number of reversible modifications that includes detyrosination, glutamylation/glycylation and acetylation (Hammond *et al.*, 2008).

Detyrosination involves the removal of the gene-encoded C-terminal tyrosine of α-tubulin in microtubule polymers by a still unidentified carboxypeptidase, this detyrosination being important for kinesin-1 preferential binding to microtubules (Reed *et al.*, 2006) and the consequent transport of vimentin filaments (Liao *et al.*, 1998). On the other hand, glutamylation/glycylation involve the addition of glutamate or glycine residues onto glutamate residues in the C-terminal tails of both α-and β-tubulin (Verhey *et al.*, 2007). While glycylation is mainly limited to tubulin incorporated into axonemes (cilia and flagella), glutamylation is more prevalent in neuronal cells, centrioles, and the mitotic spindle. Both modifications have been found on the same tubulin C-terminal end and it has been proposed that there are cross-talks between the α- and β-tubulin tails that regulate the type and extent for each post-translational modification (Duan *et al.*, 2002).

It is accepted that most tubulin modifications occur on polymerized microtubules, whereas the substrate for the reversal of these modifications is the free tubulin heterodimer (Westermann *et al.*, 2003). Nonetheless, this heuristic model still deserves more conclusive experimental support. Some of the tubulin-modifying enzymes might compete or cooperate with other factors involved in regulating microtubule function. Thus, the class II histone-deacetylase HDAC6, known to function as a potent α-tubulin deacetylase, binds to several microtubule-binding proteins such as plus-end marker proteins (Zilberman *et al.*, 2009). Recent evidence suggests that tubulin acetylation is involved in neurodegenerative diseases (Creppe *et al.*, 2009). The list of HDAC6 partners with which the enzyme can interact has expanded considerably during last several years. Unquestionably, the regulation of HDAC6 activity by Hsp90 is one of the most striking (Kovacs *et al.*, 2005; Murphy *et al.*, 2005), the chaperone itself being substrate of the enzyme. When HDAC6 was knocked-down, Hsp90 was hyperacetylated and much less of the Hsp90 cochaperone p23 was recovered associated to Hsp90. The glucocorticoid receptor becomes defective in ligand binding, nuclear translocation, and gene activation. The acetylated Hsp90 has a lower affinity for ATP than the wild-type Hsp90, and the dynamic the glucocorticoid receptor•Hsp90 heterocomplex assembly/disassembly that occurs in HDAC6 knockdown cells is manifest as a ~100-fold shift to the right *versus* wild-type cells in the dose response of steroid for transcriptional activation. However, the receptor itself is not affected by HDAC6.

Several other targets besides tubulin and Hsp90 were shown to be deacetylated by HDAC6. Among them, cortactin, an actin-binding protein promoting Arp2/3-dependent actin nucleation (Ammer *et al.*, 2008), whereas Hsp90 can regulate actin polymerization by interacting with and enhancing the function of the Neural Wiskott-Aldrich syndrome protein, N-WASP (Park *et al.*, 2005). Involvement of Hsp90 in HDAC6-dependent actin remodeling was also recently reported (Gao *et al.*, 2007). Because assembly of tubulin is affected by Hsp90, both targets of HDAC6, cortactin and Hsp90, could mediate the effect of HDAC6 inhibition on microtubules acting, directly or indirectly, via remodeling of the actin cytoskeleton. The putative participation of Hsp90 or cortactin in the effect of HDAC6 inhibition on microtubule dynamics requires the physical presence of HDAC6, rather than just increased acetylation of Hsp90 or cortactin, since HDAC6 knockdown, which enhances such acetylation, is not sufficient for alteration of microtubule dynamics.

A significant fraction of Hsp90 is also associated to microfilaments along with the cochaperone p23 (Czar *et al.*, 1996). When cells are treated with colcemid under conditions that eliminate microtubules, the fibrils condense into bright rope-like bundles located in the immediate perinuclear area and extending toward the cell periphery. Identical images are observed with an antibody against vimentin and the pattern seen after colcemid treatment is classical for IFs. Heat-shock causes the spatial arrangement of IFs to change from a fine thread-like network to a large, perinuclear aggregate, whereas in thermotolerant cells, integrity of the IF network is enhanced, and maintenance of this organization is required for establishment of increased heat resistance (Lee *et al.*, 1995). When brain tumor cells are treated before heat-shock with angulatin A, a drug that disrupts vimentin intermediate filaments, they do not acquire thermotolerance, while those treated after receiving a heat-shock gain protection. The vimentin intermediate filaments in cells exposed to the drug prior to heating are arranged in tight perinuclear aggregates unable to return to their normal distribution upon recovery. In contrast, IFs in cells treated with the drug subsequent to thermal stress resume a dispersed organization. Hsc70 colocalizes with vimentin during these spatial changes, most likely favoring IF stabilization.

Role of Immunophilins in the Organization of Cytoskeleton

A high percentage of prolines are common to most intrinsically disordered proteins, and tau is no exception (Elbi *et al.*). Nearly 10% of full-length tau is composed of proline residues and more than 20% of the residues between I^{151} and Q^{244} are proline. Most of the known functions of tau are mediated through microtubule binding domains distal to this proline-rich region. Recent studies have revealed that the peptidylprolyl isomerase immunophilins FKBP51 and FKBP52 interact directly with tau protein (Chambraud *et al.*, 2010; Jinwal *et al.*, 2010). It has also been shown that tau phosphorylation increases its binding affinity for FKBP52, whereas the overexpression of the immunophilin could prevent the accumulation of tau protein induced by nerve growth factor. Interestingly, it has also been reported that FKBP52 inhibits tubulin polymerization, which is triggered by tau protein in vitro (Chambraud *et al.*, 2010). On the other hand, FKBP51 stabilizes microtubules in a PPIase activity-dependent manner and also enhances the association of tau with Hsp90, although the FKBP51/tau interaction is not dependent on the chaperone (Jinwal *et al.*, 2010). In this complex, phosphorylated tau is in a *trans* configuration, and the PPIase activity of FKBP51 isomerizes tau to a *cis* configuration, which enhances dephosphorylation of tau by the immunophilin-like phosphatase PP5. This is a critical event because dephosphorylated tau is recycled to microtubules and stabilizes them. On the other hand, highly phosphorylated tau exhibits reduced microtubule binding, leading to a loss of microtubule integrity.

Because both immunophilins, FKBP51 and FKBP52, possess PPIase activity and are capable to bind phosphorylated tau, it is currently unclear why FKBP51 is selectively able to isomerize tau and promote microtubule polymerization. One possible explanation could be the fact that both TPR-domain immunophilins, as well as CyP40 and PP5, are capable to compete for a common binding-site in Hsp90 complexes such as those found in steroid receptors (Galigniana *et al.*, 2010b; Silverstein *et al.*, 1999), so it would be possible that a similar competition may take place in the phospho-tau•Hsp90 heterocomplex containing a TPR-domain cochaperone. In turn, the qualitative composition of such heterocomplex may depend on the relative abundance of TPR proteins in a given cell, just like it has been postulated for steroid receptors (Banerjee *et al.*, 2008; Galigniana *et al.*, 2010b; Gross *et al.*,

2008). Moreover, structural comparison of FKBP51 and FKBP52 revealed that the TPR domains have different spatial orientations (Wu *et al.*, 2004).

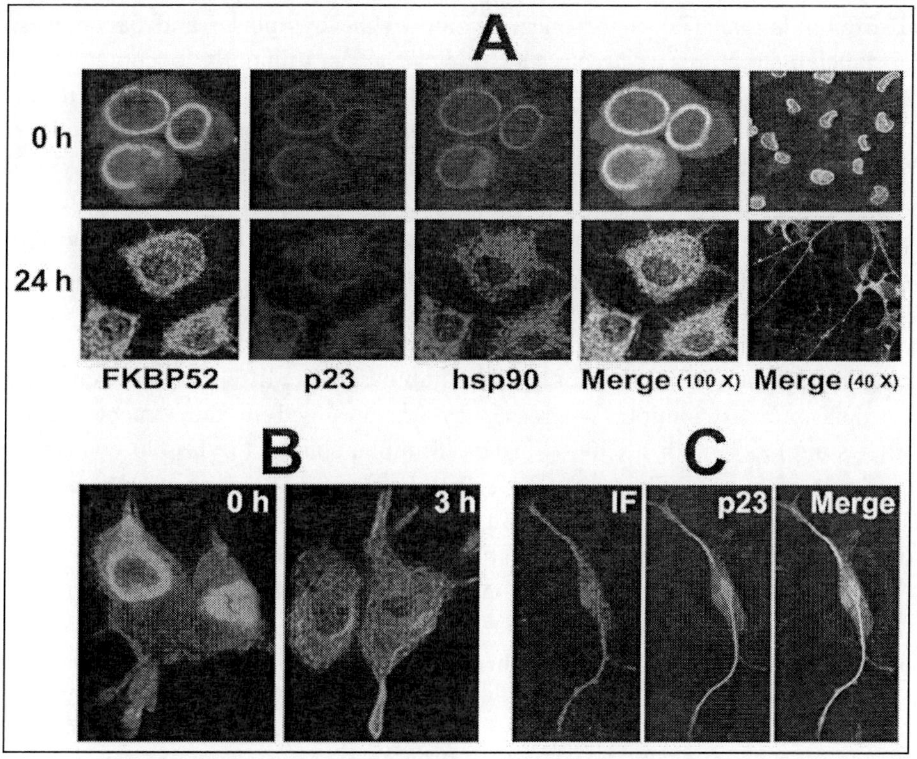

Figure 1. *Subcellular redistribution of chaperones and cytoskeleton rearrangement.* (A) Images by confocal microscopy of embryonic hippocampal neurons prior and 24 h after treatment with 1 μM FK506. Image on the right hand shows a wider field. (B) Microtubules reorganize after 3 h of differentiation. (C) Intermediate filaments (IF) of FK506-differentiated neurons recruit the Hsp90-cochaperone p23.

That FKBP52 and FKBP51 are important effectors of neurite length may explain, at least in part, the observations that the immunophilin ligand FK506 exhibits potent neurotrophic effects (Quintá *et al.*, 2010). FK506 has been shown to promote neuroprotective and neuroregenerative effects in a number of injury models. The persistence of the capability to hasten nerve outgrowth by FK506-derivatives devoid of immunosuppressive effects indicated that both effects are independent (Gold *et al.*, 2003), and experiments with mice where the *FKBP1A* gene was knocked-out clearly showed that such neurotrophic effect is not mediated by FK506•FKBP12 complexes (Gold *et al.*, 1999). Recently, it was demonstrated that the FKBP52•Hsp90•p23 complex forms a perinuclear structure in undifferentiated N2a neuroblastoma cells and hippocampal cells isolated from rat embryos (Quintá *et al.*, 2010). This peculiar structure undergoes a rapid subcellular redistribution along the cytoplasm (Fig.1A), which is detectable within the first hour of stimulation with FK506, FKBP52 being concentrated in terminal axons and arborization areas. In agreement with the potential importance of this immunophilin, knock-down experiments showed that FKBP52 plays a key role in the architecture of these nuclear rings since these structures faded in most cells (if not all of them) where the expression of the immunophilin was abrogated. Importantly, both the rate of cell differentiation and neurite outgrowth were also inhibited. Therefore, there is a direct relationship between the disassembly of the chaperone complex and the progression of neuronal differentiation. The chaperones migrate to the cytoplasm and become associated with cytoskeletal structures, whereas the nuclear areas originally occupied by them in undifferentiated cells become transcriptionally active after they redistribute. In contrast to FKBP52, FKBP51 is not induced during differentiation, it remains in the cell body and replaces FKBP52 in the annular structures of the nucleus complexed with hsp70 (Quintá *et al.*, 2010). While the overexpression of FKBP52 induces fast differentiation of N2a neuroblastoma cells and neurites were long, the opposite action is observed after knocking-down FKBP52. On the other hand, FKBP51 overexpression decreases both the neurite length and the rate of cell differentiation, and its knock-down favored neurite outgrowth. In other words, it appears that both immunophilins have antagonistic action during the very early steps of neuronal differentiation. Surprisingly, microtubules of undifferentiated neurons show a diffuse cytoplasmic distribution, whereas they are arranged in the expected filamentary pattern after stimulation with FK506 (Fig.1B). Simultaneously, the Hsp90 cochaperone p23 becomes associated to neurofilaments (Fig.1C) and seems to be dissociated from both Hsp90 and FKBP52. Importantly, after 5-6 days in culture with FK506 and no other trophic factor added to the medium, embryonic day 17 rat hippocampal neurons show a new redistribution of FKBP52 only, such that the immunophilin is concentrated back in nuclei but, in contrast to what it was observed in the undifferentiated state, in a diffuse manner without showing any particular subnuclear arrangement. All these observations suggest an active role of the immunophilin•Hsp90•p23 complex in the arrangement of the neuronal cytoskeleton during the early steps of cell differentiation. The above-described observations by Quintá et al. apparently collide with those reported by Chambroud (Chambraud *et al.*, 2007; Chambraud *et al.*, 2010) in that these studies reported that overexpression of FKBP52 reduced neurite outgrowth in response to 5 days in culture with nerve growth factor in PC12 pheochromocytoma cells always maintained in a medium with fetal serum. However there are important experimental differences between both studies, in particular those referred to the cell type, length of stimulation in culture and culture conditions.

S100A proteins constitute a large family of calcium binding proteins which are linked to many neurological disorders, inflammatory diseases, innate immunity, wound healing, and cancer processes (Wright *et al.*, 2009). Several S100A proteins can bind to the TPR domain of FKBP52 protein, which leads to the dissociation of the Hsp90•FKBP52 complex in the presence of calcium (Shimamoto *et al.*, 2010). Apart from the possible crosstalk between S100A proteins and pro- or anti-angiogenic factors, S100A4 and A13 have been reported to participate directly in the angiogenic process in other tissues. Because the common denominator able to affect the biological activity of S100A proteins in all these biological processes is calcium homeostasis, it could be implied that the function of FKBP52 can be regulated by calcium within the cell, just like tau phosphorylation and pathology in Alzheimer disease are regulated. Moreover, FKBP52 counts with a calmodulin binding domain, whose biological function is still enigmatic. Recently, FKBP52-dependent regulation of TRPC calcium channels was reported in neurons (Shim *et al.*, 2009).

The immunophilin-like protein phosphatase 5 (PP5) has three TPR units for binding to the Hsp90 chaperone, a PPIase-like domain, and the Ser/Thr phosphatase sequence (Hinds *et al.*, 2008). PP5 can bind to trafficking glucocorticoid receptor heterocomplexes, is associated to microtubules and interacts with dynein motors (Galigniana *et al.*, 2002; Galigniana *et al.*, 1999; Silverstein *et al.*, 1997). The binding to Hsp90 complex activates the PP5 enzyme which is inactive outside the complex. PP5 dephosphorylates tau protein which was phosphorylated with cAMP-dependent protein kinase and GSK-3β kinase, and this dephosphorylation of tau is connected to the aggregates formed in Alzheimer disease (Liu *et al.*, 2005). The role of this microtubule-associated immunophilin in the protection against tau pathology is largely unknown, although it is inferred that PP5, which is highly expressed in mammalian brain, possesses a great potential to dephosphorylate tau protein e.g. in Hsp90-FKBP51 complexes and enhance the binding of tau to microtubules.

Microtubule Architecture, Chaperones and Disease

The dynamic state of microtubules and their abundance may directly change the stiffness of the cytoskeleton, which in turn, influences the contractility of myocytes (Ishibashi *et al.*, 2003). Progressive increases in microtubule density have been observed in severe pressure overload cardiac myocytes (Tagawa *et al.*, 1998). It has been well established that increased density of the microtubule network is one cause of contractile dysfunction characteristic of severe pressure overload cardiac hypertrophy (Kumarapeli *et al.*, 2004). Microtubule network densification imposes a viscous load on active myofilaments, such that structural damping via intracellular frictional dissipation impedes sarcomere shortening. The physiological and pathophysiological roles of several heat-shock proteins in the heart, especially during cardiac ischemia/reperfusion injury have been further defined in recent years by transgenesis and gene targeting. The small Hsp-subfamily of Hsp27 (Hsp25 in murine) and notably aB-crystallin (CryAB) are among the most intensely studied stress proteins that associate with the muscle intermediate filaments and microtubules (Kumarapeli *et al.*, 2004). The consensus appears to be that loss-of-function of molecular chaperones compromises the ability of the heart to handle stress while their gain-of-function is cardioprotective. Thus, Hsp70 is required for late-phase, but not early-phase protection by ischemic preconditioning of the heart. Late-phase reduction of infarct size by ischemic preconditioning was completely abolished by ablation of the gene encoding for the inducible form of Hsp70.

In the several neurodegenerative disease models, overexpression of Hsp70 also improved the severities of several disease phenotypes without visibly affecting aggregate formation, suggesting that chaperones do not prevent aggregation *per se*, but rather redirect aggregates into amorphous deposits, thereby sequestering potentially toxic species from bulk solution. In Alzheimer's disease, cognitive dysfunction and neuronal loss are directly linked to the intracellular accumulation of the microtubule-associated protein tau into filamentous tangles (Koren *et al.*, 2011; Roberson *et al.*, 2011). Tau pathology is also found in approximately 15 other neurodegenerative diseases, some caused by mutations in the tau gene (Hardy *et al.*, 2006), whereas tau expression is increased in some cases of Parkinson's disease (Simon-Sanchez *et al.*, 2009). Proteins that are able to constrain the structural freedom of tau are essential for tau processing, so interactions of tau with molecular chaperones and immunophilins revealed the importance of the regulation of tau phosphorylation cycles and overall stability (Jinwal *et al.*, 2010; Luo).

Importantly, recent studies have revealed microtubule stabilization as a mechanism complementary to actin dynamics underlying neuronal polarization. Moreover, stable microtubules in the axon may form a landmark to confer identity to the axon, such that it is currently required to assess whether or not cytoskeletal effectors can affect and directly regulate neuronal polarity (Witte *et al.*, 2008a). In line with their importance for the reorganization of cellular shape in other cell types, microtubules have been found to play an active role in axon specification, complementary to the actin network. Axonal microtubules show increased stability, and microtubule stabilization is sufficient to induce axon formation in dissociated unpolarized hippocampal neurons (Witte *et al.*, 2008b). The microtubule regulator collapsing response mediator protein-2 (CRMP-2) binds tubulin heterodimers and promotes increased microtubule assembly during polarization (Fukata *et al.*, 2002). On the other hand, glycogen synthase kinase-3b (GSK-3β) regulates the activity of CRMP-2 and the microtubule affinity of microtubule associated proteins (Yoshimura *et al.*, 2005), so the pharmacological inhibition of GSK-3β promotes the formation of multiple axons by microtubule stabilization.

It is clear that the cytoskeletal proteostasis machinery will be engaged in rapidly growing tumor cells and in their response to antitubulin chemotherapeutics. In a novel anti-mitotic approach, a recent study showed that disrupting the CCT/TRiC•β-tubulin interaction with a small molecule induces apoptosis even in cancer cells resistant to other tubulin-binding chemotherapies (Lin *et al.*, 2009). CCT/TRiC has another link to cancer whereby its interaction with the von Hippel–Lindau (VHL) tumor suppressor protein is essential for assembly of the VHL E3-ubiquitin ligase complex; tumor-inducing mutations in VHL specifically disrupt CCT/TRiC binding (Feldman *et al.*, 2003). Bisphosphonates, which are used to treat cancer-induced bone loss (and which are the subject of clinical trials against breast cancer due to their effects on the migration, invasion and proliferation of cancer cells) trigger a strong up-regulation of TBCB and prominent disruption of the microtubule cytoskeleton (Bivi *et al.*, 2009). Drugs targeting tubulin folding cofactors could be beneficial because they would be tubulin-specific and could have anti-proliferative and anti-metastatic effects (Jo *et al.*, 2010). Tubulin is degraded in response to microtubule-destabilizing drugs, so a parallel increase in the concentration of soluble tubulin is observed. When the function of CCT/TRiC is impaired, tubulin does not fold efficiently and the non-native tubulin is rapidly removed by proteolysis. Accordingly, treatments of cells with the proteasome inhibitor MG132 demonstrated that α-tubulin and β-tubulin are highly ubiquitylated (Meierhofer *et al.*, 2008). Moreover, tubulin from various human cell types is targeted by cancer-preventive isothiocyanates which covalently modify tubulins and induce conformational changes that trigger proteasome-dependent tubulin degradation (Sellin *et al.*, 2008). Interestingly, the overexpression of β-tubulin is toxic in *Sacharomyces cerevisiae*, but α-tubulin overexpression is well tolerated (Burke *et al.*, 1989). However, this particular effect is not observed in mammalian cells, perhaps due to the autoregulation mechanisms for β-tubulin degradation.

Microtubules are one of the major targets with a proven record among chemotherapy drugs. Compounds such as taxans, taxoteres, vinca alkaloids, etc., bind to α- or β-tubulin and alter the organization of microtubules (Yamada *et al.*, 2010). Whereas the first two classes of drugs stabilize microtubules leading to the loss of microtubule elasticity, vinca alkaloids destabilize and dissolve microtubules. The effect on microtubule inhibition is most prominent in mitotic cells where mitotic spindles play a cardinal role in chromosome segregation, and this is why these types of drugs have been classified as anti-mitotic agents. Some chaperone factors such as survivin are known to localize to the mitotic spindle by interaction with tubulin during mitosis and regulate mitosis. Survivin is known to be expressed across most tumour cell types and is at the same time absent in normal non-malignant cells, which transforms this chaperone in a potential target. The naturally occurring compound resveratrol (a polyphenol found in grapes and red wine) can be used as a sensitizer for anticancer drug-induced apoptosis by the action of causing cell cycle arrest (Fulda *et al.*, 2004). This cell cycle arrest causes a dramatic decline in survivin levels in the cells. Survivin is a dual functioning protein overexpressed in many types of cancer (Sah *et al.*, 2006). It inhibits the apoptosis of cancer cells by inhibiting caspases, and also promotes cancer cell growth by stabilizing microtubules during mitosis. Importantly, Hsp90 binds and stabilizes surviving (Fortugno *et al.*, 2003), so down-regulation of survivin could be one of the important therapeutic functions of Hsp90 inhibitors such as geldanamycin derivatives. Hsp90 associated with survivin is overexpressed in cancers with roles in mitotic control and apoptosis inhibition. The cytoprotection mechanism of this survivin•Hsp90 association is centred on the mitochondrial pathway, where surviving has a role in the regulation of mitochondrial apoptosis specifically in tumours (Fortugno *et al.*, 2003). The disruption of the survivin•Hsp90 complex destabilizes surviving leading to mitochondrial apoptosis and ultimately cell growth suppression. However, Hsp90 interaction with survivin also enables stabilization of cofactors such as AKT, Erb-2 and HIF-1α, which can lead to tumour progression (Cheung *et al.*, 2010).

The association of Hsp90 with microtubules shows several implications. Many proteins responsible for malignant progression within tumor cells are Hsp90-dependent and dozens of oncogenic substrates have been identified to date. Hsp90 fulfils the six hallmarks of cancer (Donnelly *et al.*, 2008). Therefore, targeting Hsp90 can offer a unified mechanism for chemotherapy. In addition, Hsp90 is overexpressed in malignant cells, and its expression correlates with the proliferation of these cells (Csermely *et al.*, 1998). Hsp90 inhibitors represent a unique class of compounds that demonstrate high differential selectivity for malignant versus normal cells at concentrations that are well tolerated by humans (Neckers, 2002; Sausville *et al.*, 2003; Trepel *et al.*, 2010). There are currently twenty Hsp90-targeted clinical trials in progress and many more inhibitors are in preclinical development, in particular, geldanamycin-derivatives. Recently, it was shown that Hsp90 inhibits tubulin aggregation and that tubulin polymerization is slower in the presence of Hsp90 than in its absence (Weis *et al.*, 2010), indicating that the polymerization state of microtubules depends on the cell type and the stress conditions. Hsp90 sequesters soluble (i.e. unstable) tubulin dimers to prevent their irreversible aggregation and to maintain them in a conformational active state for several hours. Once the stress conditions have ended, tubulin is able to form microtubules again, thus, restoring their cellular functions. As it is expected, inhibitors of Hsp90 prevent this action and unstabilize the cell due to microtubule disassembly.

Taxol is a well-known drug for the treatment of cancer that is responsible for the stabilization of microtubules and inhibition of mitosis. Interestingly, taxol induces transcription factors and kinase activation, mimicking the effect of bacterial lipopolysaccharide (LPS), an attribute unrelated to its tubulin-binding properties (Ding *et al.*, 1990). Actually, a significant amount of evidence suggests that the LPS-mimetic activity of taxol is independent of β-tubulin binding. Affinity chromatography experiments with lysates from both mouse brain and macrophage cell lines led to affinity purification of two chaperones, Hsp70 and Hsp90 (Byrd *et al.*, 1999). In contrast to typical Hsp90-binding drugs, taxol exhibits a stimulatory response, mediating the activation of macrophages and exerting the LPS-mimetic effects observed. The geldanamycin derivative 17-AAG behaves synergistically with taxol-induced apoptosis due to sensitization of tumor cells by suppression of Akt (Solit *et al.*, 2003), a Ser/Thr protein kinase that is stabilized by Hsp90 as well as its activating kinase PDK1 (Basso *et al.*, 2002). Therefore, the potential use of Hsp90 inhibitors in combination with proapoptotic therapies represents an exciting new strategy for chemotherapy.

Envoy

Clearly, molecular chaperones perform different but cooperating roles in the regulation of cytoskeleton function. To date, many questions concerning the interplay of molecular chaperones and the cytoskeleton remain to be elucidated. Clarification of these issues requires a better understanding of our knowledge for both groups of proteins molecular chaperones and cytoskeletal elements. Nevertheless, this is not an easy issue. Within cells, the protein synthesis and folding take place at the ribosome surface. Both processes are rapid and occur in a protein-rich milieu that exerts different influences on the final product. As a case in point, it is not always possible to differentiate in vivo the individual effects of molecular chaperones that act cooperatively and/or sequentially on nascent proteins, so the problem was partially overcome by analysis of cells with mutated chaperones; however, this type of approach is limited to those organisms for which genetic manipulation is possible.

On the other hand, while studies in vitro allow defined conditions, they suffer from the limitations imposed by the artificial environment. Protein concentration and diversity in vitro are often much lower than in the cytoplasm and only one of many cooperative reactions influencing protein folding may be under study. Thus, a molecular chaperone might seem unrelated to a particular cytoskeletal component only because a cochaperone is missing. Conversely, molecular chaperones tend to recognize hydrophobic regions exposed to the aqueous environment, a generalized mechanism providing ample opportunity for non-specific interactions. This concern is especially pertinent in an in vitro situation, where relatively high levels of a limited number of proteins dissolved in low salt non-physiological buffers may drive inappropriate reactions. Artifactual protein-protein associations are also encountered when searching cells for microtubule-associated proteins, an example which provides a strong cautionary message to the study of chaperone-substrate interactions.

In spite of all these problems associated with their study, the available evidence proves univocally that molecular chaperones not only assist in the formation of cytoskeletal proteins, but also they influence the structural and functional organization of microtubules, microfilaments and intermediate filaments.

Acknowledgments

The experimental work described in this chapter was supported by grants from Agencia Nacional de Promoción Científica y Tecnológica (PICT 2010-1170) and Universidad de Buenos Aires (UBACyT Grant).

References

Adachi H, Katsuno M, Minamiyama M, Sang C, Pagoulatos G, Angelidis C, *et al.* (2003). Heat shock protein 70 chaperone overexpression ameliorates phenotypes of the spinal and bulbar muscular atrophy transgenic mouse model by reducing nuclear-localized mutant androgen receptor protein. *J Neurosci* 23(6): 2203-2211.

Albanese V, Yam AY, Baughman J, Parnot C, Frydman J (2006). Systems analyses reveal two chaperone networks with distinct functions in eukaryotic cells. *Cell* 124(1): 75-88.

Ammer AG, Weed SA (2008). Cortactin branches out: roles in regulating protrusive actin dynamics. *Cell Motil Cytoskeleton* 65(9): 687-707.

Aravind L, Koonin EV (1999). Gleaning non-trivial structural, functional and evolutionary information about proteins by iterative database searches. *J Mol Biol* 287(5): 1023-1040.

Aylett CH, Wang Q, Michie KA, Amos LA, Lowe J (2010). Filament structure of bacterial tubulin homologue TubZ. *Proc Natl Acad Sci U S A* 107(46): 19766-19771.

Bailey CK, Andriola IF, Kampinga HH, Merry DE (2002). Molecular chaperones enhance the degradation of expanded polyglutamine repeat androgen receptor in a cellular model of spinal and bulbar muscular atrophy. *Hum Mol Genet* 11(5): 515-523.

Banerjee A, Periyasamy S, Wolf IM, Hinds TD, Jr., Yong W, Shou W, *et al.* (2008). Control of glucocorticoid and progesterone receptor subcellular localization by the ligand-binding domain is mediated by distinct interactions with tetratricopeptide repeat proteins. *Biochemistry* 47(39): 10471-10480.

Basso AD, Solit DB, Chiosis G, Giri B, Tsichlis P, Rosen N (2002). Akt forms an intracellular complex with heat shock protein 90 (Hsp90) and Cdc37 and is destabilized by inhibitors of Hsp90 function. *J Biol Chem* 277(42): 39858-39866.

Bechtold DA, Brown IR (2003). Induction of Hsp27 and Hsp32 stress proteins and vimentin in glial cells of the rat hippocampus following hyperthermia. *Neurochem Res* 28(8): 1163-1173.

Bercovich B, Stancovski I, Mayer A, Blumenfeld N, Laszlo A, Schwartz AL, *et al.* (1997). Ubiquitin-dependent degradation of certain protein substrates in vitro requires the molecular chaperone Hsc70. *J Biol Chem* 272(14): 9002-9010.

Bivi N, Romanello M, Harrison R, Clarke I, Hoyle DC, Moro L, *et al.* (2009). Identification of secondary targets of N-containing bisphosphonates in mammalian cells via parallel competition analysis of the barcoded yeast deletion collection. *Genome Biol* 10(9): R93.

Bjorkdahl C, Sjogren MJ, Zhou X, Concha H, Avila J, Winblad B, *et al.* (2008). Small heat shock proteins Hsp27 or alphaB-crystallin and the protein components of neurofibrillary tangles: tau and neurofilaments. *J Neurosci Res* 86(6): 1343-1352.

Bork P, Sander C, Valencia A (1992). An ATPase domain common to prokaryotic cell cycle proteins, sugar kinases, actin, and hsp70 heat shock proteins. *Proc Natl Acad Sci U S A* 89(16): 7290-7294.

Brackley KI, Grantham J (2010). Subunits of the chaperonin CCT interact with F-actin and influence cell shape and cytoskeletal assembly. *Exp Cell Res* 316(4): 543-553.

Broadley SA, Hartl FU (2009). The role of molecular chaperones in human misfolding diseases. *FEBS Lett* 583(16): 2647-2653.

Burke D, Gasdaska P, Hartwell L (1989). Dominant effects of tubulin overexpression in Saccharomyces cerevisiae. *Mol Cell Biol* 9(3): 1049-1059.

Byrd CA, Bornmann W, Erdjument-Bromage H, Tempst P, Pavletich N, Rosen N, *et al.* (1999). Heat shock protein 90 mediates macrophage activation by Taxol and bacterial lipopolysaccharide. *Proc Natl Acad Sci U S A* 96(10): 5645-5650.

Cabeen MT, Jacobs-Wagner C (2010). The bacterial cytoskeleton. *Annu Rev Genet* 44: 365-392.

Capetanaki Y, Bloch RJ, Kouloumenta A, Mavroidis M, Psarras S (2007). Muscle intermediate filaments and their links to membranes and membranous organelles. *Exp Cell Res* 313(10): 2063-2076.

Colo GP, Rubio MF, Nojek IM, Werbajh SE, Echeverria PC, Alvarado CV, *et al.* (2008). The p160 nuclear receptor co-activator RAC3 exerts an anti-apoptotic role through a cytoplasmatic action. *Oncogene* 27(17): 2430-2444.

Cox MB, Riggs DL, Hessling M, Schumacher F, Buchner J, Smith DF (2007). FK506-binding protein 52 phosphorylation: a potential mechanism for regulating steroid hormone receptor activity. *Mol Endocrinol* 21(12): 2956-2967.

Creppe C, Malinouskaya L, Volvert ML, Gillard M, Close P, Malaise O, *et al.* (2009). Elongator controls the migration and differentiation of cortical neurons through acetylation of alpha-tubulin. *Cell* 136(3): 551-564.

Csermely P, Schnaider T, Soti C, Prohaszka Z, Nardai G (1998). The 90-kDa molecular chaperone family: structure, function, and clinical applications. A comprehensive review. *Pharmacol Ther* 79(2): 129-168.

Cuellar J, Martin-Benito J, Scheres SH, Sousa R, Moro F, Lopez-Vinas E, *et al.* (2008). The structure of CCT-Hsc70 NBD suggests a mechanism for Hsp70 delivery of substrates to the chaperonin. *Nat Struct Mol Biol* 15(8): 858-864.

Czar MJ, Owens-Grillo JK, Yem AW, Leach KL, Deibel MR, Jr., Welsh MJ, *et al.* (1994). The hsp56 immunophilin component of untransformed steroid receptor complexes is localized both to microtubules in the cytoplasm and to the same nonrandom regions within the nucleus as the steroid receptor. *Mol Endocrinol* 8(12): 1731-1741.

Czar MJ, Welsh MJ, Pratt WB (1996). Immunofluorescence localization of the 90-kDa heat-shock protein to cytoskeleton. *Eur J Cell Biol* 70(4): 322-330.

Chadli A, Bouhouche I, Sullivan W, Stensgard B, McMahon N, Catelli MG, *et al.* (2000). Dimerization and N-terminal domain proximity underlie the function of the molecular chaperone heat shock protein 90. *Proc Natl Acad Sci U S A* 97(23): 12524-12529.

Chambraud B, Belabes H, Fontaine-Lenoir V, Fellous A, Baulieu EE (2007). The immunophilin FKBP52 specifically binds to tubulin and prevents microtubule formation. *Faseb J* 21(11): 2787-2797.

Chambraud B, Sardin E, Giustiniani J, Dounane O, Schumacher M, Goedert M, *et al.* (2010). A role for FKBP52 in Tau protein function. *Proc Natl Acad Sci U S A* 107(6): 2658-2663.

Chan HY, Warrick JM, Gray-Board GL, Paulson HL, Bonini NM (2000). Mechanisms of chaperone suppression of polyglutamine disease: selectivity, synergy and modulation of protein solubility in Drosophila. *Hum Mol Genet* 9(19): 2811-2820.

Cheung CH, Chen HH, Cheng LT, Lyu KW, Kanwar JR, Chang JY (2010). Targeting Hsp90 with small molecule inhibitors induces the over-expression of the anti-apoptotic molecule, survivin, in human A549, HONE-1 and HT-29 cancer cells. *Mol Cancer* 9: 77.

Dabir DV, Trojanowski JQ, Richter-Landsberg C, Lee VM, Forman MS (2004). Expression of the small heat-shock protein alphaB-crystallin in tauopathies with glial pathology. *Am J Pathol* 164(1): 155-166.

Davies TH, Sanchez ER (2005). Fkbp52. *Int J Biochem Cell Biol* 37(1): 42-47.

Defeu Soufo HJ, Graumann PL (2004). Dynamic movement of actin-like proteins within bacterial cells. *EMBO Rep* 5(8): 789-794.

DeZwaan DC, Freeman BC (2010). HSP90 manages the ends. *Trends Biochem Sci* 35(7): 384-391.

Dickey CA, Kamal A, Lundgren K, Klosak N, Bailey RM, Dunmore J, *et al.* (2007). The high-affinity HSP90-CHIP complex recognizes and selectively degrades phosphorylated tau client proteins. *J Clin Invest* 117(3): 648-658.

Ding AH, Porteu F, Sanchez E, Nathan CF (1990). Shared actions of endotoxin and taxol on TNF receptors and TNF release. *Science* 248(4953): 370-372.

Donnelly AC, Mays JR, Burlison JA, Nelson JT, Vielhauer G, Holzbeierlein J, *et al.* (2008). The design, synthesis, and evaluation of coumarin ring derivatives of the novobiocin scaffold that exhibit antiproliferative activity. *J Org Chem* 73(22): 8901-8920.

Dou F, Netzer WJ, Tanemura K, Li F, Hartl FU, Takashima A, *et al.* (2003). Chaperones increase association of tau protein with microtubules. *Proc Natl Acad Sci U S A* 100(2): 721-726.

Duan J, Gorovsky MA (2002). Both carboxy-terminal tails of alpha- and beta-tubulin are essential, but either one will suffice. *Curr Biol* 12(4): 313-316.

Dunn AY, Melville MW, Frydman J (2001). Review: cellular substrates of the eukaryotic chaperonin TRiC/CCT. *J Struct Biol* 135(2): 176-184.

Echeverria PC, Picard D (2010). Molecular chaperones, essential partners of steroid hormone receptors for activity and mobility. *Biochim Biophys Acta* 1803(6): 641-649.

Elbi C, Walker DA, Lewis M, Romero G, Sullivan WP, Toft DO, *et al.* (2004). A novel in situ assay for the identification and characterization of soluble nuclear mobility factors. *Sci STKE* 2004(238): pl10.

Elliott E, Tsvetkov P, Ginzburg I (2007). BAG-1 associates with Hsc70.Tau complex and regulates the proteasomal degradation of Tau protein. *J Biol Chem* 282(51): 37276-37284.

Erickson HP (2007). Evolution of the cytoskeleton. *Bioessays* 29(7): 668-677.

Feldman DE, Spiess C, Howard DE, Frydman J (2003). Tumorigenic mutations in VHL disrupt folding in vivo by interfering with chaperonin binding. *Mol Cell* 12(5): 1213-1224.

Fisher BR, Heredia DJ, Brown KM (1996). Heat-induced alterations in embryonic cytoskeletal and stress proteins precede somite malformations in rat embryos. *Teratog Carcinog Mutagen* 16(1): 49-64.

Fisher BR, Heredia DJ, Brown KM (1995). Induction of hsp72 in heat-treated rat embryos: a tissue-specific response. *Teratology* 52(2): 90-100.

Fortugno P, Beltrami E, Plescia J, Fontana J, Pradhan D, Marchisio PC, *et al.* (2003). Regulation of survivin function by Hsp90. *Proc Natl Acad Sci U S A* 100(24): 13791-13796.

Fukata Y, Itoh TJ, Kimura T, Menager C, Nishimura T, Shiromizu T, *et al.* (2002). CRMP-2 binds to tubulin heterodimers to promote microtubule assembly. *Nat Cell Biol* 4(8): 583-591.

Fulda S, Debatin KM (2004). Sensitization for anticancer drug-induced apoptosis by the chemopreventive agent resveratrol. *Oncogene* 23(40): 6702-6711.

Galigniana MD, Echeverría PC, Erlejman AG, Piwien-Pilipuk G (2010a). Role of molecular chaperones and TPR-domain proteins in the cytoplasmic transport of steroid receptors and their passage through the nuclear pore. *Nucleus* 1: 299-308.

Galigniana MD, Erlejman AG, Monte M, Gomez-Sanchez C, Piwien-Pilipuk G (2010b). The hsp90-FKBP52 Complex Links the Mineralocorticoid Receptor to Motor Proteins and Persists Bound to the Receptor in Early Nuclear Events. *Mol Cell Biol* 30(5): 1285-1298.

Galigniana MD, Harrell JM, Murphy PJ, Chinkers M, Radanyi C, Renoir JM, *et al.* (2002). Binding of hsp90-associated immunophilins to cytoplasmic dynein: direct binding and in vivo evidence that the peptidylprolyl isomerase domain is a dynein interaction domain. *Biochemistry* 41(46): 13602-13610.

Galigniana MD, Harrell JM, O'Hagen HM, Ljungman M, Pratt WB (2004a). Hsp90-binding immunophilins link p53 to dynein during p53 transport to the nucleus. *J Biol Chem* 279(21): 22483-22489.

Galigniana MD, Housley PR, DeFranco DB, Pratt WB (1999). Inhibition of glucocorticoid receptor nucleocytoplasmic shuttling by okadaic acid requires intact cytoskeleton. *J Biol Chem* 274(23): 16222-16227.

Galigniana MD, Morishima Y, Gallay PA, Pratt WB (2004b). Cyclophilin-A is bound through its peptidylprolyl isomerase domain to the cytoplasmic dynein motor protein complex. *J Biol Chem* 279(53): 55754-55759.

Galigniana MD, Radanyi C, Renoir JM, Housley PR, Pratt WB (2001). Evidence that the peptidylprolyl isomerase domain of the hsp90-binding immunophilin FKBP52 is involved in both dynein interaction and glucocorticoid receptor movement to the nucleus. *J Biol Chem* 276(18): 14884-14889.

Gallo LI, Ghini AA, Pilipuk GP, Galigniana MD (2007). Differential recruitment of tetratricorpeptide repeat domain immunophilins to the mineralocorticoid receptor influences both heat-shock protein 90-dependent retrotransport and hormone-dependent transcriptional activity. *Biochemistry* 46(49): 14044-14057.

Gao YS, Hubbert CC, Lu J, Lee YS, Lee JY, Yao TP (2007). Histone deacetylase 6 regulates growth factor-induced actin remodeling and endocytosis. *Mol Cell Biol* 27(24): 8637-8647.

Gold BG, Densmore V, Shou W, Matzuk MM, Gordon HS (1999). Immunophilin FK506-binding protein 52 (not FK506-binding protein 12) mediates the neurotrophic action of FK506. *J Pharmacol Exp Ther* 289(3): 1202-1210.

Gold BG, Villafranca JE (2003). Neuroimmunophilin ligands: the development of novel neuroregenerative/ neuroprotective compounds. *Curr Top Med Chem* 3(12): 1368-1375.

Goldbaum O, Riedel M, Stahnke T, Richter-Landsberg C (2009). The small heat shock protein HSP25 protects astrocytes against stress induced by proteasomal inhibition. *Glia* 57(14): 1566-1577.

Goldfarb SB, Kashlan OB, Watkins JN, Suaud L, Yan W, Kleyman TR, *et al.* (2006). Differential effects of Hsc70 and Hsp70 on the intracellular trafficking and functional expression of epithelial sodium channels. *Proc Natl Acad Sci U S A* 103(15): 5817-5822.

Grad I, McKee TA, Ludwig SM, Hoyle GW, Ruiz P, Wurst W, *et al.* (2006). The Hsp90 cochaperone p23 is essential for perinatal survival. *Mol Cell Biol* 26(23): 8976-8983.

Grantham J, Brackley KI, Willison KR (2006). Substantial CCT activity is required for cell cycle progression and cytoskeletal organization in mammalian cells. *Exp Cell Res* 312(12): 2309-2324.

Green KJ, Bohringer M, Gocken T, Jones JC (2005). Intermediate filament associated proteins. *Adv Protein Chem* 70: 143-202.

Gross KL, Westberry JM, Hubler TR, Sadosky PW, Singh RJ, Taylor RL, *et al.* (2008). Androgen resistance in squirrel monkeys (Saimiri spp.). *Comp Med* 58(4): 381-388.

Guillozet-Bongaarts AL, Garcia-Sierra F, Reynolds MR, Horowitz PM, Fu Y, Wang T, *et al.* (2005). Tau truncation during neurofibrillary tangle evolution in Alzheimer's disease. *Neurobiol Aging* 26(7): 1015-1022.

Hammond JW, Cai D, Verhey KJ (2008). Tubulin modifications and their cellular functions. *Curr Opin Cell Biol* 20(1): 71-76.

Hardy J, Pittman A, Myers A, Fung HC, de Silva R, Duckworth J (2006). Tangle diseases and the tau haplotypes. *Alzheimer Dis Assoc Disord* 20(1): 60-62.

Hartl FU, Hayer-Hartl M (2009). Converging concepts of protein folding in vitro and in vivo. *Nat Struct Mol Biol* 16(6): 574-581.

Herrmann H, Strelkov SV, Burkhard P, Aebi U (2009). Intermediate filaments: primary determinants of cell architecture and plasticity. *J Clin Invest* 119(7): 1772-1783.

Hinds TD, Jr., Sanchez ER (2008). Protein phosphatase 5. *Int J Biochem Cell Biol* 40(11): 2358-2362.

Horwich AL, Farr GW, Fenton WA (2006). GroEL-GroES-mediated protein folding. *Chem Rev* 106(5): 1917-1930.

Ishibashi Y, Takahashi M, Isomatsu Y, Qiao F, Iijima Y, Shiraishi H, *et al.* (2003). Role of microtubules versus myosin heavy chain isoforms in contractile dysfunction of hypertrophied murine cardiocytes. *Am J Physiol Heart Circ Physiol* 285(3): H1270-1285.

Jakob U, Muse W, Eser M, Bardwell JC (1999). Chaperone activity with a redox switch. *Cell* 96(3): 341-352.

Jana NR, Tanaka M, Wang G, Nukina N (2000). Polyglutamine length-dependent interaction of Hsp40 and Hsp70 family chaperones with truncated N-terminal huntingtin: their role in suppression of aggregation and cellular toxicity. *Hum Mol Genet* 9(13): 2009-2018.

Jinwal UK, Koren J, 3rd, Borysov SI, Schmid AB, Abisambra JF, Blair LJ, *et al.* (2010). The Hsp90 cochaperone, FKBP51, increases Tau stability and polymerizes microtubules. *J Neurosci* 30(2): 591-599.

Jo H, Loison F, Hattori H, Silberstein LE, Yu H, Luo HR (2010). Natural product Celastrol destabilizes tubulin heterodimer and facilitates mitotic cell death triggered by microtubule-targeting anti-cancer drugs. *PLoS One* 5(4): e10318.

Kadota Y, Shirasu K, Guerois R (2010). NLR sensors meet at the SGT1-HSP90 crossroad. *Trends Biochem Sci* 35(4): 199-207.

Kampinga HH, Craig EA (2010). The HSP70 chaperone machinery: J proteins as drivers of functional specificity. *Nat Rev Mol Cell Biol* 11(8): 579-592.

Kang CB, Hong Y, Dhe-Paganon S, Yoon HS (2008). FKBP family proteins: immunophilins with versatile biological functions. *Neurosignals* 16(4): 318-325.

Karagoz GE, Duarte AM, Ippel H, Uetrecht C, Sinnige T, van Rosmalen M, *et al.* (2011). N-terminal domain of human Hsp90 triggers binding to the cochaperone p23. *Proc Natl Acad Sci U S A* 108(2): 580-585.

Kaverina I, Krylyshkina O, Beningo K, Anderson K, Wang YL, Small JV (2002). Tensile stress stimulates microtubule outgrowth in living cells. *J Cell Sci* 115(Pt 11): 2283-2291.

Khatau SB, Kim DH, Hale CM, Bloom RJ, Wirtz D (2010). The perinuclear actin cap in health and disease. *Nucleus* 1(4): 337-342.

Klucken J, Shin Y, Masliah E, Hyman BT, McLean PJ (2004). Hsp70 Reduces alpha-Synuclein Aggregation and Toxicity. *J Biol Chem* 279(24): 25497-25502.

Koren J, 3rd, Jinwal UK, Davey Z, Kiray J, Arulselvam K, Dickey CA (2011). Bending Tau into Shape: The Emerging Role of Peptidyl-Prolyl Isomerases in Tauopathies. *Mol Neurobiol*.

Kovacs JJ, Murphy PJ, Gaillard S, Zhao X, Wu JT, Nicchitta CV, *et al.* (2005). HDAC6 regulates Hsp90 acetylation and chaperone-dependent activation of glucocorticoid receptor. *Mol Cell* 18(5): 601-607.

Kumarapeli AR, Wang X (2004). Genetic modification of the heart: chaperones and the cytoskeleton. *J Mol Cell Cardiol* 37(6): 1097-1109.

Lee YC, Lai YK (1995). Integrity of intermediate filaments is associated with the development of acquired thermotolerance in 9L rat brain tumor cells. *J Cell Biochem* 57(1): 150-162.

Liao G, Gundersen GG (1998). Kinesin is a candidate for cross-bridging microtubules and intermediate filaments. Selective binding of kinesin to detyrosinated tubulin and vimentin. *J Biol Chem* 273(16): 9797-9803.

Lin YF, Tsai WP, Liu HG, Liang PH (2009). Intracellular beta-tubulin/chaperonin containing TCP1-beta complex serves as a novel chemotherapeutic target against drug-resistant tumors. *Cancer Res* 69(17): 6879-6888.

Liu F, Grundke-Iqbal I, Iqbal K, Gong CX (2005). Contributions of protein phosphatases PP1, PP2A, PP2B and PP5 to the regulation of tau phosphorylation. *Eur J Neurosci* 22(8): 1942-1950.

Luo W, Dou F, Rodina A, Chip S, Kim J, Zhao Q, *et al.* (2007). Roles of heat-shock protein 90 in maintaining and facilitating the neurodegenerative phenotype in tauopathies. *Proc Natl Acad Sci U S A* 104(22): 9511-9516.

Luo W, Sun W, Taldone T, Rodina A, Chiosis G (2010). Heat shock protein 90 in neurodegenerative diseases. *Mol Neurodegener* 5: 24.

Mayer MP (2010). Gymnastics of molecular chaperones. *Mol Cell* 39(3): 321-331.

Meierhofer D, Wang X, Huang L, Kaiser P (2008). Quantitative analysis of global ubiquitination in HeLa cells by mass spectrometry. *J Proteome Res* 7(10): 4566-4576.

Michie KA, Lowe J (2006). Dynamic filaments of the bacterial cytoskeleton. *Annu Rev Biochem* 75: 467-492.

Mogilner A, Oster G (2003). Polymer motors: pushing out the front and pulling up the back. *Curr Biol* 13(18): R721-733.

Morimoto RI (2008). Proteotoxic stress and inducible chaperone networks in neurodegenerative disease and aging. *Genes Dev* 22(11): 1427-1438.

Muchowski PJ, Wacker JL (2005). Modulation of neurodegeneration by molecular chaperones. *Nat Rev Neurosci* 6(1): 11-22.

Murphy PJ, Morishima Y, Kovacs JJ, Yao TP, Pratt WB (2005). Regulation of the dynamics of hsp90 action on the glucocorticoid receptor by acetylation/deacetylation of the chaperone. *J Biol Chem* 280(40): 33792-33799.

Neckers L (2002). Hsp90 inhibitors as novel cancer chemotherapeutic agents. *Trends Mol Med* 8(4 Suppl): S55-61.

Omary MB (2009). "IF-pathies": a broad spectrum of intermediate filament-associated diseases. *J Clin Invest* 119(7): 1756-1762.

Pallari HM, Eriksson JE (2006). Intermediate filaments as signaling platforms. *Sci STKE* 2006(366): pe53.

Park SJ, Suetsugu S, Takenawa T (2005). Interaction of HSP90 to N-WASP leads to activation and protection from proteasome-dependent degradation. *EMBO J* 24(8): 1557-1570.

Perrot-Applanat M, Cibert C, Geraud G, Renoir JM, Baulieu EE (1995). The 59 kDa FK506-binding protein, a 90 kDa heat shock protein binding immunophilin (FKBP59-HBI), is associated with the nucleus, the cytoskeleton and mitotic apparatus. *J Cell Sci* 108 (Pt 5): 2037-2051.

Pratt WB, Galigniana MD, Harrell JM, DeFranco DB (2004a). Role of hsp90 and the hsp90-binding immunophilins in signalling protein movement. *Cell Signal* 16(8): 857-872.

Pratt WB, Galigniana MD, Morishima Y, Murphy PJ (2004b). Role of molecular chaperones in steroid receptor action. *Essays Biochem* 40: 41-58.

Pratt WB, Toft DO (2003). Regulation of signaling protein function and trafficking by the hsp90/hsp70-based chaperone machinery. *Exp Biol Med (Maywood)* 228(2): 111-133.

Pratt WB, Toft DO (1997). Steroid receptor interactions with heat shock protein and immunophilin chaperones. *Endocr Rev* 18(3): 306-360.

Quintá HR, Maschi D, Gomez-Sanchez C, Piwien Pilipuk G, Galigniana MD (2010). Subcellular rearrangement of hsp90-binding immunophilins accompanies neuronal differentiation and neurite outgrowth. *J Neurochem* 115: 716-734.

Reed NA, Cai D, Blasius TL, Jih GT, Meyhofer E, Gaertig J, *et al.* (2006). Microtubule acetylation promotes kinesin-1 binding and transport. *Curr Biol* 16(21): 2166-2172.

Richter K, Haslbeck M, Buchner J (2010). The heat shock response: life on the verge of death. *Mol Cell* 40(2): 253-266.

Roberson ED, Halabisky B, Yoo JW, Yao J, Chin J, Yan F, *et al.* (2011). Amyloid-beta/Fyn-induced synaptic, network, and cognitive impairments depend on tau levels in multiple mouse models of Alzheimer's disease. *J Neurosci* 31(2): 700-711.

Rycyzyn MA, Reilly SC, O'Malley K, Clevenger CV (2000). Role of cyclophilin B in prolactin signal transduction and nuclear retrotranslocation. *Mol Endocrinol* 14(8): 1175-1186.

Sah NK, Khan Z, Khan GJ, Bisen PS (2006). Structural, functional and therapeutic biology of survivin. *Cancer Lett* 244(2): 164-171.

Sahara N, Maeda S, Yoshiike Y, Mizoroki T, Yamashita S, Murayama M, *et al.* (2007). Molecular chaperone-mediated tau protein metabolism counteracts the formation of granular tau oligomers in human brain. *J Neurosci Res* 85(14): 3098-3108.

Sausville EA, Elsayed Y, Monga M, Kim G (2003). Signal transduction--directed cancer treatments. *Annu Rev Pharmacol Toxicol* 43: 199-231.

Sellin ME, Holmfeldt P, Stenmark S, Gullberg M (2008). Op18/Stathmin counteracts the activity of overexpressed tubulin-disrupting proteins in a human leukemia cell line. *Exp Cell Res* 314(6): 1367-1377.

Shih YL, Rothfield L (2006). The bacterial cytoskeleton. *Microbiol Mol Biol Rev* 70(3): 729-754.

Shim S, Yuan JP, Kim JY, Zeng W, Huang G, Milshteyn A, *et al.* (2009). Peptidyl-prolyl isomerase FKBP52 controls chemotropic guidance of neuronal growth cones via regulation of TRPC1 channel opening. *Neuron* 64(4): 471-483.

Shimamoto S, Kubota Y, Tokumitsu H, Kobayashi R (2010). S100 proteins regulate the interaction of Hsp90 with Cyclophilin 40 and FKBP52 through their tetratricopeptide repeats. *FEBS Lett* 584(6): 1119-1125.

Shimura H, Miura-Shimura Y, Kosik KS (2004). Binding of tau to heat shock protein 27 leads to decreased concentration of hyperphosphorylated tau and enhanced cell survival. *J Biol Chem* 279(17): 17957-17962.

Silverstein AM, Galigniana MD, Chen MS, Owens-Grillo JK, Chinkers M, Pratt WB (1997). Protein phosphatase 5 is a major component of glucocorticoid receptor.hsp90 complexes with properties of an FK506-binding immunophilin. *J Biol Chem* 272(26): 16224-16230.

Silverstein AM, Galigniana MD, Kanelakis KC, Radanyi C, Renoir JM, Pratt WB (1999). Different regions of the immunophilin FKBP52 determine its association with the glucocorticoid receptor, hsp90, and cytoplasmic dynein. *J Biol Chem* 274(52): 36980-36986.

Simon-Sanchez J, Schulte C, Bras JM, Sharma M, Gibbs JR, Berg D, *et al.* (2009). Genome-wide association study reveals genetic risk underlying Parkinson's disease. *Nat Genet* 41(12): 1308-1312.

Sinars CR, Cheung-Flynn J, Rimerman RA, Scammell JG, Smith DF, Clardy J (2003). Structure of the large FK506-binding protein FKBP51, an Hsp90-binding protein and a component of steroid receptor complexes. *Proc Natl Acad Sci U S A* 100(3): 868-873.

Solit DB, Basso AD, Olshen AB, Scher HI, Rosen N (2003). Inhibition of heat shock protein 90 function down-regulates Akt kinase and sensitizes tumors to Taxol. *Cancer Res* 63(9): 2139-2144.

Spiess C, Meyer AS, Reissmann S, Frydman J (2004). Mechanism of the eukaryotic chaperonin: protein folding in the chamber of secrets. *Trends Cell Biol* 14(11): 598-604.

Sreedhar AS, Kalmar E, Csermely P, Shen YF (2004). Hsp90 isoforms: functions, expression and clinical importance. *FEBS Lett* 562(1-3): 11-15.

Szeverenyi I, Cassidy AJ, Chung CW, Lee BT, Common JE, Ogg SC, *et al.* (2008). The Human Intermediate Filament Database: comprehensive information on a gene family involved in many human diseases. *Hum Mutat* 29(3): 351-360.

Tagawa H, Koide M, Sato H, Zile MR, Carabello BA, Cooper Gt (1998). Cytoskeletal role in the transition from compensated to decompensated hypertrophy during adult canine left ventricular pressure overloading. *Circ Res* 82(7): 751-761.

Tanioka T, Nakatani Y, Semmyo N, Murakami M, Kudo I (2000). Molecular identification of cytosolic prostaglandin E2 synthase that is functionally coupled with cyclooxygenase-1 in immediate prostaglandin E2 biosynthesis. *J Biol Chem* 275(42): 32775-32782.

Thomas M, Harrell JM, Morishima Y, Peng HM, Pratt WB, Lieberman AP (2006). Pharmacologic and genetic inhibition of hsp90-dependent trafficking reduces aggregation and promotes degradation of the expanded glutamine androgen receptor without stress protein induction. *Hum Mol Genet* 15(11): 1876-1883.

Thulasiraman V, Yang CF, Frydman J (1999). In vivo newly translated polypeptides are sequestered in a protected folding environment. *EMBO J* 18(1): 85-95.

Toivola DM, Strnad P, Habtezion A, Omary MB (2010). Intermediate filaments take the heat as stress proteins. *Trends Cell Biol* 20(2): 79-91.

Toivola DM, Tao GZ, Habtezion A, Liao J, Omary MB (2005). Cellular integrity plus: organelle-related and protein-targeting functions of intermediate filaments. *Trends Cell Biol* 15(11): 608-617.

Trepel J, Mollapour M, Giaccone G, Neckers L (2010). Targeting the dynamic HSP90 complex in cancer. *Nat Rev Cancer* 10(8): 537-549.

Verhey KJ, Gaertig J (2007). The tubulin code. *Cell Cycle* 6(17): 2152-2160.

Vogel J, Drapkin B, Oomen J, Beach D, Bloom K, Snyder M (2001). Phosphorylation of gamma-tubulin regulates microtubule organization in budding yeast. *Dev Cell* 1(5): 621-631.

Weaver CL, Espinoza M, Kress Y, Davies P (2000). Conformational change as one of the earliest alterations of tau in Alzheimer's disease. *Neurobiol Aging* 21(5): 719-727.

Weis F, Moullintraffort L, Heichette C, Chretien D, Garnier C (2010). The 90-kDa heat shock protein Hsp90 protects tubulin against thermal denaturation. *J Biol Chem* 285(13): 9525-9534.

Westermann S, Weber K (2003). Post-translational modifications regulate microtubule function. *Nat Rev Mol Cell Biol* 4(12): 938-947.

Wiesner S, Helfer E, Didry D, Ducouret G, Lafuma F, Carlier MF, *et al.* (2003). A biomimetic motility assay provides insight into the mechanism of actin-based motility. *J Cell Biol* 160(3): 387-398.

Wirtz D, Khatau SB (2010). Protein filaments: Bundles from boundaries. *Nat Mater* 9(10): 788-790.

Witte H, Bradke F (2008a). The role of the cytoskeleton during neuronal polarization. *Curr Opin Neurobiol* 18(5): 479-487.

Witte H, Neukirchen D, Bradke F (2008b). Microtubule stabilization specifies initial neuronal polarization. *J Cell Biol* 180(3): 619-632.

Wright NT, Cannon BR, Zimmer DB, Weber DJ (2009). S100A1: Structure, Function, and Therapeutic Potential. *Curr Chem Biol* 3(2): 138-145.

Wu B, Li P, Liu Y, Lou Z, Ding Y, Shu C, *et al.* (2004). 3D structure of human FK506-binding protein 52: implications for the assembly of the glucocorticoid receptor/Hsp90 /immunophilin heterocomplex. *Proc Natl Acad Sci U S A* 101(22): 8348-8353.

Yamada HY, Rao CV (2010). Genes that modulate the sensitivity for anti-microtubule drug-mediated chemotherapy. *Curr Cancer Drug Targets* 10(6): 623-633.

Yoshimura T, Kawano Y, Arimura N, Kawabata S, Kikuchi A, Kaibuchi K (2005). GSK-3beta regulates phosphorylation of CRMP-2 and neuronal polarity. *Cell* 120(1): 137-149.

Zhai Y, Kronebusch PJ, Simon PM, Borisy GG (1996). Microtubule dynamics at the G2/M transition: abrupt breakdown of cytoplasmic microtubules at nuclear envelope breakdown and implications for spindle morphogenesis. *J Cell Biol* 135(1): 201-214.

Zhao W, Zhong L, Wu J, Chen L, Qing K, Weigel-Kelley KA, *et al.* (2006). Role of cellular FKBP52 protein in intracellular trafficking of recombinant adeno-associated virus 2 vectors. *Virology* 353(2): 283-293.

Zilberman Y, Ballestrem C, Carramusa L, Mazitschek R, Khochbin S, Bershadsky A (2009). Regulation of microtubule dynamics by inhibition of the tubulin deacetylase HDAC6. *J Cell Sci* 122(Pt 19): 3531-3541.

In: Tubulin: Structure, Functions and Roles in Disease
Editors: Wei Yamauchi and Antonio Sokić
ISBN: 978-1-62100-141-6

Chapter III

Trypanosome Tubulin Molecules as Targets for Drug and Vaccine Development for Protozoan Infections

Barbara A. Katzenback[1], Debbie A. Plouffe[2] and Miodrag Belosevic*[*,1,3]

[1] Department of Biological Sciences, University of Alberta, Edmonton, Alberta, Canada
[2] 10 AquaBounty Canada Inc., Prince Edward Island, Canada
[3] School of Public Health, University of Alberta, Edmonton, Alberta, Canada

Abstract

Parasites from the kinetoplastid group (*Trypanosoma* spp., *Leishmania* spp.) are of considerable importance because they are responsible for serious disease and mortality in humans and economically important livestock. The identification of specific parasite molecules responsible for the induction of protective antibodies in the host are of great interest since they represent potential candidates for treatment of, or vaccination against, infections of the host. Despite the high degree of sequence conservation of tubulin molecules among different host and parasite species and the sequestered intracellular location of tubulin, a number of studies involving trypanosomatids have identified parasite tubulin molecules as key targets for drug chemotherapy or for use in vaccination studies to selectively target the parasite while limiting toxicity to the host. These observations were not limited to mammalian trypanosomatids tubulin, as similar observations for trypanosomes of ectothermic vertebrates such as fish have been reported. The combination of these studies suggests that immunization of multiple hosts with parasite tubulin from one trypanosome species may provide protection against homologous and heterologous challenge. In this review, we focus on the use of parasite tubulin as a chemotherapeutic target in mammalian systems, and as a vaccine target in vertebrate hosts, the binding specificity of anti-tubulin antibodies generated during host

[*] E-mail address: mike.belosevic@ualberta.ca; Tel: (780) 492-1266; Fax: (780) 492-2216; Address: CW-405 Biological Sciences Building, Department of Biological Sciences, University of Alberta, Edmonton, AB, T6G 2E9 Canada (Corresponding author).

vaccination and their potential mechanism of action to prevent parasite growth and survival.

1. Introduction

Microtubules are of critical importance in eukaryotic cells as they are involved in processes such as cell division, intracellular vesicle transport, and maintaining cell structure. Similarly, tubulin plays a key role in the cell biology of the kinetoplastid protozoa, which include the members of the genera *Trypanosoma* spp. and *Leishmania* spp. [1, 2]. Tubulins heterodimerize to comprise the subpellicular lattice (shape and structure of the organism), and are important for motility (flagellar axoneme) and cell division (mitotic spindle) of trypanosomatids [3]. The trypanosomatid group of parasites is responsible for a number of serious diseases resulting in mortality of humans and economically important livestock. Currently, the available drugs used for treatment of these parasitic infections are highly toxic to the vertebrate host, particularly during pregnancy, and often require the administration of multiple doses. Vaccines for these parasitic infections are not available. Therefore, the identification of a target parasite molecule that could be used in the development of non-toxic chemotherapeutic or preventative vaccine strategies is of great interest. While there have been many parasite molecules that have been identified as possible targets for drug chemotherapy or for use in vaccination studies, in this review we will focus on the role of parasite tubulin and how this molecule can be targeted using chemotherapy or used as an immunogen to elicit a protective response in the vertebrate host.

2. Kinetoplastid Protozoa

Trypanosoma brucei is a blood borne protozoan parasite transmitted through the blood meal of the the tsetse fly and is the causative agent of African trypanosomiasis [4]. *T. brucei rhodesiense* and *T. brucei gambiense* are transmitted in humans, while *T. brucei brucei* is transmitted between domestic animals such as cattle. Regardless of the subspecies, *T. brucei* infections often result in mortality of the host. This is due to the elaborate immune evasion mechanism that *T. brucei* posses-the variable surface glycoprotein (VSG) coat [5]. This parasite avoids elimination from the host by undergoing antigenic variation, where the parasite sequentially presents a single variable surface glycoprotein (VSG) on its surface. However, just as the vertebrate host immune system begins to generate antibodies to this VSG coat, the parasites synchronously express a different VSG on their surface thereby evading the host immune response. *T. brucei* has more than 1000 genes that encode different VSGs. For recent reviews on African trypanosomes and VSGs see [6, 7].

Trypanosoma cruzi, the causative agent of Chagas disease, is transmitted by the Reduviidae family of hematophagus bugs. Upon entering the vertebrate host, *T. cruzi* invades a number of different cell types where it takes up intracellular residence, but prefers skeletal, cardiac and muscle cells [8]. *T. cruzi* infection triggers the excessive activation of the immune system and often results in the enlargement of vital organs such as the spleen and heart. The over-activation of the immune system and chronic inflammation often leads to death of the host, reviewed in [8].

Trypanosoma carassii is a natural blood-borne pathogen of freshwater fish, such as carp (*Cyprinus carpio*), tench (*Tinca tinca*), and eel (*Anguilla* spp.) and can be experimentally infective to fish such as the tinfoil barb (*Barbus conhus*) and goldfish (*Carassius auratus* L.) [9, 10]. Unlike the blood-borne *T. brucei*, *T. carassii* does not appear to possess antigentic variation. It is believed that fish develop antibody-mediated immunity towards *T. carassii* because (a) recovered fish develop long-lasting resistance to re-infection [11]; (b) passive transfer of immune serum or purified IgM to naïve individuals confers protection upon homologous challenge [12-14]; (c) antibodies from infected hosts are able to recognize parasite cell lysates [12]; and (d) resistance to re-infection in recovered hosts can be suppressed by the administration of corticosteroids [11].

Leishmania spp. cause the disease known as leishmaniasis that collectively encompasses a spectrum of self-healing cutaneous leishmaniasis to visceral leishmaniasis, which is usually fatal. (Leishmaniasis is found in a number of tropical and subtropical regions and is transmitted from host to host by the sandfly vector. Once inside the vertebrate host, the parasite resides primarily within the phagolysosomal vacuoles of the phagocytic macrophage. For a recent review on the disease, life cycle, and interaction and evasion mechanisms of *Leishmania* spp. in the vertebrate host response see [15].

The various immune evasion mechanisms of these parasites, such as antigenic variation, the intracellular nature of the parasites, and the modulation of the immune system, presents a challenge for identifying a parasite target molecule(s) that can be used effectively in protection against infection. One promising parasite molecule that has been identified is tubulin.

3. Trypanosomatid Tubulin Structure

Six different types of tubulins have been identified in kinetoplastid parasites and include alpha, beta [3], gamma [16, 17], epsilon, delta, and zeta tubulins [18]. *T. brucei* alpha- and beta-tubulin genes are encoded by long tandem repeats in the parasite genome and the corresponding proteins are abundant [19, 20]. In *Leishmania* spp., the α- and β-tubulin genes are also arranged in arrays, but are located on different chromosomes, a derived characteristic from the ancestral alternating arrays of tubulins [21]. Alpha- and β-tubulins form heterodimers which polymerize to form microtubules. The α-/β- tubulin heterodimers can undergo a number of post-translational modifications including acetylation, phosphorylation, tyrosination, and glutamylation. These post-translational modifications are often both tubulin-dependant and site-specific, with many of the modifications occurring in the acidic carboxy-terminal end of the tubulin molecule [22]. Gamma-tubulin, which appears to be only a single copy in the genome, is much lower in abundance compared to α- and β-tubulins, but appears to be constitutively expressed throughout the life cycle of the trypanosome [16]. Based on immunofluorescence studies using a polyclonal γ-tubulin antibody, γ-tubulin was found to be in close association with the nucleolus, the basal bodies, the very anterior tip of the parasites, and the flagellum attachment zone in trypanosomes [16]. In *L. tropica*, γ-tubulin shows a complex distribution pattern based on the stage of the *Leishmania* life cycle [17]. *L. tropica* had a similar distribution of γ-tubulin to that observed in trypanosomes, however, γ-tubulin appears to be associated with the basal bodies, the flagella, and the subpellicular corset of *Leishmania* cells in interphase [17]. Gamma-tubulin, unlike α- and β-tubulin, does not have an acidic carboxy terminus. Based on sequence structure, it is hypothesized that the post-translational modifications of γ-tubulin are most likely different from those of α-and β-tubulins [16]. The function of delta, epsilon and zeta tubulins and their modifications in kinetoplastids have not been investigated in parasites. However, despite the presence of additional tubulin subunits, the alpha- and beta-tubulin subunits are the most abundant in trypanosomatid parasites. The use of RNAi has been used to examine the functions of tubulin genes in *T. brucei* and knock-down of α-tubulin, induced a "FAT" cell phenotype in which the characteristic shape of the trypanosomes was disrupted as a result of the loss of tubulin molecules required for the generation of microtubules necessary for maintaining parasite structure [23, 24].

4. Tubulin as a Drug Target

Trypanosomatid infections have been classified as neglected diseases, and to date, there are no highly effective chemotherapeutic agents to control these diseases. Many of the existing drug treatments are toxic to the host and often involve multiple regimented treatment schedules that are impractical to administer in developing countries in which these diseases are endemic [25, 26]. The current anti- *Trypanosoma* chemotherapeutic drugs target a number of different parasite molecules. These targets include parasite DNA, proteases, or proteins and enzymes involved in processes such as sugar metabolism or polyamine biosynthesis [25, 26]. While select molecules or enzymes are unique to the parasite, many have high homology to their mammalian counterparts, resulting in toxicity to the host. Furthermore, parasites are developing resistance to certain drugs, because they have the ability to mutate the target molecule(s) [27, 28]. Thus, there is an urgent need to develop more effective and less toxic drugs for control of these parasitic infections. The development of effective drugs must address two important issues: (a) the identification of compounds that are specific for the parasite molecule only, with no or limited reactivity with the host molecule; and (b) that the target molecules are of critical importance to the parasite survival in order to prevent the development of drug resistance.

Parasite tubulin molecules are now being examined as potential drug targets. Tubulins, as mentioned above, play an essential role within a eukaryotic cell and would prove difficult for the parasite to modify them to gain drug resistance. Additionally, despite their highly conserved nature in eukaryotic cells, there are significant differences in susceptibility of tubulin molecules in different organisms to anti-mitotic agents. Although a number of dugs such as colchicines and some benzimidazoles [29] can be used for treatment in a number of *Trypanosoma* infections due to their greater affinity for trypanosome tubulin compared to host tubulin [30], we will focus on a new group of compounds, the dinitroanilines, that have been identified as a group of anti-mitotic compounds with activity against protozoan parasites [25, 31-33].

Dinitroanilines were initially used as herbicides due to their ability to selectively bind and inhibit plant tubulin assembly, reviewed in [34]. Through a series of studies it was demonstrated that the dinitroanilines not only selectively bound to plant tubulins, but also to protozoan tubulins, yet they did not exhibit the same activity against vertebrate and fungi microtubules [35-39]. Dinitroanilines are capable of acting on a number of different protozoan parasites including *Trypanosoma* spp. and *Leishmania* spp., which are the focus of this review, as well as *Cryptosporidium parvum*, *Plasmodium falciparum*, *Entamoeba* spp. and *Toxoplasma gondii* [32, 36, 38, 40-45]. The selective activity of dinitroanilines on protozoan tubulin generated a number of interesting questions that researchers have tried to address: (a) what is the mechanism of action of dinitroanilines? (b) how are protozoan and vertebrate tubulins different from each other? (c) where do the dinitroanilines bind the protozoan tubulin molecule? (d) can dinitroaniline derivatives be generated that will have greater potency against protozoan tubulins? (e) can parasites acquire resistance to dinitroanilines?, and (f) what does this tell us about tubulin biology?

Over twenty years ago, the dinitroanilines oryzalin [40] and trifluralin [36] were first shown to have anti-mitotic activity against *Leishmania.* These compounds inhibited the polymerization of purified *Leishmania* tubulin *in vitro* [31, 40, 46], and possessed selective anti-kinetoplastid activity by inhibiting *Leishmania* spp. and *Trypanosoma* spp. growth and differentiation *in vitro* [42, 45-47]. Since these initial studies with oryzalin and trifluralin, a number of dinitroaniline derivatives have been developed and tested as potential chemotherapeutic agents for trypanosomatids [46-50]. An oryzalin analog, GB-II-5, was generated that was different from the parent compound in the substitution of the sulfonamide with a phenyl ring at the N1 position. GB-II-5 was tested *in vitro* against *L. donovani* and *T. brucei brucei* parasites, and was found to be the most active compound, as determined by a cell viability assay against these kinetoplastid parasites. GB-II-5 was found to be more active against *T. brucei brucei* than against *L. donovani* [47]. Furthermore, GB-II-5 was able to inhibit leishmanial tubulin polymerization in a dose-dependent manner and was considerably more potent than oryzalin [46, 47]. Despite the potent activity of GB-II-5 against kinetoplastids and their tubulin, GB-II-5 had minimal effects on mammalian cell viability or purified mammalian brain tubulin polymerization [47]. The action of GB-II-5 on tubulin molecules was further supported using cell cycle analysis of *L. donovani* and *T. brucei brucei* parasites treated with GB-II-5. When treated with GB-II-5, a population of both *L. donovani* and *T. brucei brucei* parasites contained four times the amount of DNA compared to non-treated parasite controls. This suggested that the action of GB-II-5 on tubulin molecules was inhibiting the parasites from dividing and completing their cell cycle [47]. Based on these promising results using GB-II-5, over 30 derivatives were generated based on GB-II-5 and tested for anti-kinetoplastid activity [48]. From this study, it was noted that many of the compounds generated were more active against *T. brucei* than *L. donovani*, and the compounds that possessed the greatest activity were those compounds that had an aryl substitution on the sulfonamide nitrogen, with a meta-substitution being more potent than a para substitution. One compound, GB-II-150, was found to have potent activity with similar biological effects on kinetoplastids as GB-II-5, and suggested that the N^1-aryl-N^4-dialkyl-3,5-dinitrosulfanilamides may be an excellent candidate for use in treatment of leishmaniasis and trypanosomiasis [48]. However, upon administering these compounds to mice *in vivo* at a dose of 20 mg kg-1, infected mice did not survive any longer than the control animals. The lack of *in vivo* activity was later attributed to the rapid metabolism of GB-II-150, as well as many other compounds tested [47, 48, 51, 52]. Therefore, the modification of dinitroaniline compounds to minimize degradation of dinitroanilines by metabolism, while maintaining potent anti-parasite activity and limiting toxicity to the host tubulin, is a daunting endeavor, yet essential for the successful use of these compounds as therapeutic drugs against kinetoplastid diseases. However, the lack of toxicity observed in mice injected with GB-II-5 or GB-II-150 was encouraging and prompted further studies into the modification of these compounds in attempts to find a drug with greater stability *in vivo* [50, 51]. Despite the identification of a number of new compounds with apparent greater stability *in vivo*, there have yet to be any reports on *in vivo* dinitroaniline activity against trypanosomatids [50, 51].

Another barrier to the potency of the dinitroaniline compounds and their derivatives is the host cell membrane, as is the case in intracellular *T. cruzi* and *Leishmania* spp. infections [51]. The investigations into how these potential therapeutic compounds get taken up and accumulated by host cells will be important in developing new dinitroaniline derivatives.

To facilitate the design of new dinitroanilines with potential anti-parasitic activity it is important to determine the dinitroaniline target and the binding site within this target. Much of the work to elucidate the mechanism of action has been performed using the *Toxoplasma gondii* model system [32, 38]. Although the exact mechanism of action of dinitroanilines is not known, it is believed that they act by inducing disassembly of microtubules [49]. Microtubule destabilization is known to occur though the binding of dinitroanilines to α-tubulin subunits in *T. gondii* [38, 49], *L. major* [49] and *T. brucei* [53], which is in contrast to numerous other anti-mitotic compounds which bind and interact with β-tubulin molecules, as reviewed in [54]. The selective binding of dinitroanilines to parasite α-tubulin appears to be due to small differences in the protein structure between protozoan and mammalian tubulin [49]. The prediction of the dinitroaniline parasite α-tubulin binding site was determined primarily through computer docking and simulation studies [49] and the generation of point mutations in α-tubulin molecule to examine how these mutations affected dinitroaniline binding [38, 55-57]. From studies using oryzalin, a representative dinitroaniline, a binding site mapping to a number of residues beneath the N-loop of the α-tubulin molecule was proposed [38]. Another computation analysis study using oryzalin, trifluralin, GB-II-5 and six analogues of GB-II-5 confirmed the dinitroaniline consensus binding site to be beneath the H1-S2 loop of the α-tubulin molecule, resulting in restriction of the flexibility of this loop and loss of inter-protofilament contacts leading to microtubule disassembly [49]. Subsequently, the mutation of a number of amino acid residues in the dinitroaniline binding site of the *T. gondii* α-tubulin subunit conferred parasite resistance to oryzalin [55-57]. However, these mutations in *T. gondii* α-tubulin did not occur without a cost to normal tubulin function [55, 57]. While these studies provided strong evidence that dinitroanilines bound to parasite α-tubulin, they also reinforce the potential for protozoan parasites to develop resistance to dinitroanilines through mutation of their α-tubulin molecules.

It is important to note that protozoan tubulin molecules are not identical; there are small differences in the amino acid sequences and potentially in the post-translational modifications of tubulin molecules amongst parasite species that may influence drug efficacy. For example, these differences are manifested in the varying toxicity observed from incubation with a particular dinitroaniline using a panel of parasites. Oryzalin, was shown to be more effective in inhibiting *T. gondii* replication and less effective in preventing *Leishmania* spp. replication [49]. Another example is the oryzalin derivative GB-II-5 that is more effective against trypanosomatids, and less effective against the apicomplexans such as *T. gondii* [56]. These different inhibitory effects of dinitroanilines on kinetoplastids and apicomplexans are present despite the 86% amino acid identity of their tubulin molecules [56]. Similarly, despite the high amino acid identity of trypanosomal and leishmanial α- and β-tubulins, 94.5% and

95.2%, respectively [53], varying degrees of anti-parasite effects were observed using dinitroaniline derivative. Examples of this include GB-II-5, GB-150 [48], and a number of their analogues [50]. These oryzalin derivates displayed increased inhibitory activity compared to oryzalin when used for *in vitro* assays against *T. brucei* (cell viability, proliferation, cell division, and tubulin polymerization) compared to that of *L. donovani* [48, 50]. These observations stress the importance of using multiple protozoan model systems to examine the development of anti-parasite compounds as chemotherapeutic agents. It is likely that the design of unique dinitroaniline derivatives will be required to treat the specific parasitic infection.

5. Trypanosome Tubulin and Host Immunity

5.1. Introduction

Microtubules are of critical importance in trypanosome biology. They comprise the subpellicular lattice (shape and structure of the organism), are important for motility (flagellar axoneme) and cell division (mitotic spindle) [3]. Despite the high degree of sequence conservation of tubulin among different species and the sequestered intracellular location of the protein, a number of studies involving trypanosomatids have identified tubulin as an immunogen. In this section we focus on the role of *Trypanosoma* (*T. brucei*, *T. cruzi*, and *T. carassii)* and *Leishmania* spp. tubulin as an immunogen in their respective vertebrate host.

5.2. Trypanosoma brucei

The VSG coat is the major immune evasion mechanism of *T. brucei*. The unpredictable nature of expression of the 1000s of VSG genes contained within the parasite genome makes it impossible to produce an effective vaccine preparation targeting a single VSG as the parasites may not express that VSG at the time of infection, and may never express that particular VSG. Therefore, the identification of an invariant and essential parasite molecule as a target for antibody-mediated immunity, such as trypanosome tubulin, is a superior vaccination strategy. The first immunization studies in cattle using a flagellar pocket antigen preparation from *T. brucei rhodesiense* afforded partial protection against heterologous challenge with *T. vivax* or *T. congolense*. Trypanosome infections were reduced by approximately 50% in cattle immunized with the *T. brucei* flagellar pocket preparation, resulting in partial protection of hosts. While the protective antigen responsible for the resistance was not determined, the abundance and presence of tubulin in the flagellar pocket was suggestive for tubulin to be the target antigen [58].

Studies by Lubega et al. [59] demonstrated that mice immunized with a purified and renatured β-tubulin preparation mount a protective immune response against *T. brucei*, *T. rhodesiense*, or *T. congolense* [59]. Sixty to 75% of the immunized mice were protected against homologous or heterologous challenge, and those mice that were not protected exhibited partial protection as demonstrated by the longer pre-patent period and delayed time to death when compared to control mice. However, mice immunized with denatured tubulin, β-tubulin C-terminal synthetic peptides, or rat tubulin did not demonstrate significant antibody production or protection from homologous challenge [59]. These were the first studies that demonstrated that purified renatured parasite tubulin could be used as a protective immunogen. It was also demonstrated that incubation of *T. brucei* with the anti-tubulin antibodies *in vitro* (anti-renatured tubulin, anti-denatured tubulin, or anti-synthetic tubulin peptides) inhibited the growth of cultured parasites in a dose-dependent fashion. However, the antibodies that were most effective in preventing parasite growth were antibodies generated towards the native, renatured tubulin [60] while anti-rat tubulin antibodies had no effect on trypanosome growth. Anti-renatured tubulin antibodies were uniformly detected in the cytoplasm of permeabilized parasites [60]. The in *vivo* and *in vitro* studies with anti-*T. brucei* β-tubulin antibodies suggest a potential mechanism of action for protective antibodies in immunized animals [60].

Due to the conserved nature of trypanosomatid tubulins, >95% amino acid identity, it was hypothesized that a tubulin subunit from a single trypanosome species could be used to immunize an animal, and confer protection against many trypanosome species [59]. Upon performing these studies, mice immunized with β-tubulin from *T. evansi* were found to be resistant to both homologous challenge and to heterologous challenge with *T. equiperdum* or *T. brucei* [61]. Greater than 70% of immunized mice showed no detectable levels of trypanosomes in their blood, and those that became infected showed a delay in the day of parasite detection and day of host death. Furthermore, anti-β-tubulin antibodies from mouse or rabbit serum specifically bound parasite, but not host tubulin, and were capable of inhibiting the growth of all three trypanosome species *in vitro* [61].

5.3. Trypanosoma cruzi

The role of the innate immune system in combating *T. cruzi* during the acute stage of infection, and in controlling the levels of parasites within the host in the chronic stage of disease, has been reviewed previously [8] and will not be discussed here. Due to the intracellular nature of this trypanosome species, the effectiveness of acquired host antibodies are limited. The humoral response of the host has a limited window in which it can directly access the parasites and this interaction can occur when trypanosomes lyse the host cell and enter the extracellular fluid or blood prior to invasion of additional host cells. To date, there are no reports of anti-tubulin antibodies elicited as a result of immunization being effective in limiting or controlling the progression of this infection. However, there are a few studies that suggest that host anti-tubulin antibodies may play a role in limiting the intensity of *T. cruzi* infections. One study identified parasite tubulin in the urine of *T. cruzi*-infected patients suggesting that tubulin is present in the extracellular environment, perhaps as a result of parasite lysis during the course of infection, leading to an improved chance of immune recognition [62]. Furthermore, low levels of natural antibodies to tubulin within naïve hosts can increase in titre following infection with *Leishmania* spp. and *T. cruzi* or, following immunization with *T. cruzi* exoantigens [63-65]. In many cases, the anti-tubulin antibodies generated are not parasite specific and react with host and parasite tubulin. These non-specific anti-tubulin antibodies are believed to contribute to the destruction of host tissue in the chronic stage of Chagas disease. Interestingly, the pre-existing anti-tubulin antibodies and the infection-induced antibodies do not recognize common epitopes nor do they possess the same affinities for their respective binding sites [66]. Another possible explanation is the production of molecules by the parasites that are tubulin-like, or that contain epitopes that cross-react with parasite tubulin such as those that have been identified for *T. cruzi* [67]. Antigenic mimicry is a common strategy employed by parasites, including trypanosomes, to evade host immune defenses [68]. In fact, mimicry of host molecules has also been hypothesized to be associated with the autoimmune-type pathology induced by *T. cruzi* during the chronic stages of the infection [69].

The role of anti-tubulin antibodies during *T. cruzi* infection appears to be controversial. While it may be possible that these anti-tubulin antibodies play a role in the humoral immune response involved in controlling *T. cruzi* infection, it seems more likely that the presence of these antibodies may just be another strategy of the parasite to induce an over-reactive immune response.

5.4. Trypanosoma carassii *(syn* T. danilewskyi*)*

Recovered fish hosts develop resistance to re-infection with *T. carassii,* and this resistance is antibody mediated [11-14]. Further studies related to the nature of the protective antigens revealed that immunization of fish with *T. carassii* excretory/secretory (ES) products prior to challenge with the parasite decreased prevalence of infection and reduced parasite levels in the fish blood by three logs [70]. Serum from survivors recognized parasite ES products in a western blot, and two of these molecules were identified as α- and β-tubulin [71].

T. carassii α- and β-tubulin sequences share >95% similarity to the deduced sequences for related trypanosome species and ~80% identity to fish tubulin molecules [71]. Furthermore, *T. carassii* tubulin molecules retain a number of conserved residues that are commonly associated with post-translational modifications and binding properties of other tubulin molecules [3, 71]. It appears that *T. carassii* α-tubulin shares a lower degree of similarity to α-tubulin from other trypanosomatids and vertebrates, particularly in the C-terminus, when compared to the high degree of similarity shared among the β-tubulin sequences. It has been previously reported that one of the most antigenic portions of the trypanosome tubulin molecule lies in the C-terminus of both subunits in which there is the greatest amount of sequence heterogeneity between species. In fact, the species specificity of anti-tubulin antibodies is likely due to the high immunogenicity of the C-terminus [60, 72].

An affinity purified rabbit polyclonal antibody generated towards the renatured recombinant α-tubulin molecule recognized the recombinant α-and β-tubulin molecules, as well as the native *T. carassii* tubulin molecule in parasite cell lysates; it did not recognize tubulin in goldfish cell lysates [71]. Immunofluorescence studies with the affinity-purified anti-α-tubulin IgG revealed uniform, specific, labeling of permeabilized trypanosomes, and incubation of live *T. carassii* with 40 μg, 80 μg or 160 μg of the anti-α-tubulin IgG inhibited parasite growth in vitro in a dose dependent manner compared to isotype controls [14, 71]. Goldfish immunization studies involving administration of 40 μg and 80 μg of recombinant β-tubulin prior to challenge with *T. carassii* resulted in a ≤2.5 log decrease in the abundance of parasites in the early stages of the challenge, thus conferring partial protection to fish against *T. carassii* infection. Furthermore, immunization of fish with β-tubulin enhanced the levels of anti-tubulin antibodies present in the serum when compared to non-immunized controls [14]. These studies demonstrate the immunogenic nature of *T. carassii* tubulin in poikilothermic vertebrates.

While the precise mode of action of anti-tubulin antibodes towards trypanosomes has not been determined, it is known that anti-tubulin antibodies are present in the flagellar pocket of mammalian trypanosomes, and these anti-tubulin antibodies recognize molecules inside membrane-permeabilized parasites [60]. Interestingly, rabbit anti-recombinant *T. carassii* α-tubulin IgG could also be detected intracellularly in non-permeabilized parasites that were pre-incubated with the antibody. The specific internal structures that were observed when the parasites were incubated with 80μg of IgG could not be positively identified, but they might represent the trypanosome nucleus or membrane-bound compartments formed as part of the endocytic pathway. Internal accumulation of the antibody was even more striking at higher IgG concentrations since heavy staining of whole parasites was observed. In other eukaryotic cells, monoclonal anti-tubulin antibodies have been shown to inhibit motility of the flagellum, and to prevent the assembly of microtubules *in vitro* [73, 74]. Therefore, like *T. brucei*, internalization of anti-tubulin antibodies might play a significant role in limiting the proliferation of *T. carassii in vitro* [60]. Parasite tubulin represents a good immunotherapeutic target for mammalian trypanosomes, as well as for trypanosomes of cold-blooded lower vertebrates such as fish [60].

5.5. Leishmania

Leishmaniasis ranges from a self-limiting cutaneous lesion to visceral leishmaniasis in which the infection with the parasite is systemic and infects many organs resulting in a lethal infection [75]. Individuals that recover from leishmaniasis are normally protected by a strong immune response elicited by the infection. The production of a strong host cell-mediated immune response towards the pathogen, as well as the large number of conserved antigens between Leishmania species, has suggested that the development of a vaccine against Leishmania preparation may be promising. However, despite this, there are currently no vaccines that have been generated to combat this disease [75].

There are a number of candidate *Leishmania* antigens that have been found to elicit an immune response in the mouse model system. One of the antigens that has shown potential as a vaccine candidate are the tubulin molecules, although studies examining tubulins as potential vaccine targets are limited. In one study, the authors isolated *Leishmania* specific T-cells from a host infected with *Leishmania*, and used these T-cells to screen a *Leishmania* amastigote cDNA expression library. Interestingly, the *Leishmania* specific T-cells recognized both amastigote α- and β-tubulin in the cDNA expression library. However, the parasite epitopes recognized by the T-cells appeared to be divergent from that of eukaryotic tubulins, despite the high degree of sequence conservation [76]. Furthermore, recombinant *Leishmania* tubulin, but not host (mouse) tubulin was able to induce a proliferative response in the *Leishmania* specific T-cells. These data suggest that *Leishmania* α- and β-tubulins are important T-cell stimulating antigens, and despite the sequence conservation, possess unique epitopes that are specific to parasite tubulin. These properties are important for the induction of a cell-mediated immune response for the clearance of the parasite while limiting an autoimmune response to the host tubulin molecules.

Another study examined the ability of *Leishmania* tubulin to be used as a vaccine candidate. Mice were immunized with Leishmania antigens from *L. donovani* promastigotes using cationic liposomes for antigen delivery. One of the antigens from the promastigotes mix that appeared to elicit an immune response, termed LD51 due to its 51kDa molecular weight, corresponded to *Leishmania* β-tubulin. Subsequently, mice were immunized with LD51 and challenged with *L. donovani* AG83, a strain of *L. donovani* that is known to induce a progressive infection in the liver and spleen of mice. The immunization of mice with LD51 prior to challenge reduced parasite burden in the liver by 72% and by 75% in the spleen compared to controls. Furthermore, a significantly higher delayed type hypersensitivity response (DTH) and splenocyte proliferative response was observed in mice immunized with LD51 compared to control mice immunized with cationic liposomes alone, indicative of an enhanced cell-mediated immune response in LD51 immunized mice. Splenocytes from LD51-immunized mice were also capable of producing significantly higher levels of IL-4, IL-12 and interferon gamma when re-stimulated with LD51 compared to splenocytes from control animals. The magnitude of cytokine release appeared to correlate with the decreased parasite burden observed in the tissues. Not only were the splenocytes from LD51-immunized mice capable of higher cytokine production, but also increased production of antibodies, specifically IgG21 and IgG1, most likely as a result of antibody class switching induced by the high levels of IL-4 and interferon gamma in LD51 immunized mice [77].

These studies suggest that *Leishmania* β-tubulin is a target of the cell-mediated immune response, however, the detection of anti-tubulin antibodies present in the sera of dogs infected with *L. donovani* [76, 78] also suggests that soluble parasite tubulin is present within the host and is a target of the host antibody response. However, the mechanism of the action of these host antibodies remains to be elucidated. We speculate that, as in the case with the intracellular parasite *T. cruzi*, that some *Leishmania* are destroyed upon entry of the host, releasing intracellular contents such as tubulin that may then be targeted by the host antibody response. These antibodies could then play an important role in interfering with tubulin polymerization if they managed to be taken up by extracellular *Leishmania* following lysis of the macrophage host cells.

Conclusion

Parasite tubulins are emerging as a candidate target for chemotherapeutic and vaccination strategies. Dinitroanilines represent an emerging group of chemotherapeutic agents for use against trypanosomatids, although it is likely that a number of unique dinitroanilines will need to be produced depending on the protozoan infection. However, there are many questions that still remain. What is the mechanism of dinitroaniline action? How does the modification of these compounds influence their activity *in vitro* and *in vivo*? Can a dinitroaniline be designed that is metabolically stable, and if so, can this compound work *in vivo*? Lastly, what can these studies tell us about the biology of tubulins? The studies reviewed in this paper suggest that β-tubulin is a vaccine candidate for extracellular trypanosomes of both endothermic and ectothermic vertebrates. However, the mechanism of antibody uptake and the mechanism of how these antibodies function to eliminate parasites are unclear. The use of trypanosomatid tubulin molecules as a strategy to eliminate these infections is a novel and exciting field. Future studies will yield insight into the mechanism of action and advance our ability to use tubulin molecules as a trypanosomatid target.

Acknowlegments

This research was supported by a grant from the Natural Sciences and Engineering Research Council of Canada (NSERC) Discovery grant to MB and by an NSERC and AIF doctoral scholarships to BAK.

References

[1] Simpson, AG; Lukes, J; Roger, AJ. The evolutionary history of kinetoplastids and their kinetoplasts. *Mol Biol Evol,* 2002 19, 2071-2083.

[2] Stevens, JR; Gibson, W. The molecular evolution of trypanosomes. *Parasitol Today,* 1999 15, 432-437.

[3] Gull, K. The cytoskeleton of trypanosomatid parasites. *Annu Rev Microbiol,* 1999 53, 629-655.

[4] Baral, TN. Immunobiology of African trypanosomes: need of alternative interventions. *J Biomed Biotechnol,* 2010 2010, 389153.

[5] Borst, P; Rudenko, G. Antigenic variation in African trypanosomes. *Science,* 1994 264, 1872-1873.

[6] Sternberg, JM; Maclean, L. A spectrum of disease in human African trypanosomiasis: the host and parasite genetics of virulence. *Parasitology,* 2010 137, 2007-2015.

[7] Horn, D; McCulloch, R. Molecular mechanisms underlying the control of antigenic variation in African trypanosomes. *Curr Opin Microbiol,* 2010 13, 700-705.

[8] Junqueira, C; Caetano, B; Bartholomeu, DC; Melo, MB; Ropert, C; Rodrigues, MM; Gazzinelli, RT. The endless race between *Trypanosoma cruzi* and host immunity: lessons for and beyond Chagas disease. *Expert Rev Mol Med,* 2010 12, e29.

[9] Lom, J. 1973. Experimental infection of goldfish with blood flagellates. In *International Congress on Protozoology.*

[10] Woo, PT; Black, GA. *Trypanosoma danilewskyi*: host specificity and host's effect on morphometrics. *J Parasitol,* 1984 70, 788-793.

[11] Islam, AK; Woo, PT. *Trypanosoma danilewskyi* in Carassius auratus: the nature of protective immunity in recovered goldfish. *J Parasitol,* 1991 77, 258-262.

[12] Overath, P; Haag, J; Mameza, MG; Lischke, A. Freshwater fish trypanosomes: definition of two types, host control by antibodies and lack of antigenic variation. *Parasitology,* 1999 119 (Pt 6), 591-601.

[13] Woo, PT. Acquired immunity against *Trypanosoma danilewskyi* in goldfish, Carassius auratus. *Parasitology,* 1981 83, 343-346.

[14] Katzenback, BA; Plouffe, DA; Haddad, G; Belosevic, M. Administration of recombinant parasite beta-tubulin to goldfish (*Carassius auratus* L.) confers partial

protection against challenge infection with *Trypanosoma danilewskyi* Laveran and Mesnil, 1904. *Vet Parasitol,* 2008 151, 36-45.

[15] Mougneau, E; Bihl, F; Glaichenhaus, N. Cell biology and immunology of *Leishmania. Immunol Rev,* 2011 240, 286-296.

[16] Scott, V; Sherwin, T; Gull, K. gamma-tubulin in trypanosomes: molecular characterisation and localisation to multiple and diverse microtubule organising centres. *J Cell Sci,* 1997 110 (Pt 2), 157-168.

[17] Libusova, L; Sulimenko, T; Sulimenko, V; Hozak, P; Draber, P. gamma-Tubulin in Leishmania: cell cycle-dependent changes in subcellular localization and heterogeneity of its isoforms. *Exp Cell Res,* 2004 295, 375-386.

[18] Vaughan, S; Attwood, T; Navarro, M; Scott, V; McKean, P; Gull, K. New tubulins in protozoal parasites. *Curr Biol,* 2000 10, R258-259.

[19] Seebeck, T; Whittaker, PA; Imboden, MA; Hardman, N; Braun, R. Tubulin genes of *Trypanosoma brucei*: a tightly clustered family of alternating genes. *Proc Natl Acad Sci U S A,* 1983 80, 4634-4638.

[20] Thomashow, LS; Milhausen, M; Rutter, WJ; Agabian, N. Tubulin genes are tandemly linked and clustered in the genome of *Trypanosoma brucei. Cell,* 1983 32, 35-43.

[21] Jackson, AP; Vaughan, S; Gull, K. Evolution of tubulin gene arrays in Trypanosomatid parasites: genomic restructuring in Leishmania. *BMC Genomics,* 2006 7, 261.

[22] Schneider, A; Plessmann, U; Weber, K. Subpellicular and flagellar microtubules of *Trypanosoma brucei* are extensively glutamylated. *J Cell Sci,* 1997 110 (Pt 4), 431-437.

[23] Best, A; Handoko, L; Schluter, E; Goringer, HU. In vitro synthesized small interfering RNAs elicit RNA interference in african trypanosomes: an in vitro and in vivo analysis. *J Biol Chem,* 2005 280, 20573-20579.

[24] Ngo, H; Tschudi, C; Gull, K; Ullu, E. Double-stranded RNA induces mRNA degradation in *Trypanosoma brucei. Proc Natl Acad Sci U S A,* 1998 95, 14687-14692.

[25] Gonzalez, M; Cerecetto, H. Novel compounds to combat trypanosomatid infections: a medicinal chemical perspective. *Expert Opin Ther Pat,* 2011 21, 699-715.

[26] Steverding, D. The development of drugs for treatment of sleeping sickness: a historical review. *Parasit Vectors,* 2010 3, 15.

[27] Werbovetz, KA. Promising therapeutic targets for antileishmanial drugs. *Expert Opin Ther Targets,* 2002 6, 407-422.

[28] Burchmore, RJ; Ogbunude, PO; Enanga, B; Barrett, MP. Chemotherapy of human African trypanosomiasis. *Curr Pharm Des,* 2002 8, 256-267.

[29] Katiyar, SK; Gordon, VR; McLaughlin, GL; Edlind, TD. Antiprotozoal activities of benzimidazoles and correlations with beta-tubulin sequence. *Antimicrob Agents Chemother,* 1994 38, 2086-2090.

[30] Ochola, DO; Prichard, RK; Lubega, GW. Classical ligands bind tubulin of trypanosomes and inhibit their growth in vitro. *J Parasitol,* 2002 88, 600-604.

[31] Chan, MM; Grogl, M; Chen, CC; Bienen, EJ; Fong, D. Herbicides to curb human parasitic infections: in vitro and in vivo effects of trifluralin on the trypanosomatid protozoans. *Proc Natl Acad Sci U S A,* 1993 90, 5657-5661.

[32] Stokkermans, TJ; Schwartzman, JD; Keenan, K; Morrissette, NS; Tilney, LG; Roos, DS. Inhibition of *Toxoplasma gondii* replication by dinitroaniline herbicides. *Exp Parasitol,* 1996 84, 355-370.

[33] Benbow, JW; Bernberg, EL; Korda, A; Mead, JR. Synthesis and evaluation of dinitroanilines for treatment of cryptosporidiosis. *Antimicrob Agents Chemother,* 1998 42, 339-343.

[34] Morejohn, LC; Fosket, DE. The biochemistry of compounds with anti-microtubule activity in plant cells. *Pharmacol Ther,* 1991 51, 217-230.

[35] Morgan, RE; Werbovetz, KA. Selective lead compounds against kinetoplastid tubulin. *Adv Exp Med Biol,* 2008 625, 33-47.

[36] Chan, MM; Fong, D. Inhibition of leishmanias but not host macrophages by the antitubulin herbicide trifluralin. *Science,* 1990 249, 924-926.

[37] Hugdahl, JD; Morejohn, LC. Rapid and Reversible High-Affinity Binding of the Dinitroaniline Herbicide Oryzalin to Tubulin from Zea mays L. *Plant Physiol,* 1993 102, 725-740.

[38] Morrissette, NS; Mitra, A; Sept, D; Sibley, LD. Dinitroanilines bind alpha-tubulin to disrupt microtubules. *Mol Biol Cell,* 2004 15, 1960-1968.

[39] Fennell, BJ; Naughton, JA; Dempsey, E; Bell, A. Cellular and molecular actions of dinitroaniline and phosphorothioamidate herbicides on *Plasmodium falciparum*: tubulin as a specific antimalarial target. *Mol Biochem Parasitol,* 2006 145, 226-238.

[40] Chan, MM; Triemer, RE; Fong, D. Effect of the anti-microtubule drug oryzalin on growth and differentiation of the parasitic protozoan *Leishmania mexicana. Differentiation,* 1991 46, 15-21.

[41] Edlind, T; Li, J; Katiyar, S. Expression of *Cryptosporidium parvum* beta-tubulin sequences in yeast: potential model for drug development. *J Eukaryot Microbiol,* 1996 43, 86S.

[42] Armson, A; Kamau, SW; Grimm, F; Reynoldson, JA; Best, WM; MacDonald, LM; Thompson, RC. A comparison of the effects of a benzimidazole and the dinitroanilines against *Leishmania infantum. Acta Trop,* 1999 73, 303-311.

[43] Makioka, A; Kumagai, M; Ohtomo, H; Kobayashi, S; Takeuchi, T. Effect of dinitroaniline herbicides on the growth of *Entamoeba histolytica. J Parasitol,* 2000 86, 607-610.

[44] Makioka, A; Kumagai, M; Ohtomo, H; Kobayashi, S; Takeuchi, T. Effect of the antitubulin drug oryzalin on the encystation of *Entamoeba invadens. Parasitol Res,* 2000 86, 625-629.

[45] Traub-Cseko, YM; Ramalho-Ortigao, JM; Dantas, AP; de Castro, SL; Barbosa, HS; Downing, KH. Dinitroaniline herbicides against protozoan parasites: the case of *Trypanosoma cruzi. Trends Parasitol,* 2001 17, 136-141.

[46] Bhattacharya, G; Salem, MM; Werbovetz, KA. Antileishmanial dinitroaniline sulfonamides with activity against parasite tubulin. *Bioorg Med Chem Lett,* 2002 12, 2395-2398.

[47] Werbovetz, KA; Sackett, DL; Delfin, D; Bhattacharya, G; Salem, M; Obrzut, T; Rattendi, D; Bacchi, C. Selective antimicrotubule activity of N1-phenyl-3,5-dinitro-N4,N4-di-n-propylsulfanilamide (GB-II-5) against kinetoplastid parasites. *Mol Pharmacol,* 2003 64, 1325-1333.

[48] Bhattacharya, G; Herman, J; Delfin, D; Salem, MM; Barszcz, T; Mollet, M; Riccio, G; Brun, R; Werbovetz, KA. Synthesis and antitubulin activity of N1- and N4-substituted 3,5-dinitro sulfanilamides against African trypanosomes and Leishmania. *J Med Chem,* 2004 47, 1823-1832.

[49] Mitra, A; Sept, D. Binding and interaction of dinitroanilines with apicomplexan and kinetoplastid alpha-tubulin. *J Med Chem,* 2006 49, 5226-5231.

[50] George, TG; Endeshaw, MM; Morgan, RE; Mahasenan, KV; Delfin, DA; Mukherjee, MS; Yakovich, AJ; Fotie, J; Li, C; Werbovetz, KA. Synthesis, biological evaluation, and molecular modeling of 3,5-substituted-N1-phenyl-N4,N4-di-n-butylsulfanilamides as antikinetoplastid antimicrotubule agents. *Bioorg Med Chem,* 2007 15, 6071-6079.

[51] George, TG; Johnsamuel, J; Delfin, DA; Yakovich, A; Mukherjee, M; Phelps, MA; Dalton, JT; Sackett, DL; Kaiser, M; Brun, R; Werbovetz, KA. Antikinetoplastid antimitotic activity and metabolic stability of dinitroaniline sulfonamides and benzamides. *Bioorg Med Chem,* 2006 14, 5699-5710.

[52] Wu, D; George, TG; Hurh, E; Werbovetz, KA; Dalton, JT. Pre-systemic metabolism prevents in vivo antikinetoplastid activity of N1,N4-substituted 3,5-dinitro sulfanilamide, GB-II-150. *Life Sci,* 2006 79, 1081-1093.

[53] Giles, NL; Armson, A; Reid, SA. Characterization of trifluralin binding with recombinant tubulin from *Trypanosoma brucei. Parasitol Res,* 2009 104, 893-903.

[54] Kavallaris, M. Microtubules and resistance to tubulin-binding agents. *Nat Rev Cancer,* 2010 10, 194-204.

[55] Ma, C; Li, C; Ganesan, L; Oak, J; Tsai, S; Sept, D; Morrissette, NS. Mutations in alpha-tubulin confer dinitroaniline resistance at a cost to microtubule function. *Mol Biol Cell,* 2007 18, 4711-4720.

[56] Ma, C; Tran, J; Gu, F; Ochoa, R; Li, C; Sept, D; Werbovetz, K; Morrissette, N. Dinitroaniline activity in *Toxoplasma gondii* expressing wild-type or mutant alpha-tubulin. *Antimicrob Agents Chemother,* 2010 54, 1453-1460.

[57] Lyons-Abbott, S; Sackett, DL; Wloga, D; Gaertig, J; Morgan, RE; Werbovetz, KA; Morrissette, NS. alpha-Tubulin mutations alter oryzalin affinity and microtubule assembly properties to confer dinitroaniline resistance. *Eukaryot Cell,* 2010 9, 1825-1834.

[58] Mkunza, F; Olaho, WM; Powell, CN. Partial protection against natural trypanosomiasis after vaccination with a flagellar pocket antigen from *Trypanosoma brucei rhodesiense. Vaccine,* 1995 13, 151-154.

[59] Lubega, GW; Byarugaba, DK; Prichard, RK. Immunization with a tubulin-rich preparation from *Trypanosoma brucei* confers broad protection against African trypanosomosis. *Exp Parasitol,* 2002 102, 9-22.

[60] Lubega, GW; Ochola, DO; Prichard, RK. *Trypanosoma brucei*: anti-tubulin antibodies specifically inhibit trypanosome growth in culture. *Exp Parasitol,* 2002 102, 134-142.

[61] Li, SQ; Fung, MC; Reid, SA; Inoue, N; Lun, ZR. Immunization with recombinant beta-tubulin from *Trypanosoma evansi* induced protection against T. evansi, T. equiperdum and T. b. brucei infection in mice. *Parasite Immunol,* 2007 29, 191-199.

[62] Bertot, GM; Corral, RS; Fresno, M; Rodriguez, C; Katzin, AM; Grinstein, S. *Trypanosoma cruzi* tubulin eliminated in the urine of the infected host. *J Parasitol,* 1998 84, 608-614.

[63] Gruppi, A; Pistoresi-Palencia, MC; Ordonez, P; Cerban, F; Vottero-Cima, E. Enhancement of natural antibodies in mice immunized with exoantigens of pI 4.5 from *Trypanosoma cruzi. Immunol Lett,* 1994 42, 151-159.

[64] Howard, MK; Gull, K; Miles, MA. Antibodies to tubulin in patients with parasitic infections. *Clin Exp Immunol,* 1987 68, 78-85.

[65] Kahn, S; Kahn, M; Eisen, H. Polyreactive autoantibodies to negatively charged epitopes following *Trypanosoma cruzi* infection. *Eur J Immunol,* 1992 22, 3051-3056.

[66] Ternynck, T; Bleux, C; Gregoire, J; Avrameas, S; Kanellopoulos-Langevin, C. Comparison between autoantibodies arising during *Trypanosoma cruzi* infection in mice and natural autoantibodies. *J Immunol,* 1990 144, 1504-1511.

[67] Alcina, A; Fresno, M. A tubulin-related 55 kilodalton surface antigen recognized by different *Trypanosoma cruzi* stage-specific monoclonal antibodies from infected mice. *Mol Biochem Parasitol,* 1988 29, 181-190.

[68] Abu-Shakra, M; Buskila, D; Shoenfeld, Y. Molecular mimicry between host and pathogen: examples from parasites and implication. *Immunol Lett,* 1999 67, 147-152.

[69] Tarleton, RL. Chagas disease: a role for autoimmunity? *Trends Parasitol,* 2003 19, 447-451.

[70] Bienek, DR; Plouffe, DA; Wiegertjes, GF; Belosevic, M. Immunization of goldfish with excretory/secretory molecules of *Trypanosoma danilewskyi* confers protection against infection. *Dev Comp Immunol,* 2002 26, 649-657.

[71] Plouffe, DA; Belosevic, M. Antibodies that recognize alpha- and beta-tubulin inhibit in vitro growth of the fish parasite *Trypanosoma danilewskyi*, Laveran and Mesnil, 1904. *Dev Comp Immunol,* 2006 30, 685-697.

[72] de la Vina, S; Andreu, D; Medrano, FJ; Nieto, JM; Andreu, JM. Tubulin structure probed with antibodies to synthetic peptides. Mapping of three major types of limited proteolysis fragments. *Biochemistry,* 1988 27, 5352-5365.

[73] Draber, P; Vater, W; Bohm, KJ; Kuklova, E; Unger, E. Inhibition of microtubule assembly in vitro by anti-tubulin monoclonal antibodies. *FEBS Lett,* 1990 262, 209-211.

[74] Oka, MT; Nakajima, Y; Obika, M; Arai, T; Nakayama, Y; Hamaguchi, Y. The inhibition of motility of demembranated spermatozoa by anti-tubulin antibodies. *Cell Struct Funct,* 1997 22, 455-463.

[75] Modabber, F. Leishmaniasis vaccines: past, present and future. *Int J Antimicrob Agents,* 2010 36 Suppl 1, S58-61.

[76] Probst, P; Stromberg, E; Ghalib, HW; Mozel, M; Badaro, R; Reed, SG; Webb, JR. Identification and characterization of T cell-stimulating antigens from *Leishmania* by CD4 T cell expression cloning. *J Immunol,* 2001 166, 498-505.

[77] Bhowmick, S; Ali, N. Identification of novel *Leishmania donovani* antigens that help define correlates of vaccine-mediated protection in visceral leishmaniasis. *PLoS One,* 2009 4, e5820.

[78] Pateraki, E; Portocala, R; Labrousse, H; Guesdon, JL. Antiactin and antitubulin antibodies in canine visceral leishmaniasis. *Infect Immun,* 1983 42, 496-500.

In: Tubulin: Structure, Functions and Roles in Disease
Editors: Wei Yamauchi and Antonio Sokić

ISBN: 978-1-62100-141-6

Chapter IV

Tubulin: Structure, Function and Role in Disease

Ikuru Yazawa*

Laboratory of Research Resources, Research Institute for Longevity Sciences, National Center for Geriatrics and Gerontology, Aichi, Japan

Abstract

Microtubules are one of the components that make up the filamentous cytoskeleton in the nervous system. A microtubule is formed from subunits of tubulin, which is a heterodimer formed from α-tubulin and β-tubulin. The formation of protein filaments in microtubules allows for regulated filament assembly and disassembly, resulting in the reshaping of the cytoskeleton. Microtubules are dynamic structures that are maintained by a rapid and continuous exchange of tubulins between the soluble and free form and the filamentous form. Neurons regulate the length and stability of microtubules, as well as their number and geometry. Microtubule-associated proteins (MAPs) bind along the sides of microtubules and participate in the regulation of microtubule assembly and function. Drugs that have an effect on microtubule polymerization can be useful for examining the role of microtubules in disease processes. Motor proteins bind to microtubules and transport organelles and particles in neuronal processes. Deficits in microtubule and MAP function contribute to the pathogenesis of multiple diseases, including Alzheimer's disease and hereditary peripheral neuropathy. Multiple system atrophy (MSA), a nonhereditary neurodegenerative disease affecting oligodendrocytes and neurons in the human central nervous system, is characterized by abnormal accumulations of filamentous α-synuclein inclusions. Recent studies in MSA mouse models have identified that microtubule β-III tubulin binds to α-synuclein. In addition, the binding of α-synuclein to β-III tubulin promoted the accumulation of insoluble α-synuclein, leading to neuronal dysfunction and degeneration. Furthermore, α-synuclein accumulation was suppressed by inhibiting microtubule polymerization. Thus, the binding of α-synuclein to β-III tubulin is a key process in neurodegeneration in MSA. Important insights into

* E-mail address: yazawa@nils.go.jp; Tel: +81-562-46-2311; Fax: +81-562-46-8319; Address: Laboratory of Research Resources, Research Institute for Longevity Sciences, National Center for Geriatrics and Gerontology, 35 Gengo, Morioka-cho, Obu-shi, Aichi 474-7511, Japan.

neurodegeneration in MSA and therapeutic targets have emerged from the MSA mouse model.

Introduction

Understanding the structure and function of cytoskeletal filaments in the nervous system forms a foundation for understanding the pathophysiology of neurodegenerative diseases. Alternatively, understanding the pathophysiology of neurodegenerative diseases contributes to understanding the structure and function of the cytoskeleton. Therefore, knowledge of the neuronal cytoskeletal structure is crucial for understanding brain function and the mechanisms of neurodegeneration. In order to accomplish this, recent developments that have uncovered aspects of the normal role of microtubules in brain function are discussed. Further discussion will concentrate on the pathophysiological mechanisms of neurodegenerative diseases in which microtubules are implicated, including multiple system atrophy (MSA) and its cellular mechanisms of neurodegeneration. In the human central nervous system (CNS), MSA affects oligodendrocytes and neurons, and the interactions between them. Recent studies on neurodegeneration in MSA using animal and cellular models have provided evidence that the microtubule subunits of tubulin are involved in the disease, and these findings have contributed toward the development of a therapeutic strategy to treat neurodegeneration in MSA. A treatment that prevents this disease process in a mouse model has been further tested as a potential therapeutic treatment for MSA.

Structure of Microtubules

A neuron contains a cytoskeleton that consists of various filamentous structures. The three main filaments are microtubules, intermediate filaments (neurofilaments), and actin filaments. The arrangement of these three filaments gives neurons their characteristic shape. These filaments interact with each other and with other neuronal structures. They are dynamic and unstable rather than passive and stable. Each of these cytoskeletal filaments has a unique function essential for a functional nervous system, particularly for synaptic plasticity and axonal transport.

Microtubules are ubiquitous cytoskeletal components in eukaryotic cells and are abundant in the nervous system. Microtubules are formed from tubulin subunits (Figure 1). Tubulin itself is a heterodimer formed from closely related globular proteins, called α-tubulin and β-tubulin, which are tightly bound together by noncovalent bonds [1]. A microtubule is a hollow cylindrical structure built from 13 parallel protofilaments that are each composed of alternating α-tubulin and β-tubulin [1]. In electron microscopic preparations, microtubules appear as long, tubular elements with a diameter of 20–26 nm [2]. The formation of protein filaments in microtubules allows for regulated filament assembly and disassembly, which reshape the cytoskeleton [3, 4]. In addition, microtubules are dynamic structures that are maintained by rapid and continuous exchange of tubulin between the soluble and free form and the filamentous form (Figure 1). This tubulin flux is necessary for neuronal function. Addition and loss of subunits occur almost exclusively at the microtubule ends. The dynamic end of a microtubule, where growth and shrinkage are fast, is called the plus end, while the other end is called the minus end [1]. In living cells, where the minus ends of microtubules are tightly anchored at the centrosome, microtubules exchange subunits by polymerization and depolymerization at their plus ends [5]. The cytoplasmic microtubules emanate in an astral conformation from the centrosome, which is located near the nucleus [6, 7].

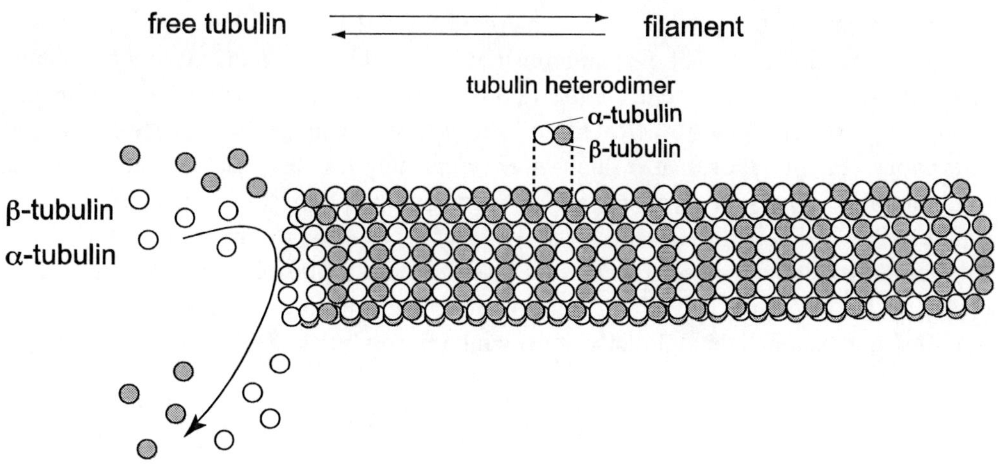

Figure 1. Microtubule structure with its tubulin subunits. A microtubule, a hollow cylindrical structure made of 13 protofilaments, is composed of tubulin subunits. Each protofilament of a microtubule is formed from tubulin heterodimers, which are tightly bound pairs of alternating α-tubulin and β-tubulin. The microtubule is a dynamic structure that is maintained by continuous exchange of tubulin between the soluble and free form and the filamentous form.

Tubulin

Tubulin exists in multiple forms in eukaryotic cells. Genes for these multiple forms are differentially expressed according to cell type and developmental stage. Most tubulin genes are expressed in the nervous system, and some are enriched or specific to neurons [8]. For example, β-III tubulin is specifically expressed in neurons [9, 10]. The different forms of tubulin are very similar, but they can have distinct locations in a cell and perform different functions [8]. The process of joining tubulin, which is a small protein, to form a long-stranded microtubule is called polymerization. Polymerization and depolymerization of microtubules result in the formation of the shape of a neuron, and both processes can be regulated by neuronal signals [11]. In addition, various drugs can have an effect on microtubule polymerization. For instance, one class of antineoplastic drugs, vincristine and other vinca alkaloids, binds to tubulin and prevents microtubule polymerization [11]. Colchicine or nocodazole binds to and stabilizes free tubulin, causing microtubule depolymerization [12-15]. In contrast, taxol binds to and stabilizes microtubules, causing a net increase in tubulin polymerization [16, 17]. These drugs can be useful in examining the role of microtubules in the disease processes underlying neurodegeneration.

Regulation of Microtubule Dynamics

Microtubules are dynamic cytoskeletal filaments in neurons. Neurons regulate the length and stability of microtubules as well as their number and geometry. Studies on microtubule dynamics *in vitro* have shown two principal mechanisms of subunit exchange known as treadmilling [18] and dynamic instability [19] (Figure 2). Treadmilling involves the addition of tubulin subunits to the plus end of a microtubule and the simultaneous loss of tubulin subunits from the minus end [5]. Dynamic instability involves the gain and loss of subunits at the same end (either the plus end or the minus end) of a microtubule during growth and shortening [5]. Microtubules nucleated by centrosomes can grow transiently at tubulin concentrations below those at which free microtubules are stable [20]. Nucleated microtubules that coexist in growing and shrinking populations rarely interconvert to free microtubules [19]. Dynamic instability is a general property of microtubules (Figure 2). Microtubule formation involves the addition of guanosine triphosphate (GTP)-bound tubulin molecules to the growing end of the microtubule. Recent studies have shown that small tubulin oligomers can be added directly to growing microtubules [21].

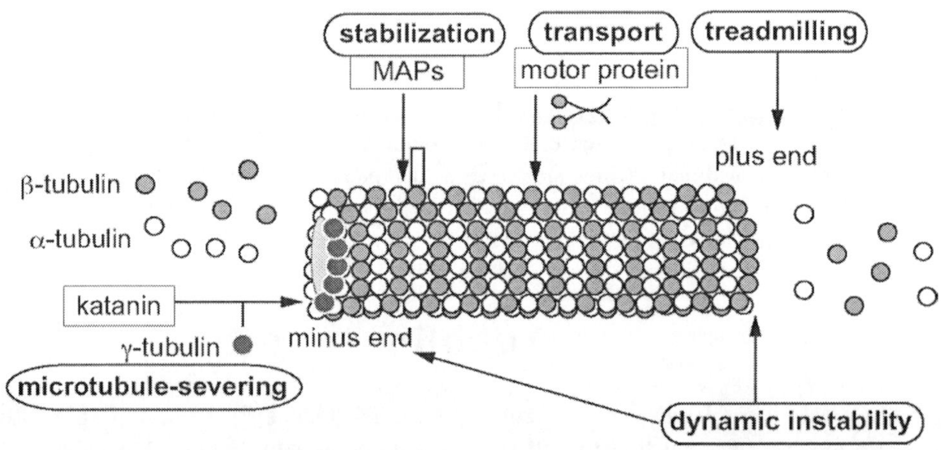

Figure 2. Regulation of microtubule dynamics. A microtubule is a dynamic cytoskeletal filament. Microtubule dynamics are regulated by mechanisms of subunit exchange: treadmilling, dynamic instability, or microtubule-severing by katanin. Microtubule-associated proteins (MAPs) stabilize or destabilize the microtubule. Motor proteins bind to the microtubule to transport organelles and particles in neuronal processes.

In addition to microtubule treadmilling and dynamic instability, there exists a third mechanism by which microtubule dynamics can be affected. In this mechanism, microtubules can be broken or severed along their length [22]. In many cells, cytoplasmic microtubules have their minus ends anchored near the centrosome and their plus ends oriented toward the periphery. Microtubules are generally nucleated from a specific intracellular location known as the microtubule-organizing center (MTOC) [1, 7, 23, 24]. MTOCs orient microtubules to grow at the plus ends (Figure 2). While α-tubulin and β-tubulin are the regular building subunits of microtubules, γ-tubulin, another type of tubulin, has a specialized role in the nucleation of microtubule growth [25]. The biochemistry of MTOCs is not fully understood, but γ-tubulin functions as a nucleating microtubule protein [26-28]. In contrast, neuronal dendrites and axons require a robust supply of noncentrosomal microtubules for structural support and material transport [22, 29]. Dendritic and axonal microtubules are nucleated at the MTOC but are subsequently released for delivery to the appropriate compartment [8]. Microtubule release from the MTOC involves the microtubule-severing protein katanin [30]. Microtubule severing by katanin is essential for regulating the length of microtubules after their release (Figure 2).

An important question is how neurons regulate their microtubules. Direct covalent modification of microtubules regulates some properties, but regulation is mostly performed by a large array of accessory proteins that bind to either microtubules or free tubulins. Proteins that bind along the sides of microtubules may alter the stability and mechanical properties of microtubules (Figure 2). The accessory proteins, collectively called microtubule-associated proteins (MAPs), participate in the regulation of microtubule assembly and function. MAPs in the nervous system fall into two heterogeneous groups: tau proteins and high-molecular weight MAPs. MAPs can either stabilize or destabilize microtubules [31-33]. A subset of MAPs can also mediate the interaction of microtubules with other cellular components. This subset is prominent in neurons where stabilized microtubules bundle from the core of the axons and dendrites that extend from the cell body [1]. These MAPs have at least one domain that binds to the microtubule surface and another that projects outward. Cells overexpressing tau, a MAP with a short projecting domain, form bundles of many closely packed microtubules [34]. Thus, MAP activities regulate the structural changes in neuronal microtubule dynamics. Tau binding to microtubules can also regulate the transport of membrane-enclosed organelles driven by molecular motors.

Function of Molecular Motors

Microtubules are responsible for the significant cellular function that involves the transport of organelles and particles in neuronal processes. Transport from the cell body toward the nerve terminals is called anterograde axonal transport [35]. Alternatively, transport from the nerve terminals toward the cell body is called retrograde axonal transport [35]. One microtubule can transport in both directions depending on the kind of motor proteins to which a particle binds (Figure 2). Kinesin is responsible for anterograde movement in microtubules. Dynein is responsible for the bulk of the retrograde transport, although some kinesins also contribute. Kinesin structurally consists of two heavy chains and two light chains [36]. It has a rod-like structure that is composed of two globular heads, a stalk, and a fan-like end with a total length of 80 nm [36]. The globular heads are composed of heavy chains that bind to microtubules and adenosine triphosphate (ATP) [37], and the light chains constitute the fan-like end [36]. All kinesins have a homologous catalytic core that binds to ATP and microtubules, and the catalytic core bears a striking structural similarity to the nucleotide-binding core of the actin-based motor myosin [38]. The neck linker, a 15-amino-acid segment in the carboxyl terminal to the catalytic core, is immobilized and extends toward the plus end of the microtubule when kinesin binds to microtubules and ATP [38]. This conformational change explains the direction of kinesin motion and the movement by kinesin dimer [38]. The tail of the motor protein binds to the cargo that they carry. Dynein is a complex of between one and three heavy chains with several intermediate and light chains [39]. Each heavy chain contains sites of ATP hydrolysis and microtubule binding, and thus constitutes the fundamental motor unit [40]. The heavy chain folds to form a globular head with two elongated structures, the stalk and the stem, which emerge from it [41, 42]. These two structures bind to the microtubule track and cargo, respectively [43]. The stem and stalk are both flexible, and they emerge separately from the head [44]. The conformational changes are fundamental for the power stroke of dynein [44].

Many different kinds of cytoskeletal motor proteins generally coexist in eukaryotic cells. Motor proteins are grouped by protein superfamilies or families. Actin-based motor proteins are members of the myosin superfamily [1]. The microtubule motor proteins are members of either the kinesin superfamily or the dynein family [36]. Most motor proteins share a similar structure involving the motor domain in the head and the tail domain that attaches to the cargo. A major function of microtubule motors is the transport and positioning of membrane-enclosed organelles. A clear example of the effect of microtubules and microtubule motors on the behavior of intracellular membranes is their role in organizing the endoplasmic reticulum (ER) and the Golgi apparatus [1]. The ER aligns with microtubules and extends to the edge of the cell, while the Golgi apparatus is located near the centrosome [1]. When cells are treated with a drug that depolymerizes microtubules, such as colchicine or nocodazole, the ER collapses into the center of the cell, and the Golgi apparatus fragments and disperses throughout the cytoplasm [1]. Thus, the motor proteins transport membrane-enclosed organelles along microtubules to their appropriate locations in the cells.

Role of Microtubules and MAPs in Disease

Growing evidence suggests that deficits in the function of microtubules and MAPs contribute to the pathogenesis of multiple diseases. A famous example of a disease-related histopathological feature that involves microtubules is the neurofibrillary tangles. Neurofibrillary tangles are composed of phosphorylated tau, which is an axonal MAP, and are a core feature of Alzheimer's disease (AD). MAPs participate in the regulation of microtubule assembly and anchor microtubules to each other and to other parts of neurons. Tau stabilizes microtubules and suppresses microtubule dynamics, thus providing structural support for axons [45]. Tau protein forms its own helical filaments when they are present at sufficiently high concentrations [1]. The accumulation of filamentous tau independent of tubulin is a well-established hallmark of several neurodegenerative diseases, collectively known as tauopathies [45]. Tauopathies comprise a heterogeneous group of age-dependent cognitive disorders, including AD and frontotemporal dementia and parkinsonism linked to chromosome 17 (FTDP-17) [46-49]. Although much attention has been focused on the extracellular deposits of amyloid β-protein in AD, the neuronal pathological changes in tau have been implicated in the dementia that accompanies the disease. The discovery of mutations in tau associated with inherited FTDP-17 has led to unequivocal evidence that aberrant tau can produce neurodegenerative diseases in humans [45].

The ability of neurons to maintain a highly specialized morphology depends on the continuous transport of proteins and organelles. Microtubules and motor proteins are two major components of the molecular motors that are involved in axonal transport. Impaired axonal transport may be a primary contributor to disease processes. The kinesin superfamily motor protein 1B (KIF1B) has been shown to transport mitochondria [50]. Studies in *KIF1B* knockout mice revealed a defect in the transportation of synaptic vesicle precursors and progressive muscular weakness [51]. One form of hereditary peripheral neuropathy, Charcot-Marie-Tooth (CMT) disease type 2A, was mapped to an interval containing the *KIF1B* gene [52]. CMT disease, which is clinically characterized by distal muscle weakness and atrophy, is divided into two forms: the demyelinating or hypertrophic form, or CMT type 1, and the axonal or neuronal form, or CMT type 2. Histologically, CMT type 2 displays chronic axonal loss and no features of hypertrophy. One form of the CMT type 2, CMT2A, is caused by a point mutation in *KIF1B* which is involved in the transport of synaptic vesicle precursors down the axon [51, 53]. Axonal transport is disturbed by the mutated motor protein kinesin in human hereditary neuropathies.

The primary disorders associated with ciliary dysfunction, termed primary ciliary dyskinesia (immotile cilia syndrome), are inherited in an autosomal recessive manner [54], and several genes encoding dynein have been implicated [55]. Numerous defects are considered under this category, including structural abnormalities of the dynein arms, radial spokes, and microtubules [54]. In addition, mutations in the heavy and intermediate chains of dynein have been described [55]. The cilia become dyskinetic, their coordinated propulsive action is diminished, and bacterial clearance is implicated [55]. The clinical effects include upper and lower respiratory tract infections, such as sinusitis and bronchiectasis, male infertility, and Kartagener's syndrome, in which situs inversus accompanies respiratory infections.

Role of Tubulin in MSA

MSA is a nonhereditary neurodegenerative disease in which oligodendrocytes and neurons in the CNS are affected. MSA represents a group of disorders that present with diverse clinical and pathological manifestation. Clinical symptoms include parkinsonism, ataxia, and autonomic failure [56]. MSA onset is usually in the sixth decade, and it is followed by progression and death, which occurs after an average disease duration of nine years [57]. The term MSA encompasses a wide variety of cases with clinical and pathological features of striatonigral degeneration, olivopontocerebellar involvement, and autonomic failure [58]. The cellular mechanisms underlying neurodegeneration in MSA are not fully understood, and unfortunately, no therapeutic approaches have yet had an immediate impact on neurodegeneration in MSA. Three significant neuropathological features characterize MSA histologically: glial cytoplasmic inclusions (GCIs), neuronal inclusions, and neuropil threads [59]. All three are composed of α-synuclein. GCIs, the first neuropathological manifestation to be described in MSA, are oligodendrocytic inclusions [60-62]. The presence of GCIs is a consistent and unifying histopathological feature observed in all the various clinical and pathological manifestations of MSA [58]. In studies on GCIs, filaments isolated from the CNS of MSA patients were identified using antibodies that recognized α-synuclein [63], the major component of the inclusions in MSA [64, 65]. GCIs may be the primary lesion that eventually compromises nerve cell function and viability in MSA [66]. However, the relevance of oligodendrocytic α-synuclein accumulation in neurodegeneration in MSA remains unknown. The vulnerability of oligodendrocytes probably underlies the susceptibility to neurodegeneration in MSA.

To understand the complexity of degeneration in the two types of CNS cells, transgenic (Tg) mouse models, in which human wild-type α-synuclein was overexpressed in oligodendrocytes in the mouse CNS, were generated under the control of different promoters [67-69]. The Tg mouse models showed abnormal aggregation of overexpressed α-synuclein in oligodendrocytes. In addition, two of the three mouse lines showed that α-synuclein accumulation in GCIs led to neuronal degeneration in the mouse CNS [68, 69]. We generated a Tg mouse model that expressed human wild-type α-synuclein in oligodendrocytes in the mouse CNS under the control of the 2′,3′-cyclic nucleotide 3′-phosphodiesterase (CNP) promoter (CNP-Tg mouse) [70]. Our study of the CNP-Tg mouse model demonstrated that the formation of oligodendrocytic α-synuclein inclusions resulted in neuronal degeneration that was exemplified by a phenotype of motor impairment, brain atrophy that was macroscopically apparent, and a histological decrease in the number of neurons with gliosis [68]. α-Synuclein has been detected in the insoluble fraction of the mouse CNS and the brain tissue of MSA patients, whereas it is typically distributed in the soluble fraction of the brain tissue of healthy humans [68, 71]. Furthermore, α-synuclein modulation by neuronal activity was disturbed in the mouse cells, and the insolubility of α-synuclein may cause abnormal regulation of α-synuclein synaptic accumulation, thus disturbing the normal function of α-synuclein [72]. Oligodendrocytic α-synuclein accumulation induced secondary neuronal dysfunction and degeneration, and a similar disease process is hypothesized to underlie MSA. Consequently, a novel pathological mechanism of neuronal degeneration was elucidated in the CNP-Tg mouse model. β-III tubulin, a type of neuron-specific tubulin in microtubules, was identified as a protein that binds to α-synuclein in the mouse CNS [72]. Binding of the protein to α-synuclein promoted α-synuclein accumulation, leading to neuronal dysfunction in the CNP-Tg mouse model [72]. Moreover, α-synuclein accumulation was suppressed by the inhibition of microtubule polymerization. These results indicate that the binding of α-synuclein to β-III tubulin is a key process in the development of neurodegeneration in the MSA mouse model.

Binding of α-synuclein to β-III Tubulin Is a Therapeutic Target for MSA

Tg mouse models have provided important evidence for an increased understanding of the pathophysiology of neurodegeneration in MSA and have led to hypotheses postulating the involvement of three major pathological processes underlying neurodegeneration in MSA (Figure 3). The first hypothesized process involves the formation of GCIs in oligodendrocytes, which increase α-synuclein expression in neurons. Using primary culture cells derived from the CNP-Tg mouse brain, the inhibition of glial cell proliferation by treatment with arabinofuranosyl cytidine was shown to prevent insoluble α-synuclein accumulation in the CNP-Tg mouse neurons [72]. Because reciprocal communication between neurons and oligodendrocytes is essential for CNS development [73], the results from the study using the CNP-Tg mouse model suggest that degenerated oligodendrocytes regulate α-synuclein protein expression in neurons. Moreover, when the primary culture cells derived from non-Tg mouse brain were treated with conditioned media derived from CNP-Tg mouse cell cultures, insoluble α-synuclein accumulated in the non-Tg mouse cells [72]. These results indicate that the conditioned media of CNP-Tg mouse cell cultures contained oligodendrocyte-derived signals that induced neuronal α-synuclein accumulation. Thus, abnormal regulation of α-synuclein expression by degenerated oligodendrocytes may be a trigger for the onset of neurodegeneration in MSA.

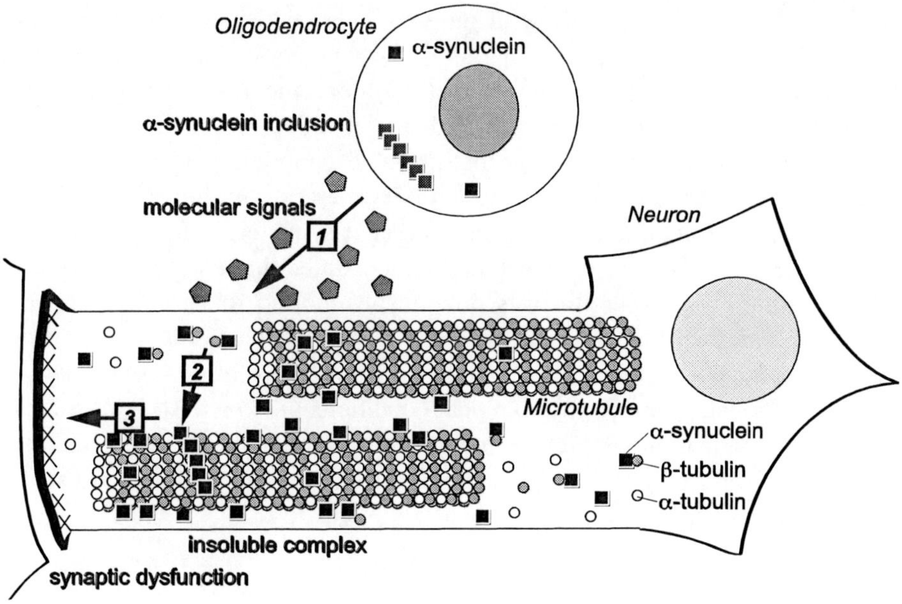

Figure 3. Schematic representation of the three cellular mechanisms of neurodegeneration in MSA. First, an oligodendrocyte degenerates because of α-synuclein accumulation, forming an oligodendrocytic inclusion (arrow *1*). Subsequently, the oligodendrocyte releases molecular signals that increase α-synuclein expression in a neuron, and endogenous neuronal α-synuclein accumulates. Endogenous α-synuclein binds to β-III tubulin in microtubules and forms insoluble protein complexes in the neuron (arrow *2*). Finally, α-synuclein modulation by neuronal activity is disturbed, and α-synuclein accumulation causes neuronal dysfunction (arrow *3*), leading to neurodegeneration.

The second hypothesized process involves the binding of α-synuclein to β-III tubulin, which induces the accumulation of insoluble protein complexes in neurons (Figure 3). The third hypothesized process involves the modulation of α-synuclein by neuronal activity, which is disturbed in MSA. Neuronal activity controls α-synuclein synaptic accumulation, which disperses from the nerve terminal in response to neuronal activity [74]. Assessment of primary culture cells revealed that the insolubility of α-synuclein caused abnormal regulation of α-synuclein synaptic accumulation, disturbing the normal function of α-synuclein. Because α-synuclein localizes to nerve terminals [75-77], it functions as a negative regulator of synaptic vesicle fusion and neurotransmitter release at the synapse when expressed at physiological levels [78]. The total number of vesicle fusion events in the α-synuclein-overexpressing cells was significantly decreased [79]. Because endogenous mouse α-synuclein expression was increased in the presynaptic terminals of CNP-Tg mouse neurons [68], the total number of synaptic vesicle fusion events may be significantly decreased, resulting in neuronal dysfunction and neurodegeneration.

Although the mechanisms of GCI development are not fully understood, the data described above strongly suggest that a therapeutic target for MSA involves the binding of α-synuclein with β-III tubulin, which results in the accumulation of insoluble complexes. The microtubule component β-III tubulin was identified as the protein that interacts with α-synuclein in the CNP-Tg mouse model [68]. Previous studies have shown that β-tubulin was located in GCIs in the brain tissues of MSA patients [62, 80-82]. Ultrastructural studies have demonstrated that GCIs are composed of oligodendrocytic microtubules with diameters of 20–30 nm [80]. Aged CNP-Tg mice with motor impairment demonstrated distinct axonal accumulation of α-synuclein and the formation of α-synuclein inclusions in the degenerating axons and axonal terminals [68]. The axonal terminal inclusions were ultrastructurally composed of disordered tubular structures with diameters of approximately 20–35 nm [68]. In addition, the formation and accumulation of insoluble complexes of α-synuclein and β-III tubulin were important disease processes of neurodegeneration in the CNP-Tg mouse model [72]. The physiological role of the interaction between α-synuclein and β-tubulin has not been completely clarified, but growing evidence has demonstrated that a definite interaction exists between the two proteins. β-Tubulin was previously reported to be coimmunoprecipitated with α-synuclein in normal human brain extracts [83, 84]. Previous reports also showed that α-synuclein interacts with heterodimeric tubulin [83] and that tubulin initiates and promotes α-synuclein fibril formation under physiological conditions *in vitro* [84]. In the mouse model, microtubule depolymerization suppressed α-synuclein accumulation [72]. Thus, the binding of α-synuclein to β-III tubulin and the formation of insoluble protein complexes are important processes of neurodegeneration in MSA.

Conclusion

Microtubules are essential to the structure and function of neurons. One cause of neuronal dysfunction and degeneration is microtubule impairment. Further understanding of the pathophysiology of microtubules in neurodegeneration will provide new treatments for intractable neurodegenerative disease. Further studies on MSA are needed to address aspects of the pathological mechanisms of neurodegeneration and the physiological function of tubulin using animal models and primary culture cells, and thereby to develop effective therapeutic approaches for MSA.

Acknowledgments

This study was supported by Research Funding for Longevity Sciences (23-11) from the National Center for Geriatrics and Gerontology, Japan and Grants-in-Aid (22500326 and 22700386) for Scientific Research from the Japan Society for the Promotion of Science.

References

[1] Alberts, B., Johnson, A., Lewis, J., Raff, M., Roberts, K., & Walter, P. (2008). *Molecular Biology of the Cell* (5th ed., pp.1047-1050). NY: Garland Science.

[2] Peters, A., Palay, S.L., & Webster, H.D. (1991). *The Fine Structure of the Nervous System, Neurons and Their Supporting Cells* (3rd ed., pp.14-137). NY: Oxford University Press.

[3] Soltys, B.J. & Borisy, G.G. (1985). Polymerization of tubulin *in vivo*: direct evidence for assembly onto microtubule ends and from centrosomes. *J. Cell Biol.*, 100, 1682-1689.

[4] Crossin, K.L. & Carney, D.H. (1981). Evidence that microtubule depolymerization early in the cell cycle is sufficient to initiate DNA synthesis. *Cell*, 23, 61-71.

[5] Rodionov, V.I. & Borisy, G.G. (1997).Microtubule treadmilling *in vivo*. *Science* 275, 215-218.

[6] Freed, J.J. & Lebowitz, M.M. (1970). The association of a class of saltatory movements with microtubules in cultured cells. *J. Cell Biol.*, 45, 334-354.

[7] Weisenberg, R.C. & Rosenfeld, A.C. (1975). *In vitro* polymerization of microtubules into asters and spindles in homogenates of surf clam eggs. *J. Cell Biol.*, 64, 146-158.

[8] Squire, L.R., Berg, D., Bloom, F.E., du Lac, S., Ghosh, A., & Spitzer, N. (2008). *Fundamental Neuroscience* (3rd ed., pp.70-86). Burlington, MA: Academic Press.

[9] Lee, M.K., Tuttle, J.B., Rebhun, L.I., Cleveland, D.W., & Frankfurter, A. (1990). The expression and posttranslational modification of a neuron-specific beta-tubulin isotype during chick embryogenesis. *Cell Motil. Cytoskeleton*, 17, 118-132.

[10] Alexander, J.E., Hunt, D.F., Lee, M.K., Shabanowitz, J., Michel, H., Berlin, S.C., MacDonald, T.L., Sundberg, R.J., Rebhun, L.I., & Frankfurter, A. (1991). Characterization of posttranslational modifications in neuron-specific class III beta-tubulin by mass spectrometry. *Proc. Natl. Acad. Sci. USA*, 88, 4685-4689.

[11] Bear, M.F., Connors, B.W., & Paradiso, M.A. (2007). *Neuroscience: Exploring the Brain* (3rd ed., pp.24-49). Baltimore, MD: Lippincott Williams & Wilkins.

[12] Malawista, S.E. & Bensch, K.G. (1967). Human polymorphonuclear leukocytes: demonstration of microtubules and effect of colchicines. *Science*, 28, 521-522.

[13] Borisy, G.G. & Taylor, E.W. (1967). The mechanism of action of colchicine. Colchicine binding to sea urchin eggs and the mitotic apparatus. *J. Cell Biol.,* 34, 535-548.

[14] Richardson, C.D. & Vance, D.E. (1978). The effect of colchicine and dibucaine on the morphogenesis of Semliki Forest virus. *J. Biol. Chem.*, 253, 4584-4589.

[15] De Brabander, M., De May, J., Joniau, M., & Geuens, G. (1977). Ultrastructural immunocytochemical distribution of tubulin in cultured cells treated with microtubule inhibitors. *Cell Biol. Int. Rep.*, 1, 177-183.

[16] Schiff, P.B., Fant, J., & Horwitz, S.B. (1979). Promotion of microtubule assembly *in vitro* by taxol. *Nature*, 277, 665-667.

[17] Schiff, P.B. & Horwitz, S.B. (1980). Taxol stabilizes microtubules in mouse fibroblast cells. *Proc. Natl. Acad. Sci. USA*, 77, 1561-1565.

[18] Margolis, R.L. & Wilson, L. (1981). Microtubule treadmills—possible molecular machinery. *Nature*, 293, 705-711.

[19] Mitchison, T. & Kirschner, M. (1984). Dynamic instability of microtubule growth. *Nature*, 312, 237-342.

[20] Mitchison, T. & Kirschner, M. (1984). Microtubule assembly nucleated by isolated centrosomes. *Nature*, 312, 232-237.

[21] Kerssemakers, J.W., Munteanu, E.L., Laan, L., Noetzel, T.L., Janson, M.E., & Dogterom, M. (2006). Assembly dynamics of microtubules at molecular resolution. *Nature*, 442, 709-712.

[22] Quarmby, L. (2000). Cellular Samurai: katanin and the severing of microtubules. *J. Cell Sci.*, 113, 2821-2827.

[23] Bergen, L.G., Kuriyama, R., & Borisy, G.G. (1980). Polarity of microtubules nucleated by centrosomes and chromosomes of Chinese hamster ovary cells *in vitro*. *J. Cell Biol.*, 84, 151-159.

[24] Roobol, A., Havercroft, J.C., & Gull, K. (1982). Microtubule nucleation by the isolated microtubule-organizing centre of *Physarum polycephalum* myxamoebae. *J. Cell Sci.*, 55, 365-381.

[25] Oakley, C.E. & Oakley, B.R. (1989). Identification of γ-tubulin, a new member of the tubulin superfamily encoded by *mipA* gene of *Aspergillus nidulans*. *Nature*, 338, 662-664.

[26] Stearns, T. & Kirschner, M. (1994). *In vitro* reconstitution of centrosome assembly and function: the central role of γ-tubulin. *Cell*, 76, 623-637.

[27] Zheng, Y., Wong, M.L., Alberts, B., & Mitchison, T. (1995). Nucleation of microtubule assembly by a γ-tubulin-containing ring complex. *Nature*, 378, 578-583.

[28] Aldaz, H., Rice, L.M., Stearns, T., & Agard, D.A. (2005). Insights into microtubule nucleation from the crystal structure of human γ-tubulin. *Nature*, 435, 523-527.

[29] Baas, P.W. (2002). Microtubule transport in the axon. *Int. Rev. Cytol.*, 212, 41-62.

[30] Ahmad, F.J., Yu, W., McNally, F.J., & Baas, P.W. (1999). An essential role for katanin in severing microtubules in the neuron. *J. Cell Biol.*, 145, 305-315.

[31] Greene, L.A., Liem, R.K., & Shelanski, M.L. (1983). Regulation of a high molecular weight microtubule-associated protein in PC12 cells by nerve growth factor. *J. Cell Biol.*, 96, 76-83.

[32] Serrano, L., de la Torre, J., Maccioni, R.B., & Avila, J. (1984). Involvement of the carboxyl-terminal domain of tubulin in the regulation of its assembly. *Proc. Natl. Acad. Sci. USA*, 81, 5989-5993.

[33] Brown, P.A. & Berlin, R.D. (1985). Packing volume of sedimented microtubules: regulation and potential relationship to an intracellular matrix. *J. Cell Biol.*, 101, 1492-1500.

[34] Chen, J., Kanai, Y., Cowan, N.J., & Hirokawa, N. (1992). Projection domains of MAP2 and tau determine spacings between microtubules in dendrites and axons. *Nature*, 360, 674-677.

[35] Brodal, P. (2010). *The Central Nervous System, Structure and Function* (4th ed., pp.1-18). NY: Oxford University Press.

[36] Hirokawa, N. (1998). Kinesin and dynein superfamily proteins the mechanism of organelle transport. *Science*, 279, 519-526.

[37] Howard, J., Hudspeth, A.J., & Vale, R.D. (1989). Movement of microtubules by single kinesin molecules. *Nature*, 342, 154-158.

[38] Rice, S., Lin, A.W., Safer, D., Hart, C.L., Naber, N., Carragher, B.O., Cain, S.M., Pechatnikova, E., Wilson-Kubalek, E.M., Whittaker, M., Pate, E., Cooke, R., Taylor,

E.W., Milligan, R.A., & Vale, R.D. (1999). A structural change in the kinesin motor protein that drives motility. *Nature*, 402, 778-784.

[39] King, S.M. (2000). The dynein microtubule motor. *Biochim. Biophys. Acta*, 1496, 60-75.

[40] Koonce, M.P. & Samsó, M. (1996). Overexpression of cytoplasmic dynein's globular head causes a collapse of the interphase microtubule network in Dictyostelium. *Mol. Biol. Cell*, 7, 935-948.

[41] Goodenough, U. & Heuser, J. (1984). Structural comparison of purified dynein proteins with *in situ* dynein arms. *J. Mol. Biol.*, 180, 1083-1118.

[42] Goodenough, U.W., Gebhart, B., Mermall, V., Mitchell, D.R., & Heuser, J.E. (1987). High-pressure liquid chromatography fractionation of Chlamydomonas dynein extracts and characterization of inner-arm dynein subunits. *J. Mol. Biol.*, 194, 481-494.

[43] King, S.M. (2000). AAA domains and organization of the dynein motor unit. *J. Cell Sci.*, 113, 2521-2526.

[44] Burgess, S.A., Walker, M.L., Sakakibara, H., Knight, P.J., & Oiwa, K. (2003). Dynein structure and power stroke. *Nature*, 421, 715-718.

[45] Garcia, M.L. & Cleveland, D.W. (2001). Going new places using an old MAP: tau, microtubules and human neurodegenerative disease. *Curr. Opin. Cell Biol.*, 13, 41-48.

[46] Poorkaj, P., Bird, T.D., Wijsman, E., Nemens, E., Garruto, R.M., Anderson, L., Andreadis, A., Wiederholt, W.C., Raskind, M., & Schellenberg, G.D. (1998). Tau is a candidate gene for chromosome 17 frontotemporal dementia. *Ann. Neurol.*, 43, 815-825.

[47] Spillantini, M.G., Murrell, J.R., Goedert, M., Farlow, M.R., Klug, A., & Ghetti, B. (1998). Mutation in the tau gene in familial multiple system tauopathy with presenile dementia. *Proc. Natl. Acad. Sci. USA*, 95, 7737-7741.

[48] Hutton, M., Lendon, C.L., Rizzu, P., Baker, M., Froelich, S., Houlden, H., Pickering-Brown, S., Chakraverty, S., Isaacs, A., Grover, A., Hackett, J., Adamson, J., Lincoln, S., Dickson, D., Davies, P., Petersen, R.C., Stevens, M., de Graaff. E., Wauters, E., van Baren, J., Hillebrand, M., Joosse, M., Kwon, J.M., Nowotny, P., Che, L.K., Norton, J., Morris, J.C., Reed, L.A., Trojanowski, J., Basun, H., Lannfelt, L., Neystat, M., Fahn, S., Dark, F., Tannenberg, T., Dodd, P.R., Hayward, N., Kwok, J.B., Schofield, P.R., Andreadis, A., Snowden, J., Craufurd, D., Neary, D., Owen, F., Oostra, B.A., Hardy, J., Goate, A., van Swieten, J., Mann, D., Lynch, T., & Heutink, P. (1998). Association of missense and 5′-splice-site mutations in tau with the inherited dementia FTDP-17. *Nature*, 393, 702-705.

[49] Hong, M., Zhukareva, V., Vogelsberg-Ragaglia, V., Wszolek, Z., Reed, L., Miller, B.I., Geschwind, D.H., Bird, T.D., McKeel, D., Goate, A., Morris, J.C., Wilhelmsen, K.C., Schellenberg, G.D., Trojanowski, J.Q., & Lee, V.M. (1998). Mutation-specific functional impairments in distinct tau isoforms of hereditary FTDP-17. *Science*, 282, 1914-1917.

[50] Nangaku, M., Sato-Yoshitake, R., Okada, Y., Noda, Y., Takemura, R., Yamazaki, H., & Hirokawa, N. (1994). KIF1B, a novel microtubule plus end-directed monomeric motor protein for transport of mitochondria. *Cell*, 79, 1209-1220.

[51] Zhao, C., Takita, J., Tanaka, Y., Setou, M., Nakagawa, T., Takeda, S., Yang, H.W., Terada, S., Nakata, T., Takei, Y., Saito, M., Tsuji, S., Hayashi, Y., & Hirokawa, N. (2001). Charcot-Marie-Tooth disease type 2A caused by mutation in a microtubule motor KIF1Bβ. *Cell*, 105, 587-597.

[52] Ben Othmane, K., Middleton, L.T., Loprest, L.J., Wilkinson, K.M., Lennon, F., Rozear, M.P., Stajich, J.M., Gaskell, P.C., Roses, A.D., Pericak-Vance, M.A., & Vance, J.M. (1993). Localization of a gene (CMT2A) for autosomal dominant Charcot-Marie-Tooth disease type 2 to chromosome 1p and evidence of genetic heterogeneity. *Genomics*, 17, 370-375.

[53] Boerkoel, C.F., Takashima, H., Garcia, C.A., Olney, R.K., Johnson, J., Berry, K., Russo, P., Kennedy, S., Teebi, A.S., Scavina, M., Williams, L.L., Mancias, P., Butler, I.J., Krajewski, K., Shy, M., & Lupski, J.R. (2002). Charcot-Marie-Tooth disease and related neuropathies: mutation distribution and genotype-phenotype correlation. *Ann. Neurol.*, 51, 190-201.

[54] Fauci, A.S., Braunwald, E., Kasper, D.L., Hauser, S.L., Longo, D.L., Jameson, J.L., & Loscalzo, J. (2008). *Harrison's Principles of Internal Medicine* (17th ed., pp.1629-1631). NY: McGraw-Hill Companies.

[55] Hirokawa, N. & Takemura, R. (2003). Biochemical and molecular characterization of diseases linked to motor proteins. *Trends Biochem. Sci.*, 28, 558-565.

[56] Graham, J.G. & Oppenheimer, D.R. (1969). Orthostatic hypotension and nicotine sensitivity in a case of multiple system atrophy. *J. Neurol. Neurosurg. Psychiatry*, 32, 28-34.

[57] Wenning, G.K., Ben-Shlomo, Y., Magalhaes, M., Daniel, S.E., & Quinn, N.P. (1994). Clinical features and natural history of multiple system atrophy. An analysis of 100 cases. *Brain*, 117, 835-845.

[58] Esiri, M. & Perl, D. (2006). *Oppenheimer's diagnostic neuropathology*. (3rd ed., pp. 312-314). London: Hodder Arnold.

[59] Ince, P.G., Clark, B., Holton, J., Revesz, T., & Wharton, S.B. (2008). Diseases of movement and system degenerations. In: Love, S., Louis, D.N., & Ellison, D.W. (Eds.), *Greenfield's Neuropathology* (8th ed., pp. 889-947). London: Arnold.

[60] Papp, M.I., Kahn, J.E., & Lantos, P.L. (1989). Glial cytoplasmic inclusions in the CNS of patients with multiple system atrophy (striatonigral degeneration, olivopontocerebellar atrophy and Shy-Drager syndrome). *J. Neurol. Sci.*, 94, 79-100.

[61] Arima, K., Ueda, K., Sunohara, N., Arakawa, K., Hirai, S., Nakamura, M., Tonozuka-Uehara, H., & Kawai, M. (1998). NACP/α-synuclein immunoreactivity in fibrillary components of neuronal and oligodendroglial cytoplasmic inclusions in the pontine nuclei in multiple system atrophy. *Acta Neuropathol. (Berl.)*, 96, 439-444.

[62] Tu, P.H., Galvin, J.E., Baba, M., Giasson, B., Tomitam, T., Leight, S., Nakajo, S., Iwatsubo, T., Trojanowski, J.Q., & Lee, V.M.Y. (1998). Glial cytoplasmic inclusions in white matter oligodendrocytes of multiple system atrophy brains contain insoluble α-synuclein. *Ann. Neurol.*, 44, 415-422.

[63] Spillantini, M.G., Crowther, R.A., Jakes, R., Cairns, N.J., Lantos, P.L., & Goedert, M. (1998). Filamentous α-synuclein inclusions link multiple system atrophy with Parkinson's disease and dementia with Lewy bodies. *Neurosci. Lett.*, 251, 205-208.

[64] Spillantini, M.G. & Goedert, M. (2000). The α-synucleinopathies: Parkinson's disease, dementia with Lewy bodies, and multiple system atrophy. *Ann. NY Acad. Sci.*, 920, 16-27.

[65] Duda, J.E., Lee, V.M.Y., & Trojanowski, J.Q. (2000). Neuropathology of synuclein aggregates. *J. Neurosci. Res.*, 61, 121-127.

[66] Goedert, M. (2001). Alpha-synuclein and neurodegenerative diseases. *Nat. Rev. Neurosci.*, 2, 492-501.

[67] Kahle, P.J., Neumann, M., Ozmen, L., Muller, V., Jacobsen, H., Spooren, W., Fuss, B., Mallon, B., Macklin, W.B., Fujiwara, H., Hasegawa, M., Iwatsubo, T., Kretzschmar, H.A., & Haass, C. (2002). Hyperphosphorylation and insolubility of α-synuclein in transgenic mouse oligodendrocytes. *EMBO Rep.*, 3, 583-588.

[68] Yazawa, I., Giasson, B.I., Sasaki, R., Zhang, B., Joyce, S., Uryu, K., Trojanowski, J.Q., & Lee, V.M.Y. (2005). Mouse model of multiple system atrophy: α-synuclein expression in oligodendrocytes causes glial and neuronal degeneration. *Neuron*, 45, 847-859.

[69] Shults, C.W., Rockenstein, E., Crews, L., Adame, A., Mante, M., Larrea, G., Hashimoto, M., Song, D., Iwatsubo, T., Tsuboi, K., & Masliah, E. (2005). Neurological and neurodegenerative alterations in a transgenic mouse model expressing human α-synuclein under oligodendrocyte promoter: implications for multiple system atrophy. *J. Neurosci.*, 25, 10689-10699.

[70] Gravel, M., Di Polo, A., Valera, P.B., & Braun, P.E. (1998). Four-kilobase sequence of the mouse CNP gene directs spatial and temporal expression of lacZ in transgenic mice. *J. Neurosci. Res.*, 53, 393-404.

[71] Dickson, D.W., Liu, W., Hardy, J., Farrer, M., Mehta, N., Uitti, R., Mark, M., Zimmerman, T., Golbe, L., Sage, J., Sima, A., D'Amato, C., Albin, R., Gilman, S., & Yen, S.H. (1999). Widespread alterations of α-synuclein in multiple system atrophy. *Am. J. Pathol.*, 155, 1241-1251.

[72] Nakayama, K., Suzuki, Y., & Yazawa, I. (2009). Microtubule depolymerization suppresses α-synuclein accumulation in a mouse model of multiple system atrophy. *Am. J. Pathol.,* 174, 1471-1480.

[73] Simons, M. & Trajkovic, K. (2006). Neuron-glia communication in the control of oligodendrocyte function and myelin biogenesis. *J. Cell Sci.*, 119, 4381-4389.

[74] Fortin, D.L., Nemani, V.M., Voglmaier, S.M., Anthony, M.D., Ryan, T.A., & Edwards, R.H. (2005). Neural activity controls the synaptic accumulation of α-synuclein. *J. Neurosci.*, 25, 10913-10921.

[75] Jakes, R., Spillantini, M.G., & Goedert, M. (1994). Identification of two distinct synucleins from human brain. *FEBS Lett.*, 345, 27-32.

[76] Iwai, A., Masliah, E., Yoshimoto, M., Ge, N., Flanagan, L., de Silva, H.A., Kittel, A., & Saitoh, T. (1995). The precursor protein of non-A beta component of Alzheimer's disease amyloid is a presynaptic protein of the central nervous system. *Neuron*, 14, 467-475.

[77] Withers, G.S., George, J.M., Banker, G.A., & Clayton, D.F. (1997). Delayed localization of synelfin (synuclein, NACP) to presynaptic terminals in cultured rat hippocampal neurons. *Brain Res. Dev. Brain Res.*, 99, 87-94.

[78] Gitler, A.D. & Shorter, J. (2007). Prime time for α-synuclein. *J. Neurosci.*, 27, 2433-2434.

[79] Larsen, K.E., Schmitz, Y., Troyer, M.D., Mosharov, E., Dietrich, P., Quazi, A.Z., Savalle, M., Nemani, V., Chaudhry, F.A., Edwards, R.H., Stefanis, L., & Sulzer, D. (2006). α-Synuclein overexpression in PC12 and chromaffin cells impairs catecholamine release by interfering with a late step in exocytosis. *J. Neurosci.*, 26, 11915-11922.

[80] Nakazato, Y., Yamazaki, H., Hirato, J., Ishida, Y., & Yamaguchi, H. (1990). Oligodendroglial microtubular tangles in olivopontocerebellar atrophy. *J. Neuropathol. Exp. Neurol.*, 49, 521-530.

[81] Abe, H., Yagishita, S., Amano, N., Iwabuchi, K., Hasegawa, K., & Kowa, K. (1992). Argyrophilic glial intracytoplasmic inclusions in multiple system atrophy: immunocytochemical and ultrastructural study. *Acta Neuropathol. (Berl.)*, 84, 273-277.

[82] Gai, W.P., Power, J.H., Blumbergs, P.C., Culvenor, J.G., & Jensen, P.H. (1999). α-Synuclein immunoisolation of glial inclusions from multiple system atrophy brain tissue reveals multiprotein components. *J. Neurochem.*, 73, 2093-2100.

[83] Payton, J.E., Perrin, R.J., Clayton, D.F., & George, J.M. (2001). Protein-protein interactions of alpha-synuclein in brain homogenates and transfected cells. *Brain Res. Mol. Brain Res.*, 95, 138-145.

[84] Alim, M.A., Hossain, M.S., Arima, K., Takeda, K., Izumiyama, Y., Nakamura, M., Kaji, H., Shinoda, T., Hisanaga, S., & Ueda, K. (2002). Tubulin seeds α-synuclein fibril formation. *J. Biol. Chem.*, 277, 2112-2117.

In: Tubulin: Structure, Functions and Roles in Disease
Editors: Wei Yamauchi and Antonio Sokić
ISBN: 978-1-62100-141-6

Chapter V

Targeted Disruption of the Tubulin Cytoskeletal Network Activates Expression of the Profibrotic Genes PAI-1 and CTGF: Implications for Vascular Disease

Rohan Samarakoon and Paul J. Higgins*
Center for Cell Biology and Cancer Research,
Albany Medical College, Albany, New York, US

Introduction

Internal constraints imposed by the microtubule, microfilament and intermediate filament cytoskeletal networks, junctional complexes and integrin-extracellular matrix (ECM) interactions are major determinants of cell structure/function and subject to modulation by a variety of external forces [1-3]. Cell cycle transit, entry into DNA synthetic phase and activation of cell death programs are each influenced by cellular structural dynamics [4-9]. This is particularly evident in the vascular system which is constantly subjected to a continuum of hemodynamic stimuli (e.g., shear strain, flow disturbances, mechanical or pulsile stretch) that alter microfilament and microtubule organization and associated signaling pathways. These same physical forces that influence cell shape impact gene expression programs that, in turn, modulate cell proliferation, migration and ECM synthesis/deposition resulting in the development of tissue-specific pathologies or wound-healing anomalies (e.g., vessel fibrotic disorders, atherosclerosis) [10-14]. Prominent among the fibrosis-promoting proteins implicated in vascular fibroproliferative disease are the matricellular proteins plasminogen activator inhibitor inhibitor-1 (PAI-1, SERPINE1) and connective tissue growth

* E-mail: higginp@mail.amc.edu; Tel: 518-262-5168; Fax: 518-262-5669; Address: Center for Cell Biology and Cancer Research, Albany Medical College (MC-165), 47 New Scotland Avenue, Albany, New York 12208 (Corresponding author).

factor (CTGF, CCN2) [15,16]. Most importantly, recent data clearly indicate that transcriptional controls for both genes are sensitive to cytoskeletal perturbations [16]. Dissection of pathways and mechanisms underlying vascular cell shape-deformation responses may lead to the identification of new, translationally-relevant, targets for the treatment of fibrotic disorders.

Mechanochemical Signaling in the Vaculature

Integrin activation initiates cellular responses in coordination with (a) growth factor receptors, (b) cadherin complexes and (c) signals from the ECM [10,17-21] resulting in recruitment of signaling molecules (e.g., focal adhesion kinase [FAK], non-receptor kinases) to ECM (focal) contact sites as evidenced in shear stress-induced endothelial cell migration [22]. Functional and spatial associations between certain non-receptor tyrosine kinases ($pp125^{FAK}$, $pp60^{c\text{-}src}$), for example, increase with an applied mechanical load [8,12,23,24] and transduce physical deformation stresses into specific biological responses as a consequence of this cytoskeletal "wiring" [8,25]. Indeed, mobilization of several kinases and their adaptors including FAK, $pp60^{c\text{-}src}$, and growth factor receptor-bound protein 2 or Grb2, to focal contacts in response to varying cellular tensional forces activates downstream cascades (i.e., involving mitogen-activated protein [MAP] kinases and the small GTPases Ras, Rac, and Rho) [26]. Mechanical stress clusters, or initiates conformational changes to, specific integrins resulting in recruitment of structural proteins to the focal adhesion complex. $pp125^{FAK}$, as part of this cytoskeletal "scaffold" is tyrosine phosphorylated at Tyr-397 and complexes with $pp60^{c\text{-}src}$, to effect load-dependent signal transduction [27,28]. The proximity of $pp60^{c\text{-}src}$ with $pp125^{FAK}$ further stimulates $pp125^{FAK}$ phosphorylation at Tyr-925, creating a binding site for Grb2. The Shc adaptor protein, which is tyrosine phosphorylated in vascular cells in response to shear stress, docks with Grb2 by an SH2-dependent mechanism [29] resulting in construction of a Shc/Grb2/Sos complex and subsequent Ras GTPase activation. MAP kinase pathways in vascular smooth muscle cells (VSMC) similarly function via $pp125^{FAK}$/$pp60^{c\text{-}src}$/Grb2 interactions with Ras as a downstream target [27,30-32] while both the extracellular signal-regulated kinase (ERK) and c-*Jun*-associated kinase (JNK) pathways are mobilized in a FAK-dependent manner in mechanical stimulated endothelial cells [27,29]. Cyclic strain stimulates p38/ERK MAP kinase phosphorylation and regulates caldesmon function, impacting cellular migration/proliferation transitions and vascular wall contractility using pathways that require Ras and Rac GTPases [33,34].

Mechanical perturbation itself activates the Rho family GTPases Rho, Rac and Cdc42 [35]. Rho kinases (Rho-associated coiled-coil forming kinases; ROCK1/2, the immediate downstream targets of RhoA) and mDia, in turn, affect cytoskeletal organization, contractility, motility and gene expression [36]. They are particularly attractive, clinically-relevant, targets in the management of cardiovascular disease [37]. Rho GTPases cycle between active GTP-bound and inactive GDP-bound states regulated by guanine nucleotide exchanges factors (GEF) and GTPase-activating proteins (GAPs) [38-41]. The temporal/spatial activation of Rho GTPases appears critical for physiological adjustments to changes in mechanical forces commonly encountered in the vascular system [42] although the mechanism of mechanotransduction is complex and may vary as a function of vascular cell type. p190RhoGAP regulates capillary network formation by integrating mechanical and chemical signaling pathways, through downstream intermediates Rho kinases, mDia, LIM-kinase, PAK1 and other small GTPases [36,43]. This GTPase activating protein also functions as an integrator of Rac and RhoA cross-talk, which have opposing effects in the control of endothelial cell morphology [44]. The available evidence, in fact, supports a model wherein Rho→mDia signaling results in Rac activation via a Src-dependent mechanism with suppression of this pathway by ROCK activity [36].

Cell Shape-Dependent Responses to Microtubule Deformation

Perturbation cellular architecture (e.g., by multidirectional force application, cadherin- or integrin-interfering antibodies, cytoskeletal-active drugs, expression of mutant or cell type "unrelated" cytoskeletal elements, substrate-modulation of cell morphology) provide accessible models to probe deformation-sensitive signaling pathways and their target genes [e.g., 2,3,24,25,45-55]. Adaptation of novel "phosphotyrosine reporter" technologies, that measure fluorescence of YFP fused to multiple SH2 domains derived from the c-Src tyrosine kinase, and specific microtubule disrupting agents, moreover, confirmed the dynamic nature of microtubule-anchored signaling events (e.g., involving $pp125^{FAK}$, paxillin, vinculin and $pp60^{c\text{-}src}$) [56]. Such elegant studies highlight the impact of microtubule disruption on signal transduction, transcriptional outputs and the phenotypic response.

Distortion of cell morphology with actin/tubulin-targeting drugs is one important approach to the identification of cell shape-responsive genes as well as, in some cases, the involved signaling pathways [57-61]. Protein complex mapping revealed that the cellular signaling apparatus is interwoven with the cytoskeletal framework and, therefore, highly sensitive to shape perturbation [3]. It has become apparent in recent years, moreover, that the expression of certain genes is particularly responsive to changes in cellular configuration regardless of the specific deformation initiator. Thus, the actin (microfilament)- and tubulin (microtubule)-based cytoskeletal networks, key regulators of cell morphology, integrate and transduce a diversity of intracellular signals gleaned from cues provided by the ECM, cell-cell interactions and growth factors [1,24,62,63]. Drug-induced disruption of either the microtubule network (with colchicine or nocodazole) or cellular microfilaments (with cytochalasin D or latrunculin B), however, effectively results in a changes in gene transcription largely as a consequence of altered signaling events [16,61,64-67]. Deformation-mediated transcriptional outputs, however, are rather limited compared to the global reprogramming of gene expression that typically accompanies exposure to serum or individual growth factors [64,65,68-77]. Microarray analyses, in fact, identified relatively few shape deformation-sensitive genes in endothelial cells and VSMC exposed to steady laminar or turbulent shear stress; the most prominent of which were PAI-1 and CTGF [78,79]. CTGF, however, can be both upregulated and downregulated in a RhoA-dependent manner in flow-adapted endothelial cells depending on the extent of cytoskeletal remodeling in response to laminar shear strain and the activation of the transcription factor KLF2 (Kruppel-like factor 2) [80-82]. Induced PAI-1 gene expression in each instance, furthermore, correlated with stress-associated cytoskeletal restructuring events [45,53,54,79,83] and is closely associated with dynamic changes in cellular morphology and shape-altering physiologic processes [53,60,65,72-77,78,84-87] implicating the cytoskeleton in the signaling apparatus [60,64]. Disruption of the microtubule network with colchicine or nocodazole, in fact, stimulates several pathways and activates Rho-GEFs leading to Rho-GTP loading with subsequent effects on PAI-1 and CTGF expression [39,65,88,89] as well as SMAD phosphorylation [16,90]. The Rho-ROCK pathway utilizes SMAD3 as a transcriptional activator for PAI-1 (but not for CTGF) induction in colchicine-treated cells while SMAD3 is required for both PAI-1 and CTGF expression in TGF-β1-stimulated VSMC [16]. These findings highlight the stimulus-specific engagement of signaling pathways that regulate the expression of these two, clinically-important, profibrotic genes. RhoA-ROCK signaling is essential for upregulation of CTGF in microvascular endothelial cells by various microtubule disrupting agents, among them combretastatin A4, a tumor vessel-targeting drug [91]. Activation of FoxO transcription factors by inhibition of phosphatidyl inositol 3-kinase further increased the stimulating effect of microtubule disruption. The genomic effects of cytoskeletal deformation are, thus, modulated by additional factors. This is best exemplified by the finding that that CTGF expression was even more responsive to combretastatin A4-induced microtubule disruption when endothelial cells were cultured under hypoxic as opposed to normoxic conditions [92].

Growth Factor Receptor Transactivation and Downstream Signaling Following Microtubule Disruption

Epidermal growth factor receptor (EGFR) activation and engagement of downstream ERK1/2 MAP kinases is a common response to clinically-relevant microtubule destabilizing drugs resulting in specific changes in gene expression [16,60,61,64,77,93,94]. EGFR phosphorylation upon microtubule disruption requires generation of reactive oxygen species (ROS) [16], in sharp contrast to EGFR activation by native ligands [95]. Vascular cell shape perturbation by cytoskeletal deformation, moreover, involves engagement of at least a subset of receptor and non-receptor tyrosine kinase cascades (e.g., EGFR, Src) leading to gene reprogramming (in the case of PAI-1 and CTGF) [16,64]. The involvement of Src kinases, furthermore, is necessary for ERK1/2 activation as well as for PAI-1 and uPA transcription suggesting that cellular deformation-initiated Src signaling is a critical event in the subsequent expression of cell shape-sensitive genes [16,60,61,64,74,96-98]. ERK1/2 activation downstream of EGFR/Src induction, however, does not appear necessary (or at least play a major role) in expression of CTGF (unlike PAI-1 and uPA induction) by microtubule disruption in VSMC. These finding suggest that further bifurcation of the signaling pathway downstream of the EGFR impacts the selective expression of particular genes [16]. Despite reports that fibroblasts produce TGF-β transcripts upon cytoskeletal disruption [99], colchicine-induced PAI-1 and CTGF expression in VSMC is independent of autocrine release of TGF-β ligands or TGF-βR activity [16,61]. Indeed, microtubule disruption initiates SMAD2/3 phosphorylation, albeit in a delayed manner (2 hours) and in sharp contrast to the rapid (within 15 minutes) and robust SMAD2/3 activation following TGF-β1 stimulation This is consistent with reports that SMAD2/3 can be activated by TGF-βR-independent mechanisms (e.g., via MSP1; a component of the mitotic check point kinase) [16,90,100,101]. SMAD3, moreover, is essential for PAI-1 expression in both TGF-β1- and colchicine-treated VSMC [16] and, in most cells, for TGF-β1-induced CTGF transcription [15]. The mechanism underlying microtubule-based EGFR transactivation, however, is uncertain and may involve ligand-independent activation of growth factor receptors as a consequence of cell deformation or the release of soluble ligands in response to mechanical forces. It appears that both can processes can activate signaling events in the vasculature [102,103].

Conclusions

Mechanosensory signaling pathways function in vascular cell migration, proliferation and differentiation as well as in cardiovascular disease progression [10,13,14,21]. The "tensegrity model" of phenotypic response, for example, suggests that the plasma membrane is hardwired to the nuclear apparatus via cytoskeletal networks that, in turn, transmit signals initiated in response to mechanical stimuli originating from either from the ECM or changing tensional forces that are common in the vascular system [8,14]. Not all stress responses, however, result in equivalent outcomes. Vascular endothelial cells can distinguish between laminar shear and disturbed flow; these stimuli initiate two very different biological responses. Laminar shear-induced mechanisms are associated with less inflammation and oxidative stress leading to an atheroprotective phenotype; disturbed flow, in contrast, results in significantly greater inflammation and increased oxidative stress, thereby, acerbating the pathophysiologic development of an atherosclerotic vessel [21,112,113]. PAI-1 expression is augmented in response to mechanical forces such as shear stress. In terms of gene expression, CTGF belongs to a select group of genes which are suppressed by laminar flow but increased in areas of non-uniform shear stress [80], whereas others (e.g., endothelial NO synthetase) are regulated in the opposite direction. Since PAI-1 and CTGF contribute to the pathogenesis of cardiovascular disease and tissue fibrosis, clarifying mechanisms associated with disturbed flow leading to genomic reprogramming are crucial for identification of novel therapeutic targets. Moreover, the potential importance of morphology-linked and, in particular, microtubule-based controls on the transcription of profibrotic genes such as PAI-1 and CTGF is underscored by (a) their marked induction in mechanical- and hypoxia-stressed, as well as growth factor-stimulated, vascular cells [15,69,78,85] and (b) the obvious changes in cytoskeletal dynamics associated with these shape-altering physiological processes. It is increasingly apparent that microtubule and microfilament networks integrate signaling originating from integrins and growth factor receptors (largely due to direct interactions) and activate kinase cascades leading to alterations in gene expression. Several of the involved intermediates (e.g., SMADs, Rho-GEFs), moreover, are sequestered on the cytoskeleton in an "inactive state" and cell shape modifications or deformation due to fluctuating mechanical forces could dissociate some of these effectors from their cytoskeletal anchors thus alleviating the "repressive state" leading to the downstream signaling. Not surprisingly, drug-based tubulin cytoskeletal modifications lead cell deformation-sensitive genetic changes including PAI-1 and CTGF induction. Interestingly, comparative analysis of PAI-1 and CTGF expression in response to microtubule disruption in vascular smooth muscle cells highlights a complex network of both unique (e.g., SRF, ERK, SMAD) and common (e.g., Rho, EGFR) signaling elements in gene transcription. Given the emerging importance of these profibrogenic factors, further clarification of the involved pathways could yield novel therapeutic tools and targets to modulate the pahtophysiology of cardiovascular disease.

Acknowledgments

Supported by NIH grant GM057242.

References

[1] Fuchs, E. and Karakesisoglou, I. (2001) Bridging cytoskeletal intersections. *Genes Dev.* 15, 1-14.

[2] Ko KS and McCulloch CA (2001) Intercellular mechanotransduction: cellular circuits that coordinate tissue responses to mechanical loading. *Biochem Biophys Res Commun.* 285, 1077-1083.

[3] Forgacs, G., Yook, S.H., Janmey, P.A., Jeong, H. and Burd, C.G. (2004) Role of the cytoskeleton in signaling networks. *J. Cell Sci.* 117, 2769-2775.

[4] Folkman, J. and Moscona, A. (1978) Role of cell shape in growth control. *Nature* 273, 345-349.

[5] Ben-Ze'ev, A., Farmer, S.R., and Penman, S. (1980) Protein synthesis requires cell-surface contact while nuclear events respond to cell shape in anchorage-dependent fibroblasts. *Cell* 21, 365-372.

[6] Crossin KL, Carney DH (1981) Evidence that microtubule depolymerization early in the cell cycle is sufficient to initiate DNA synthesis. *Cell* 23, 61-71.

[7] Chou IN, Zeiger J, Solomon JA, Black PH (1981) Stimulation of plasminogen activator expression and induction of DNA synthesis by microtubule-disruptive drugs. *Biochem Biophys Res Comm* 101, 1266-1273.

[8] Wang N, Tytell JD and Ingber DE (2009) Mechanotransduction at a distance: mechanically coupling the extracellular matrix with the nucleus. Nat *Rev Mol Cell Biol* 10, 75-82

[9] Aplin AE, Howe AK and Juliano RL (1999) Cell adhesion molecules, signal transduction and cell growth. *Curr Opin Cell Biol* 11, 737-744.

[10] Lehoux S, Castier Y and Tedgui A (2006) Molecular mechanisms of the vascular responses to haemodynamic forces. *J Intern Med.* 259, 381-392.

[11] Chien S. (2007) Mechanotransduction and endothelial cell homeostasis: the wisdom of the cell. *Am J Physiol Heart Circ Physiol* 292:H1209-1224.

[12] Shyy JY and Chien S (2002) Role of integrins in endothelial mechanosensing of shear stress. *Circ Res.* 91, 769-775

[13] Lehoux, S. and Tedgui, A. (2003) Cellular mechanics and gene expression in blood vessels. *J. Biomech.* 36, 631-643.

[14] Orr AW, Helmke BP, Blackman BR and Schwartz MA (2006) Mechanisms of mechanotransduction. *Dev Cell.* 10, 11-20.

[15] Cicha I, Goppelt-Struebe M (2009) Connective tissue growth factor: context-dependent functions and mechanisms of regulation. *Biofactors*. 35, 200-208.

[16] Samarakoon R., Higgins CE, Higgins SP, Higgins PJ (2009) Differential requirement for MEK/ERK and SMAD signaling in PAI-1 and CTGF expression in response to microtubule disruption. *Cell Signal* 21, 986-995.

[17] Tzima E, Irani-Tehrani M, Kiosses WB, Dejana E, Schultz DA, Engelhardt B, Cao G, DeLisser H and Schwartz MA (2005) A mechanosensory complex that mediates the endothelial cell response to fluid shear stress. *Nature*. 437, 426-431.

[18] Chiquet M, Gelman L, Lutz R, Maier S (2009) From mechanotransduction to extracellular matrix gene expression in fibroblasts. *Biochim Biophys Acta* 1793, 911-920.

[19] Haga JH, Li YS and Chien S (2007) Molecular basis of the effects of mechanical stretch on vascular smooth muscle cells. *J Biomech*. 40, 947-960.

[20] Hahn C, Orr AW, Sanders JM, Jhaveri KA and Schwartz MA (2009) The subendothelial extracellular matrix modulates JNK activation by flow. *Circ Res* 104, 995-1003.

[21] de Cavanagh EM, Ferder M, Inserra F, Ferder L (2009) Angiotensin II, mitochondria, cytoskeletal, and extracellular matrix connections: an integrating viewpoint. *Am J Physiol Heart Circ Physio*l 296, H550-558.

[22] Li S, Butler P, Wang Y, Hu Y, Han DC, Usami S, Guan JL and Chien S (2002) The role of the dynamics of focal adhesion kinase in the mechanotaxis of endothelial cells. *Proc Natl Acad Sci U.S.A*. 99, 3546-3551.

[23] Geiger B (2006) A role for p130Cas in mechanotransduction. *Cell* 127, 879-881.

[24] Banes AJ, Tsuzaki M, Yamamoto J, Fisher T, Brigman B, Brown T, Millar L (1995) Mechanoreception at the cellular level. *Biochem Cell Biol* 73, 349-365.

[25] Chien S (2007) Mechanotransduction and endothelial cell homeostasis: the wisdom of the cell. *Am J Physiol Heart Circ Physiol*. 292, H1209-1224.

[26] Miranti CK, Brugge JS (2002) Sensing the environment: a historical perspective on integrin signal transduction. *Nature Cell Biol* 4, E83-90.

[27] Li S, Kim M, Hu YL, Jalali S, Schlaepfer DD, Hunter T, Chien S and Shyy JY (1997) Fluid shear stress activation of focal adhesion kinase. Linking to mitogen-activated protein kinases. *J Biol Chem* 272, 30455-30462

[28] Na S, Collin O, Chowdhury F, Tay B, Ouyang M, Wang Y and Wang N (2008) Rapid signal transduction in living cells is a unique feature of mechanotransduction. *Proc Natl Acad Sci U.S.A*. 105, 6626-6631

[29] Chen KD, Li YS, Kim M, Li S, Yuan S, Chien S, Shyy JY (1999) Mechanotransduction in response to shear stress. Roles of receptor tyrosine kinases, integrins, and Shc. *J Biol Chem* 274, 18393-18400.

[30] Williams B (1998) Mechanical influences on vascular smooth muscle cell function. *J Hypertension* 16, 1921-1929.

[31] Rao GN (1996) Hydrogen peroxide induces complex formation of SHC-grb2-SOS with receptor tyrosine kinase and activate signal-related protein kinases group of mitogen-activated protein kinases. *Oncogene* 13, 713-719.

[32] Iwasaki H, Shichiri M, Marumo F, Hirata Y (2000) Mechanical stretch stimulates growth of vascular smooth muscle cells via epidermal growth factor receptor. *Am J Physiol Heart Circ Physiol* 278, H521-529.

[33] Li C, Hu Y, Strum G, Wick G, Xu Q (2000) Ras/Rac-dependent activation of p38 mitogen-activated protein kinases in smooth muscle cells stimulated by cyclic strain stress. *Arterioscler Thromb vasc Biol* 20, E1-9.

[34] Adam LP, Franklin MT, Raff GJ and Hathaway DR (1995) Activation of mitogen-activated protein kinase in porcine carotid arteries. *Circ Res* 76, 183-90.

[35] Birukov KG. (2009) Small GTPases in mechanosensitive regulation of endothelial barrier. *Microvasc Res* 77, 46-52.

[36] Narumiya S, Tanji M, Ishizaki T (2009) Rho signalling, ROCK and mDia1, in transformation, metastasis and invasion. *Cancer Metastasis Rev* 28, 65-76.

[37] Shimokawa H (2002) Rho-kinase as a novel therapeutic target in treatment of cardiovascular disease. *J Cardiovasc Pharmacol* 39, 319-327.

[38] Arthur WT, Petch LA, Burridge K (2000) Integrin engagement suppesses RhoA activity via a c-Src-dependent mechanism. *Curr Biol* 10, 719-722.

[39] Ren XD, Kiosses WB and Schwartz MA (1999) Regulation of the small GTP-binding protein Rho by cell adhesion and the cytoskeleton. *EMBO J* 18, 578-585.

[40] Wojciak-Stothard B, Ridley AJ (2003) Shear stress-induced endothelial cell polarization is mediated by Rho and Rac but not Cdc42 or PI 3-kinases. *J Cell Biol* 161, 429-439.

[41] Ghosh K, Thodeti CK, Dudley AC, Mammoto A, Klagsbrun M, Ingber DE (2008) Tumor-derived endothelial cells exhibit aberrant Rho-mediated mechanosensing and abnormal angiogenesis in vitro. *Proc Natl Acad Sci U.S.A.* 105, 11305-11310.

[42] Mammoto A, Huang S, Ingber DE (2007) Filamin links cell shape and cytoskeletal structure to Rho regulation by controlling accumulation of p190RhoGAP in lipid rafts. *J Cell Sci* 120, 456-467.

[43] Mammoto A, Connor KM, Mammoto T, Yung CW, Huh D, Aderman CM, Mostoslavsky G, Smith LE, Ingber DE (2009) A mechanosensitive transcriptional mechanism that controls angiogenesis. *Nature* 457, 1103-1108.

[44] Mammoto T, Parikh SM, Mammoto A, Gallagher D, Chan B, Mostoslavsky G, Ingber DE, Sukhatme VP (2007) Angiopoietin-1 requires p190 RhoGAP to protect against vascular leakage in vivo. *J Biol Chem* 282, 23910-23918.

[45] Providence KM, Kutz SM, Higgins PJ (1999) perturbation of the actin cytoskeleton induces PAI-1 gene expresión in cultured epithelial cells independent of substrate anchorage. *Cell Motil Cytoskeleton* 42, 218-229.

[46] Sotiropoulos A, Gineitis D, Copeland J and Treisman R (2000). Signal-regulated activation of serum response factor is mediated by changes in actin dynamics. *Cell*; 98:159-69.

[47] Yang C, Patel K, Harding P, Sorokin A, Glass WF (2007) Regulation of TGF-β1/MAPK-mediated PAI-1 gene expression by the actin cytoskeleton in human mesangial cells. *Exp Cell Res* 313, 1240-1250.

[48] Charest JL, Jennings JM, King WP, Kowalczyk AP, Garcia AJ (2009) Cadherin-mediated cell-cell-contact regulates keratinocyte differentiation. *J Invest Dermatol* 129, 564-572.

[49] Wang, T-H., Wang, H.S., Ichijo, H., Giannakakou, P., Foster, J.S., Fojo, T., and Wimalasena, J. (1998) Microtubule-interfering agents activate c-Jun N-terminal kinase/stress-activated protein kinase through both Ras and apoptosis signal-regulating kinase pathways. *J Biol Chem* 273, 4928-4936.

[50] Ingber DE (2002) Mechanical signaling and the cellular response to extracellular matrix in angiogenesis and cardiovascular physiology. *Circ Res* 91, 877-887.

[51] Higgins PJ, Ryan MP (1992) Identification of the 52 kDa cytoskeletal-like protein of cytochalasin D-stimulated normal rat kidney (NRK/CD) cells as substrate-associated glycoprotein p52 [plasminogen –activator inhibitor type-1 (PAI-1)]. Expression of p52(PAI-1) in NRK/CD cells is regulated at the level of mRNA abundance. *Biochem J* 284, 433-439.

[52] Zheng G, Lyons JG, Tan TK, Wang Y, Hsu TT, Min D, Succar L, Rangan GK, Hu M, Henderson BR, Alexander SI, harris DC (2009) Disruption of E-cadherin by matrix metalloproteinase directly mediates epithelial-mesenchymal transition downstream of transforming growth factor-β1 in renal tubular epithelial cells. *Am J Path* 175, 580-591.

[53] Higgins PJ, Ryan MP, Ahmed A (1992) Cell shape-associated transcriptional activation of the p52(PAI-1) gene in rat kidney cells. *Biochem J* 288, 1017-1024.

[54] Thakar RG, Cheng Q, Patel S, Chu J, Nasir M, Liepmann D, Komvopoulos K (2009) Cell-shape regulation of smooth muscle cell proliferation. *Biophys J* 96, 3423-3432.

[55] Lee JY, Jones C, Zern MA, Revzin A (2006) Analysis of local tissue-specific gene expression in cellular micropatterns. *Anal Chem* 78, 8305-8312.

[56] Kirchner J, Kam Z, Tzur G, Bershadsky AD, Geiger B (2003) Live-cell monitoring of tyrosine phosphorylation in focal adhesions following microtubule disruption. *J Cell Sci* 116, 975-986.

[57] Kustermans G, Piette J, Legrand-Poels S (2008) Actin-targeting natural compounds as tools to study the role of actin cytoskeleton in signal transduction. *Biochem Pharmacol* 76, 1310-1322.

[58] Jordan MA, Wilson L (2004) Microtubules as a target for anticancer drugs. *Nat Rev Cancer* 4, 253-265

[59] Downing KH (2000) Structural basis for the interaction of tubulin with proteins and drugs that affect microtubule dynamics. *Annu Rev Cell Dev Biol* 16, 89-111.

[60] Irigoyen J P, Besser D, NagamineY (1997) Cytoskeleton reorganization induces the urokinase-type plasminogen activator gene via the Ras/extracellular signal-regulated kinase (ERK) signaling pathway. *J Biol Chem* 272, 1904-1909.

[61] Samarakoon R, Higgins PJ (2002) MEK/ERK pathway mediates cell-shape-dependent plasminogen activator inhibitor type 1 gene expression upon drug-induced disruption of the microfilament and microtubule networks. *J Cell Sci* 115, 3093-3103.

[62] Gundersen GG, Cook TA (1999) Microtubules and signal transduction. *Curr Opin Cell Biol* 11, 81-94.

[63] Janmey PA (1998) The cytoskeleton and cell signaling: component localization and mechanical coupling. *Physiol Rev* 78, 763-781.

[64] Samrakoon R, Higgins PJ (2003) pp60c-src mediates ERK activation/nuclear localization and PAI-1 gene expression in response to cellular deformation. *J Cell Physiol* 195, 411-420.

[65] Ott C, Iwanciw D, Graness A, Giehl K, Goppelt-Struebe M (2003) Modulation of the expression of connective tissue growth factor by alterations of the cytoskeleton. *J Biol Chem* 278, 44305-44311.

[66] Chaqour B, Yang R and Sha Q (2006). Mechanical stretch modulates the promoter activity of the profibrotic factor CCN2 through increased actin polymerization and NF-kappaB activation. *J Biol Chem* 281, 20608-20622.

[67] Muehlich S, Cicha I, Garlichs CD, Krueger B, Posern G, Goppelt-Struebe M (2007). Actin-dependent regulation of connective tissue growth factor. *Am J Physiol Cell Physiol* 292, C1732-1738.

[68] Iyer VR, Eisen MB, Ross DT, Schuler G, Moore T, Lee JC, Trent JM, Staudt LM, Hudson J, Boguski MS, Lashkari D, Shalon D, Botstein D, Brown PO (1999) The transcriptional program in the response of human fibroblasts to serum. *Science* 283, 83-87.

[69] Zavadil J, Bitzer M, Liang D, Yang YC, Massimi A, Kneitz S, Piek E, Bottinger EP (2001) Genetic Programs of epithelial plasticity directed by transforming growth factor-β. *Proc Natl Acad Sci USA* 98, 6686-6691.

[70] Qi L, Higgins SP, Lu Q, Samarakoon R, Wilkins-Port CE, Ye Q, Higgins CE, Staiano-Coico L, Higgins PJ (2008) SERPINE1 (PAI-1) is a prominent member of the early $G_0 \rightarrow G_1$ transition "wound repair" transcriptome in p53 mutant human keratinocytes. *J Invest Dermatol* 128, 749-753.

[71] Ohura N, Yamamoto K, Ichioka S, Sokabe T, Nakatsuka H, Baba A, Shibata M, Nakatsuka T, Harii K, Wada Y, Kohro T, Kodama T, Ando J. 2003. Global analysis of shear stress-responsive genes in vascular endothelial cells. *J Atheroscler Thromb* 10, 304-313.

[72] Aggeler J, Frisch SM, Werb Z (1984) Changes in cell shape correlate with collagenase gene expression in rabbit synovial fibroblasts. *J Cell Biol* 98, 1662-1671.

[73] Unemori EN, Werb Z (1986) Reorganization of polymerized actin: a possible trigger for induction of procollagenase in fibroblasts cultured in and on collagen gels. *J Cell Biol* 103, 1021-1031.

[74] Irigoyen JP, Nagamine Y (1999) Cytoskeletal reorganization leads to induction of the urokinase-type plasminogen activator gene by activating FAK and Src and subsequently the Ras/Erk signaling pathway. *Biochem Biophys Res Commun* 262, 666-670.

[75] Bayraktutan U, Jones P (1996) Expression of the human gene encoding urokinase plasminogen activator receptor is activated by disruption of the cytoskeleton. *Exp Cell Res* 221,486-495.

[76] Ailenberg M, Silverman M (1996) Cellular activation of mesangial gelatinase by cytochalasin D is accompanied by enhanced mRNA expression of both gelatinase A and its membrane-associated gelatinase A activator (MT-MMP). *Biochem J* 313, 879-884.

[77] Subbaramaiah K, Hart JC, Norton L, Dannenberg AJ (2000) Microtubule-interfering agents stimulate the transcription of cyclooxygenase-2. Evidence for involvement of ERK1/2 and p38 mitogen-activated protein kinase cascades. *J Biol Chem* 275, 14838-14845.

[78] Feng Y, Yang J-H, Huang H, Kennedy SP, Turi TG, Thompson JF, Libby P, Lee RT (1999) Transcriptional profile of mechanically induced genes in human vascular smooth muscle cells. *Circ Res* 85, 1118-1123.

[79] Garcia-CardenaG, Comander JI, Blackman BR, Anderson KR, Gimbrone MA (2001) Mechanosensitive endothelial gene expression profiles: script for the role of hemodynamics in athrogenesis? *Ann NY Acad Sci* 947, 1-6.

[80] Cicha I, Goppelt-Struebe M, Muehlich S, Yilmaz A, Raaz D, Daniel WG, Garlich CD. 2008. Pharmacological inhibition of RhoA signaling prevents connective tissue growth

factor induction in endothelial cells exposed to non-uniform shear stress. *Atherosclerosis* 196, 136-145.

[81] Mack PJ, Zhang Y, Chung S, Vickerman V, Kamm RD, Garcia-Cardena G (2009) Biomechanical Regulation of Endothelium-dependent Events Critical for Adaptive Remodeling. *J Biol Chem* 284, 8412-8420.

[82] Boon RA, Leyen TA, Fontijin RD, Fledderus JO, Baggen JM, Volger OL, van Nieuw Amerongen GP, Horrevoets AJ. 2009. KLF2-induced actin shear fibers control both alignment to flow and JNK signaling in vascular endothelium. *Blood* (in press).

[83] Hawks K, Higgins PJ (1998) Cell shape-dependent pathway of plasminogen activator inhibitor type-1 gene expression requires cytoskeletal reorganization. *J Cell Physiol* 176, 293-302.

[84] Ryan MP, Higgins PJ (1993) Growth state-regulated expression of p52(PAI-1) in normal rat kidney cells. *J Cell Physiol* 155, 376-384.

[85] Providence KM, Kutz SM, Staiano-Coico L, Higgins PJ (2000) PAI-1 gene expression is regionally induced in wounded epithelial cell monolayers and required for injury repair. *J Cell Physiol* 182, 269-280.

[86] Cicha I, Beronov K, Ramirez EL, Osterode K, Goppelt-Struebe M, Raaz D, Yilmaz A, Daniel WG, Garlichs CD. (2009) Shear stress preconditioning modulates endothelial susceptibility to circulating TNF-alpha and monocytic cell recruitment in a simplified model of arterial bifurcations. *Atherosclerosis* 207, 93-102.

[87] van Horck FP, Ahmadian MR, Haeusler LC, Moolenaar WH, Kranenburg O (2001) Characterization of p190RhoGEF, a RhoA-specific guanine nucleotide exchange factor that interacts with microtubules. *J Biol Chem* 276, 4948-4956.

[88] Chang YC, Nalbant P, Birkenfeld J, Chang ZF, Bokoch GM (2008) GEF-H1 couples nocodazole-induced microtubule disassembly to cell contractility via RhoA. *Mol Biol Cell* 19, 2147-2153.

[89] Krendel M, Zenke FT, Bokoch GM (2002) Nucleotide exchange factor GEF-H1 mediates cross-talk between microtubules and the actin cytoskeleton. *Nature Cell Biol* 4, 294-301.

[90] Dong C, Li Z, Alvarez R Jr, Feng XH, Goldschmidt-Clermont PJ (2000) Microtubule binding to Smads may regulate TGF β activity. *Mol Cell* 5, 27-34.

[91] Samarin J, Cicha I, Goppelt-Struebe M (2009) Cell type-specific regulation of CCN2 protein expression by PI3K-AKT-FoxO signaling. *J Cell Commun Signal* 3, 79-84.

[92] Samarin J, Wessel J, Cicha I, Kroening S, Warnecke C, Goppelt-Struebe M (2010) FOXO proteins mediate hypoxic induction of connective tissue growth factor (CTGF) in endothelial cells. *J Biol Chem* 285, 4328-4336.

[93] Rijken PJ, van Hal GJ, van der Heyden MA, Verkleij AJ, Boonstra J (1998) Actin polymerization is required for negative feedback regulation of epidermal growth factor signal transduction. *Exp Cell Res* 243, 254-262.

[94] Yujiri T, Fanger GR, Garrington TP, Schlesinger TK, Gibson S, Johnson GL (1999) MEK kinase 1 (MEKK1) transduces c-Jun NH_2-terminal kinase activation in response to changes in the microtubule cytoskeleton. *J Biol Chem*. 274, 12605-12610.

[95] Zandi R, Larsen AB, Andersen P, Stockhausen MT, Poulsen HS (2007) Mechanisms for oncogenic activation of the epidermal growth factor receptor. *Cell Signal* 19, 2013-2023.

[96] Faisal A, Kleiner S, Nagamine Y (2004) Non-redundant role of Shc in Erk activation by cytoskeletal reorganization. *J Biol Chem* 279, 3202-3211.

[97] Schmid-Alliana A, Menou L, Manie S, Schmid-Antomarchi H, Millet MA, Giuriato S, Ferrua B, Rossi B (1998) Microtubule integrity regulates src-like and extracellular signal-regulated kinase activities in human pro-monocytic cells. Importance for interleukin-1 production. *J Biol Chem* 273, 3394-3400.

[98] Graness A, Cicha I, Goppelt-Struebe M (2006) Contribution of Src-FAK signaling to the induction of connective tissue growth factor (CTGF) in renal fibroblasts. *Kidney Int.* 69, 1341-1349.

[99] Varedi M, Ghahary A, Scott PG, Tredget EE (1997) Cytoskeleton regulates expression of genes for transforming growth factor-β1 and extracellular matrix proteins in dermal fibroblasts. *J Cell Physiol* 172, 192-199.

[100] Zhu S, Wang W, Clarke DC, Liu X (2007) Activation of Mps1 promotes transforming growth factor-beta-independent Smad signaling. *J Biol Chem.* 282, 18327-18338.

[101] Liu X, Zhu S, Wang T, Hummers L, Wigley FM, Goldschmidt-Clermont PJ, Dong C (2005) Paclitaxel modulates TGFβ signaling in scleroderma skin grafts in immunodeficient mice. *PLoS Med* 2, e354.

[102] Adam RM, Eaton SH, Estrada C, Nimgaonkar A, Shih SC, Smithe LE, Kohane IS, Bagli D, Freeman MR (2004) Mechanical stretch is a highly selective regulator of gene expression in human bladder smooth muscle cells, *Physiol Genomics* 20, 36-44.

[103] Tschumperlin DJ, Dai G, Maly IV, Kikuchi T, Laiho LH, McVittie AK, Haley KJ, Lilly CM, So PT, Lauffenburger DA, Kamm RD, Drazen JM (2004) Mechanotransduction through growth-factor shedding into the extracellular space. *Nature* 429, 83-86.

In: Tubulin: Structure, Functions and Roles in Disease
Editors: Wei Yamauchi and Antonio Sokić ISBN: 978-1-62100-141-6

Chapter VI

Tubulin Inhibitors as Vascular Disrupting Agents (VDAS)

Jérémie Fournier-Dit-Chabert and Sylvie Ducki
Clermont Université, ENSCCF, EA 987, LCHG,
BP, F-63000 Clermont-Ferrand, France

Abstract

Vascular Disrupting Agents (VDAs) are mostly composed of tubulin inhibitors (TIs). They differ from anti-angiogenesis agents as they destroy within minutes to hours newly formed blood vessels. VDAs effectively starve the tumour to death. Deprived from vital nutrients and oxygen, the tumour goes into necrosis. A majority of cancers are solid tumours that require neovasculature to survive. VDAs are therefore emerging as a promising counterpart to conventional cytotoxic chemotherapy.

This chapter presents a new class of promising compounds, VDAs, describing their effect in contrast with anti-angiogenic drugs. Inhibitors of tubulin polymerisation and microtubule-depolymerising agents will be presented. Some of these VDAs are currently under clinical study to treat a variety of solid tumours or exploited in ophthalmology as therapies against Age-Related Macular Degeneration (AMD). We will focus on the largest group of VDAs that are tubulin-binding agents, with the combretastatins (Zybrestat, AVE-8062, Oxi-4503) as the lead compounds. New emerging VDAs will be presented, and progresses through their clinical or preclinical stage will be reviewed. The advantages and the downsides of this new chemotherapy will be explored in terms of efficiency and side-effects.

Introduction

Angiogenesis and Tumour Vasculature

Tumour angiogenesis is essential for tumour growth, progression and metastasis development. This complex physiological process was first described by Judah Folkman in the 1970s.[1] Its regulation by biomolecules such as vascular endothelial growth factor (VEGF), fibroblast growth factor (FGF) and platelet-derived growth factor (PDGF) has been largely described and its inhibition by molecules such as angiostatin or endostatin have been widely studied [2-5].

For tumours to grow over 2 mm, they require more oxygen and nutrients to develop their own blood supply. The tumour produces angiogenic stimulators (by overexpression) that initiate endothelial cells proliferation and migration of these cells lead to the formation of neovasculature. Angiogenesis has also been found to play an important role in cancer.

Established tumour vasculature is morphologically different from normal blood vessels. They proliferate at a much higher rate than normal tissues and this aggressive growth leads to the development of a disorganized and aberrant network of blood vessels (uneven shape, inconsistent diameter, disorganised, lack of hierarchy, abnormal bludges, blind ends, leaky) fundamentally different from the orderly, hierarchically organised and efficient normal vascular network.[6, 7] Tumour blood vessels also differ by their functionality as their ability to deliver oxygen and nutrients to the tumour and remove wastes is severely diminished, resulting in more permeable vasculature and hypoxic areas limiting the effectiveness of both chemotherapy and radiotherapy in cancer treatment. Hence the unique features of tumour vasculature compared to normal tissues present an opportunity for selective therapy.[8, 9]

Targeting Tumour Vasculature

A majority of cancers (90%) are solid tumours that require an intact vasculature to grow, survive and metastasize. The observation that the lack or destruction of blood vessels in solid tumours prevented metastasis and drove to tumour necrosis led to the proposal that vascular targeting agents (VTA) could become an attractive and potentially valuable strategy for anticancer therapy.

In the late 1970s, Judah Folkman suggested the use of angiogenesis inhibiting agents (AIA) to treat cancer.[1] He had observed that a tumour was unable to grow and metastasize without a blood supply which led him to postulate that this vascular network around the tumour was critical for its survival. To prove his concept, he isolated the first AIA, endostatin and angiostatin.[4, 5] Anti-angiogenic therapy targets the formation of new blood vessels by inhibiting growth factors and/or corresponding receptors. Several AIAs have been developed and are currently undergoing clinical trials or have been approved for clinical use (Avastin™, Nexavar™, Sutent™).

In the 1980s, Juliana Denekamp proved that vascular occlusion was necessary to induce tumour necrosis.[10] Vascular disrupting agents (VDAs) target already established tumour vasculature. Current approaches involve the use of biological or ligand-directed agents, which use antibodies or peptides to target toxins or pro-coagulants to the tumour endothelium and small molecules which induce vascular shutdown by exploiting the differences between normal and tumour vasculature.

VDA therapy differs conceptually from AIA therapy in their mode of action.[11] VDAs are usually more efficient on large rather than small tumours; they cause extensive tumour necrosis (up to 95% of the tumour). However the rim of the tumour, avascularised, usually survives and later grows back. In contrast, AIA therapy is more effective on the tumour rim where fast-growing well-oxygenated cells reside and where angiogenesis is ongoing. AIAs interfere with neovasculature formation; hence have a preventive action requiring chronic administration, which is more beneficial in early stage disease. On the other hand, VDAs target the established tumour vasculature, resulting in tumour necrosis and is therefore given acutely, showing immediate effects and more efficient against advanced disease. Most anticancer agents including cytotoxics are also more efficient on fast-dividing cells, and combination with VDAs could lead to additive or synergistic activity in the treatment of solid tumours [9].

Tubulin Inhibitors as Vascular Disrupting Agents

Most of the small molecules VDAs currently on clinical trials are members of the tubulin inhibitors (TIs) class of compounds and in particular agents interacting at the colchicine-binding site (CBS) of tubulin.[12] These low molecular-weight molecules present efficient and tumour-selective anti-vascular properties at non-toxic doses.[13] A quick literature survey shows that the number of publications related to both "tubulin" and "VDA" has dramatically increased in the past decade (Figure 1).

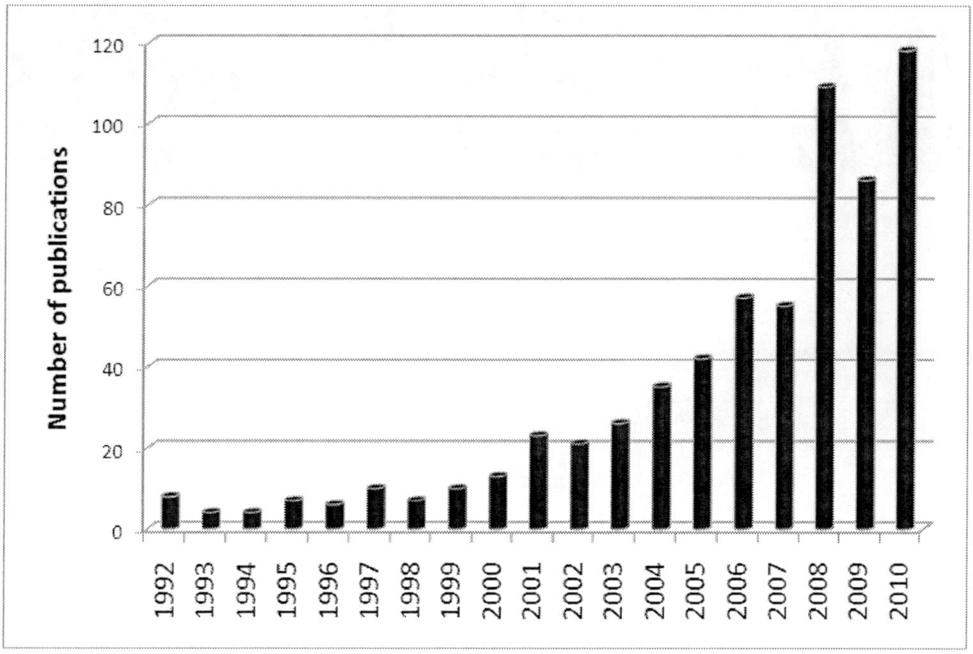

Figure 1. Literature survey (1992-2010) for "Tubulin" and "VDA" (Sciencedirect.com).

Many microtubule-interfering agents (MIAs) have been developed as anticancer agents. They are usually classed according to their site of interaction with tubulin (Figure 2). The first group contains MIAs interacting at the taxane-binding site including Paclitaxel™ and Docetaxel™ which interact with the M-loop of beta-tubulin leading to microtubule stabilisation. The second group is that of the vinca alkaloids (Vincristine™, vinblastine™, vinorelbine™), dolastatin-10 derivatives (TZT-1027, LU-103793, ILX-651)[14] and cryptophycins (ADH-1)[15] which promote microtubule depolymerisation by interacting on beta-tubulin at the interface of heterodimers. The last group is that of colchicine, where CBS agents interact with beta-tubulin within the homodimers interface, inhibiting tubulin polymerisation.[16] While taxanes and vinca alkaloids have been approved by the FDA for the treatment of various cancers, no CBS agent has yet reached the market. Given the clinical success of taxanes and vinca alkaloids in cancer treatment, research efforts have been focussed on the development of CBS agents as anticancer agents.

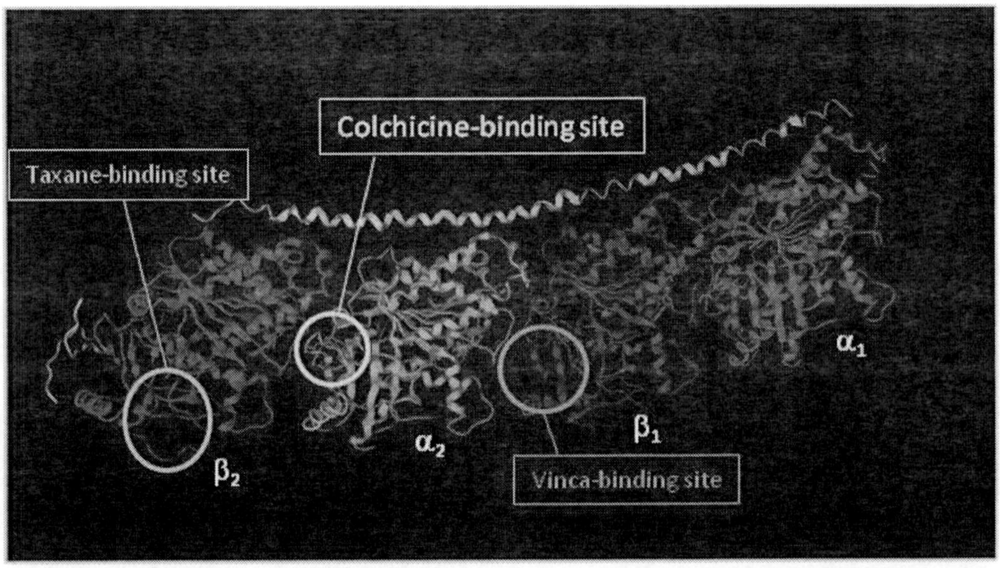

Figure 2. Structure of tubulin (PDB 1SA0) showing the three main binding sites.

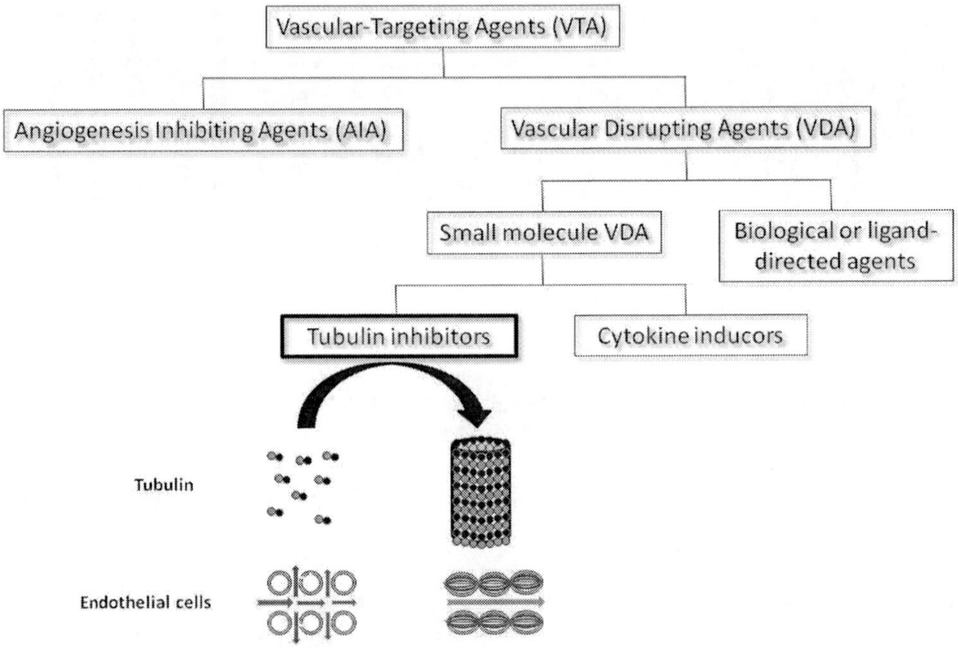

Figure 3. Summary of TI-VDA action.

The interest for CBS agent has been further raised with the discovery that these agents could act as VDAs at doses well below their maximum tolerated doses (MTD).[17] The first CBS agent, colchicine itself, isolated by Pelletier and Caventou in 1820 from an extract of *Colchicum*, has recognised tumour vascular disrupting properties but has been found too toxic for clinical use. Vincristine and vinblastine, isolated from the Madagascar periwinkle *Catharanthus roseus*, have also been shown to possess antivascular activity but only at the maximum tolerated doses limiting their clinical use as VDAs.

The search for novel and more efficient VDAs has mainly concentrated on combretastatin analogues (CA4P, CA1P, AVE-8062), currently in advanced stages of clinical development. From this discovery, other TIs emerged as potential VDAs (Table 1).

Table 1. VDAs under clinical trials as single agents

Name	Other names	Company	Core Type	Therapy	Development phase
Oxi 2021 CA4P	Zybrestat, Fosbretabulin	Oxigene	Combretastatin	Thyroid (ATC) PCV (ophta)	II (2003-2008) II (2009-2010)
Oxi-4503	CA1P	Oxigene	Combretastatin	Hepatic	II (2009-2011)
AVE-8062	Ombrabulin	Sanofi-Aventis	Combretastatin	Sarcoma	II/III (2008-2012)
BNC-105P		Bionomics	Benzofurane	Mesothelomia	II (2010)
NPI-2358	Plinabulin	Nereus	Diketo-piperazine	Solid tumours lymphoma	I (2006-2009)
MN-029	Denibulin	Medicinova	Amino-imidazole	Solid tumours lymphoma	I (2005-2007)
EPC-2407	Crinobulin	Epicept	Flavonoid	Advanced cancer	I (2006-2009)
ABT-751		Abott	Amino-pyridine	Breast Colorectal NSCLC Renal Prostate Neuroblastoma	II (2002-2004) II (2003-2005) II (2003-2006) II (2003-2006) II (2004-2007) II (2007-2012)
MPC-6827	Azixa/ verubulin	Myrexis	Quinazoline	Glioblastoma	II (2009-2011)
CYT-997		Cytopia	Amino-pyrimidine	Multiple myeloma	II (2008-2011)
ZIO-301	Indibulin	Ziopharm	Indole	Breast	I/II (2010-2011)
AEZS-112		Æterna-Zentaris	Indole	Tumours	I (2008-2009)
CKD-516		CKD Pharmac.	Phenstatin	Tumours	I (2009-2012)
ZD-6126		Astra-Zeneca	Colchicine	Kidney	II (2003-2004) [T]
CI-980			Pyrazine	Melanoma Colorectal	II (2001, failed) II (1997, failed)
T138067	Batabulin	Amgen	Sulfonamide	Liver Colorectal	II/III (2003) [T] II (1998-2001)

[T] The study has been terminated.

VDA-CBS work by interacting at the colchicine-binding site of beta-tubulin in endothelial cells, particularly reliant on the cellular cytoskeleton for shape maintenance (Figure 3). VDAs inhibit the polymerisation of tubulin into microtubules causing a rapid change in shape of endothelial cells, resulting in vascular blockade and loss of blood flow to/within the tumour. Disruption of the endothelial cell layer leads to exposure of the basement membrane, hence increasing tumour vascular permeability. This, in turn, leads to high interstitial pressures, vessel congestion and further loss of flow. Oxygen and essential nutrients are hence prevented from reaching the tumour inducing extensive tumour necrosis. The action of VDA-CBS is rapid (with maximal vessel shutdown within 1-6 hours), dose-dependent, usually sustained for 24-48 hours and selective (blood flow in normal tissues are not affected). The fragile and immature nature of tumour blood vessels is thought to be at the origin of this selectivity although it has been suggested that maturity of the tubulin could also play a role in the rapidity of the action [9].

Flavonoids (FAA, DMXAA) are simple and efficient VDAs. Their antivascular activity is dissociated with tubulin interaction as they target cytokine induction. Vadimezan™ (ASA404, DMXAA) entered phase III clinical trials in 2008 for the treatment of NSCLC but demonstrated no survival benefit so far and therefore trials were terminated [12, 18].

VDAs Undergoing Clinical Trials

Small Molecule VDAs

To date (2011), 16 β-tubulin-binding VDAs are being evaluated in clinical trials to treat a variety of solid tumours as single agents or in combination (Figure 4). The largest group of VDAs, the agents interacting at the CBS of tubulin, is composed of a variety of structures, mainly aromatics (stilbenes, N-containing heterocycles such as indole, piperazine, aminopyridine, imidazole, and O-containing heterocycles such as benzofuran, chromene) with the combretastatins (CA4P, AVE-8062, Oxi-4503) being considered as the first-in-class compounds.

Combretastatins are stilbenes, isolated from the Eastern Cape South African Bushwillow tree, *Combretum caffrum* by Pettit in 1995, which were shown in inhibit tubulin polymerization by interacting at the CBS. [19] In 1997, Chaplin reported that Combretastatin-A4 (CA4) and CA4P, the phosphate prodrug of CA4, displayed selective, rapid and extreme vascular shutdown within tumour at <10% of the MTD. [20] CA4P entered phase I clinical trials in 1998 and is currently entering phase III.

Colchicine is a tricyclic alkaloid isolated by Pelletier and Caventou in 1820 from an extract of the autumn crocus *Colchicum autumnale*. [21] The drug has been approved for its use in the treatment of gout and is also used as an anti-inflammatory drug in Bechet's disease. The derivative ZD6126 [22, 23], a phosphate prodrug of colchicine, entered clinical trials as anticancer agent in 2000 but its development has been hampered by its high toxicity, presumably due to its ability to irreversibly interact with tubulin. Trials were stopped in phase II in 2004.

BNC105P, is the phosphate prodrug of BNC105, a benzofuran analogue being developed by the Australian-based company Bionomics (bionomics.com.au). Currently in phase II clinical trials in cancer patients, the drug's therapeutic index has been shown to be 26 times greater than that of CA4. In addition to its VDA activity, the drug is cytotoxic to cancer cells which enables it to inhibit tumour growth as a single agent.

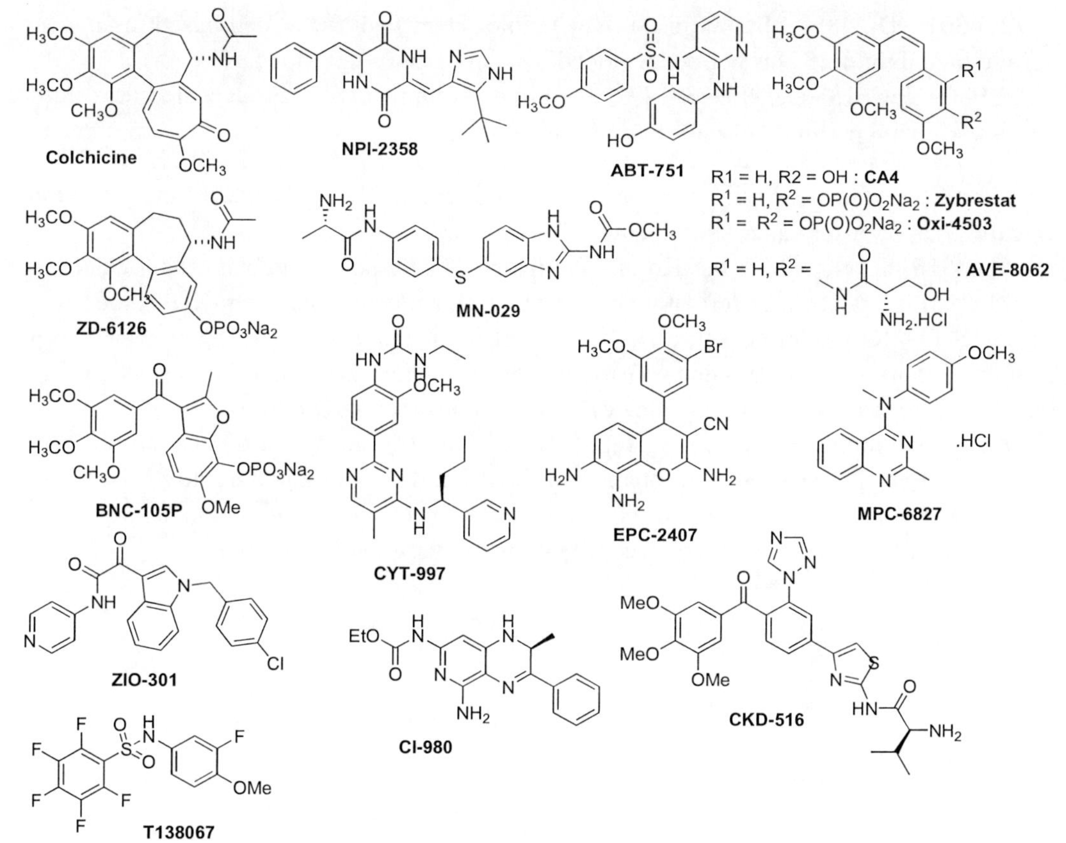

Figure 4. VDAs currently in clinical trials.

NPI2358 (Plinabulin™) is a diketopiperazine analogue of Halimide, produced by the marine microbe *Aspergillus Ustus,* and is being developed by Nereus Pharmaceuticals (www.nereuspharm.com).

The benzimidazole carbamate, MN029 (Denibulin™) developed by Angiogene Pharmaceuticals, was licensed to the Japanese company Medicinova (www.medicinova.com) in 2002 for its clinical developement. MN029 is a prodrug that is converted *in vivo* via enzymatic hydrolysis of L-alanine to the active compound, MN-022. MN-022 is subsequently N-acetylated to form MN-021.[24]

EPC2407 (Crolibulin™, Crinobulin™, MX116407), a 4-aryl-chromene analogue first identified as a caspase activator through high-throughput screening,[25-27] is being developed as a VDA by EpiCept (www.epicept.com).

ABT-751 is a CBS tubulin inhibtor (E7010) discovered through a high-throughput screening of a sulphonamide library [28] by Eisai Co. Ltd. who licensed it to Abbott Laboratories (www.abbott.com) for its clinical development.

MPC-6827 (Azixa™, Verubulin™), a quinazoline identified from a cell-based screening assay as a potent apotptosis inducer by Myriad Pharmaceuticals,[29] is being developed as a VDA by Myrexis (myrexis.com).

CYT997 is an aminopyrazine discovered by the Australian biotechnology company Cytopia Ltd (www.cytopia.com.au) through a HTS of their small molecule library.[30]

ZIO-301 (D-24851, Indibulin™), an indole identified in a cell-based assay as a microtubule inhibitor,[31] is being developed by Ziopharm (www.ziopharm.com).

CI-980 (NSC613862, NSC370147, mivobulin isethionate™) was identified as a mitotic inhibitor while preparing antifolate analogues [32, 33].

T138067 (Batabulin sodium™) is a sulphonamide that covalently binds to Cys-239 within the CBS of beta-tubulin, inhbiting tubulin polymerization [34] and is being developed clinically as a VDA for cancer treatment by Amgen.

CKD-516, developed by the Korean company CKD pharmaceuticals, (chongkun dang. en.ec21.com) is a phenstatin derivative bearing both triazole and thiazole moieties [35, 36].

AEZS-112 (chemical structure not available) is an indole derivative developed by the Canadian Æterna Zentaris (aezsinc.com). This compound is of particular interest as it exhibits three distinct activities: antiangiogenic, VDA and topoisomerase II inhibitor [37].

Other small molecules have been reported as VDAs (Figure 5) such as estradiols 2ME-2 [38-40] and ENMD-1198 [41], sulfonamide E7070 [42-46] or indole derivatives BPR0L075 [47, 48] and STA5312 (Rosabulin [49-51]) which mechanisms of action and/or tubulin binding sites are not clear and therefore will not be discussed in this chapter.

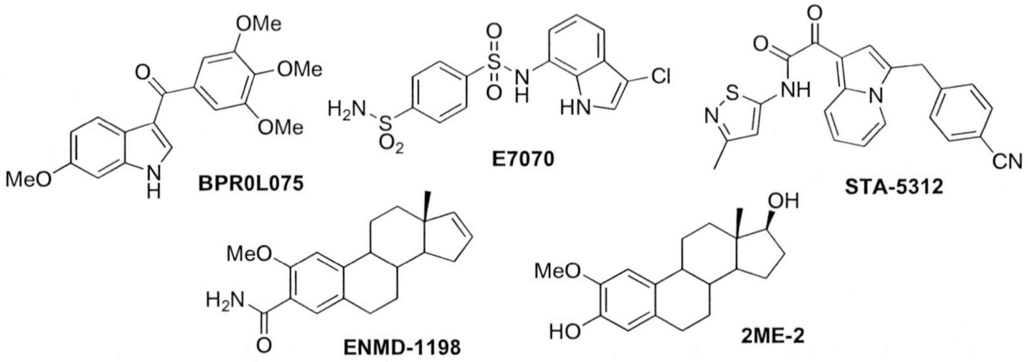

Figure 5. Other assumed VDAs.

Clinical Trials

Targeting tumour vasculature has emerged as a promising alternative to traditional cancer treatment since most solid tumours require a blood supply to survive, grow and metastasize. The rapid and selective antivascular effect of VDAs renders them attractive as alternative cancer treatment. Although no VDA targeting the CBS of tubulin has been formerly approved for clinical use, the clinical trials are proving very promising, with three NCEs having reached phase III, and eight in phase II (Table 1).

Two combretastatin analogues (AVE-8062 and CA4-P) have successfully undergone phase II trial and have reached phase III, which confirms that vascular targeting is an efficient strategy against the treatment of solid tumours. CA4-P also reached phase II clinical trials for the treatment of ophthalmologic diseases such as polypoidal choroidal vasculopathy. Five drug candidates (Oxi-4503, MPC-6827, BNC-105P, T138067 and ABT751) have entered phase II clinical trials. We note that the colchicine derivative ZD6126 and the pyrazine CI-980 trials have been abandoned due to important adverse-effects compared to the few positive responses obtained. Four candidates (CYT-997, EPC-2407, NPI-2358 and AEZS-112) have successfully undergone phase I clinical trials and results are being processed to evaluate future development plans. Finally, three molecules (ZIO-301, MN-029 and CKD-516) have recently entered a phase I trial where they are being evaluated on small groups of patients for their toxicity.

Clinical Responses and Toxicities Observed with VDA Treatment

In phase I trials, VDAs have usually been well tolerated, with few drug-related serious adverse events or dose-limiting toxicity observed and have even shown encouraging anti-tumour activity (Table 2). For instance, the leader CA4P showed much less haematological toxicity than other classes of tubulin binding anticancer drugs such as the vinca alkaloids and taxanes. In some cases, patients treated with this new therapy witnessed temporary blindness (reversible), cardiac side-effects, some pain in the tumour and the treatment is sometimes associated with nauseas and headaches.

Table 2. Dose-limiting toxicities (DLT) and frequent side-effects observed in clinical trials

Name	Cancer	Activity	DLT	Frequent side-effects
CA4P	Thyroid	1/25 TR, 6/25 SD[52] 6/18 SD.[53]	Reversible ataxia, vasovagal syncope, fatal ischemia, neuropathy.[54]	nausea, vomiting, headache,[53, 55] tumour pain,[52] increased QTc interval.[56]
Oxi-4503	Tumors	6/23 SD,[57] 2/42 PR,[58] 2/10 SD.[59]	Atrial fibrillation, shock secondary to tumor lysis syndrome, fistula formation.[57] Hypertension, visual disturbances.[58]	Hypertension, tumour pain, fatigue, myelosuppression, bowel fistula with septic shock, visual disturbance and nausea.[58]
AVE-8062	Sarcoma	3/48 PR 20/48 SD.[60][(1)]	Headache, asthenia, deep vein thrombosis	Neutropaenia, infection, nail toxicity, oesophageal fistula, fatigue, haematological toxicity, aminotransferase increase
BNC-105P	Tumors	4/21 SD	Not reported	Not Reported.[61]
NPI-2358		20% SD[62] 22% PR (vs 5%) [63][(1)]	Nausea, vomiting, fatigue, fever, tumour pain and elevations in blood pressure.[62]	Lower occurrence of neutropenia in combination with docetaxel (4% vs. 27%). [63]
MN-029	Tumors	1/20 PR,[64] 5/34 SD.[65]	transient ischemic attack, transaminitis. [65]	nausea, vomiting, diarrhoea, fatigue, headache, and anorexia.[64]

Table 2. (Continued)

Name	Cancer	Activity	DLT	Frequent side-effects
EPC-2407	Tumors	Several SD,[66] 2/11 PR.[67]	cardiac ischemia, severe pain in the tumor bed, ST depression.[67]	Increase in blood pressure, dysphoria or anxiety, hypertension.[66]
ABT-751	Pediatric[68], kidney, colorectal and NSCLC	NSA[69] 1/97 PR, 17/97 SD[70], NSA[71], 1/35 PR[72]	Fatigue, neuropathy, hypertension, vomiting, neutropenia, nausea, thrombocytopenia, dehydration, abdominal pain, constipation [73, 74]	Bone marrow suppression, gastrointestinal toxicities, sensory and motor neuropathies[73]
MPC-6827	Tumors	5/48 SD[75]	Myocardial infarction	Nausea, fatigue, flushing, and hyperglycemia
CYT-997	Tumors	18/22 SD[76, 77] 13/21 SD[78]	Prolonged QTc interval, dyspnea [76] Hypoxia, asthenia [78]	Aphasia, cerebrovascular toxicity(2)
AEZS-112	Tumors	16/22 SD[79]	Gastrointestinal problem	None reported
ZIO-301	Tumors	1/10 PR[80]	Related to lactic acid	None reported
ZD-6126	Tumors	NSA[81]	Severe cardiotoxicity[82] Phase 2 trials arrested	Abdominal pain, nausea and vomiting
CI-980	Melanoma Colorectal	NSA[83, 84] 1/25 PR[85]	Granulocytopenia[83, 84] CNS toxicity[83, 85]	nausea, vomiting, malaise, fatigue, weakness[84]. Reversible memory decline[86]
T138067	Liver, colorectal, glioma	NSA[87, 88] 2/21 PR; 9/21 SD[89] 3/18 SD[90]	Thrombocytopenia, neutropenia[91]	Hematologic, gastrointestinal toxicities[87] fatigue, nausea, neutropenia and anorexia[89, 90]

TR: total response; PR: partial response; SD: stable disease; NSA: No significant activity. (1) Studies in combination with Docetaxel. (2) Studies in combination with carboplatin.

Combretastatins: CA4P, Oxi-4503 and AVE-8062. In phase I studies, CA4P was first administered in daily perfusions once a week at doses up to 114 mg/m^2[54]. Dose-limiting toxicities (DLT), ranging from 50 to 75mg/m^2, was essentially neurological (ataxia, visual disturbance, neuropathy)[54] and cardiovascular (increased QTc interval, vasovagal syncope, hypertension, hypotension, dyspnea)[56]. In the case of CA4P, It is not clear whether the cardiotoxicity is structure- or mechanism-related. In addition, toxicity related to blood flow was reported, as one patient died from ischemia. Other adverse effects were found to be similar and realted to cytotoxic chemotherapy with nausea, vomiting, headache[55], lymphopenia, fatigue, anemia, diarrhea[54]. Tumour pain caused by the necrosis of the tumour was often reported[52]. Several patients treated in phase I trials with CA4P presented stable diseases and one patient with anaplastic thyroid cancer (ATC) even witnessed a complete recovery.[52]

The cardiac and neurological toxicities were also observed in phase I trials with the other combretastatin analogue Oxi-4503 although to a lesser extent.[58] Other toxicities included fistula formation and septic shock secondary to tumour lysis.[57] Hypertension is most likely caused by vasoconstriction of normal blood vessels. It is interesting to note that in a preclinical study in rats the use of vasodilators reduced hypertension without altering the blood flow shutdown effects [92].

Phase I trials of AVE8062 alone have not been reported but a study in combination with Docetaxel™ indicated that toxicity of this compound was more likely related to conventional cytotoxic treatments with headache, fatigue and haematological toxicity (fatal neutropenia). However, an MTD at 100mg/m^2 with grade 3 deep vein thrombosis was directly related to the antivascular treatment [60].

Overall, phase I studies showed that the combretastatins were well tolerated and the number of patients having a stable disease consecutive to the treatment was promising for the forthcoming phase II trials. Phase II clinical trials of CA4P as a single agent on ATC showed encouraging results with an increase of life expectancy (28% of the patients treated). However, combination therapy was recommended for a better outcome [53].

Other compounds. Clinical trials of VDAs have started with CI-980 in the late 1990s on melanoma and metastatic colorectal carcinoma[83, 84]. DLT was granulocytopenia and neurotoxicity (CNS), frequently combined with memory decline[86]. CI-980 was abandoned due to lack of activity in phase II studies.

In general, four primary adverse effects related to the VDA trials were reported:

1) Cardiovascular problems (ischemia, hypertension, myocardial infarction, prolonged QTc interval) were observed in MN-029[65], EPC-2407[66, 67], NPI-2358[62], ABT-751[73, 93], MPC-6827[75]. Phase 2 trials of ZD-6126 were terminated due to severe cardiotoxicity [82].
2) Neurological disorders and behavioural symptoms (neuropathy, asthenia, anxiety, anorexia), particularly with ABT-751[73] and in a lesser extent with CYT-997[78], MN-029[64], EPC-2407[66], NPI-2358[62] and TI138067 [89, 90].
3) Hematologic disorders (neutropaenia, thrombocytopenia, transaminitis) reported for MN-029[65], ABT-751[73] and TI138067[91].
4) Pain at the tumour site, provoked by tumour necrosis, or related to oral administration were also frequent: NPI-2358[62], EPC-2407[67], ABT-751[73, 74] and ZD-6126[82].

Overall, these compounds showed promising phase I results with several cases of partial responses or stable disease recorded. However, phase II trials with ABT751[69-72], CI-980[83, 84] and T138067[87-90] as single agents gave poor results with no significant increase in life expectancy. ABT-751 was recommended to pursue clinical trials in combination therapy[71].

Four molecules (ZIO-301, BNC-105P, AEZS-112, CKD-516) represent the new generation of VDAs with target, solubility and adverse effects addressed. Some of them combine several properties, e.g. AEZS-112 expressing antiangiogenic and vascular disrupting activities, as well as topoisomerase II inhibition[37]. In addition, CKD-516, AEZS-112 and ZIO-301 (Indibulin™) are orally administered and well tolerated. These drugs also showed better selectivity, hence lowering the side-effects of a clinical treatment. BNC105-P presents a “Lock-in effect”, as it is quickly cleared in normal cells but retained within tumours. In a phase I study, a weekly 16mg/m^2 dose administration showed no signs of cardiac toxicity [61]. ZIO-301 binds to a site that overlaps the CBS which is thought to be preventing neurotoxicity related to the CBS agents [94]. *In vitro* studies have shown that ZIO-301 discriminated between mature neuronal microtubules and less-modified tubulin present in immature neuronal or nonneuronal microtubules[95]. This could explain why the drug showed no signs of neurotoxicity or cardiotoxicity as DLTs in clinical trials [80]. Closely-related indole AEZS-112 showed little toxicity in phase I clinical trials [79]. Doses up to 200x3mg/week did not induce any adverse effects. In addition, AEZS-112 was shown to be metabolically stable in human plasma hence exhibiting prolonged effects as 73% of the patients treated had stable disease in various cancer types.

VDAs in Combination Therapy

There are examples where single agent VDA treatment induced tumour regression, but these tend to be exceptions rather than the norm. Considering these rather disappointing trials, most of the VDAs have been evaluated in combination chemotherapy. Indeed, the rim of the tumour, which is avascularised, usually survives to the treatment by VDAs as single agents and grows back. *In vivo* assays showed that CA4P appeared to be potent on tumour tissue with a rapid shutdown of blood flow but this effect was marginal in normal tissues and the tumour rim[96]. In clinical trials, recovery of tumour perfusion has been observed in most cases after 24 h, which has resulted in the use of an intense dose scheduling to counteract this problem[97]. Even though most VDAs exhibit potent cytotoxicities, most of them suffer from rapid clearance which prevents them from acting directly on the tumour rim. It has hence proved necessary to associate the VDAs with other therapies in order to counteract this deficiency.

Current approaches are therefore to combine VDAs with radiation therapy, antiangiogenic agents or conventional cytotoxic agents (Table 3) [98].

Combination with radiation therapy targets both the tumour vasculature and the rim of the tumour. *In vivo* studies showed that CA4P had a major effect on the radiation-resistant hypoxic tumor cells, enhancing dramatically the effectiveness of radiation therapy[99]. However, VDA administration has to be preferably given simultaneously or shortly after irradiation. Damages of radiation therapy combined with VDAs are negligible on healthy tissues such as skin[100] and also on late-responding bladder and lung cancers [101].

Preclinical studies of combretastatins combined with anti-angiogenic agents (AIA), able to prevent vasculature formation and tumour regrowth, showed encouraging synergistic activities. For example, Avastin™ plus CA4P led to a growth delay of 13 days, and 27 days for Avastin plus OXi4503[102].

Combination with chemotherapeutic agents is also of interest. VDAs target the tumor core, poorly vascularised and hypoxic, that is usually resistant to chemotherapy. Ideally, the best results will be obtained with administration of the chemotherapeutic agent prior to the VDA. It will entrap the drug in the tumour, hence exposing the cancer cells to the anticancer agent for a longer time[9].

Clinical trials in combination with CA4P and AVE8062. Clinical trials are being conducted for combretastatins, CA4P and AVE-8062, in combination with angiogenic agents (AIA) such as Avastin™ and phase I results proved enhanced efficiency. The tumour rim size and blood flow were significantly reduced when 63 mg/m^2 CA4P followed by 10 mg/kg bevacizumab were used. 9/14 patients experienced disease stabilization for >12 weeks, three patients experienced disease stabilization for >24 weeks as CA4P induced profound vascular changes which were maintained by the presence of bevacizumab [103].

Phase I trials with AVE8062 followed on day 2 by Docetaxel™ every 3 weeks showed MTDs at 35/75 mg/m² and 30/100 mg/m² respectively. 3/48 partial responses (breast cancer) and 20 stable diseases were very encouraging[60].

Several phase II trials of CA4P with Paclitaxel™ and Carboplatin™ have been conducted for ovarian, thyroid and lung cancer treatments. In solid tumors, administration of either 45 or 63 mg/m² CA4P followed by Paclitaxel™ and Carboplatin™ were well tolerated, demonstrating anti-tumor activity (3 PR and 6 SD over 13 patients) and reduction of tumor blood flow (19% vs. 46% respectively) with K_{trans} reduced by 73 and 79% 24 hours post-CA4P after 6 cycles of treatment[107].

The same combination proved effective on ovarian cancer with 6/17 PR in a phase I dose escalation study[110]. A phase II trial in patients with platinum-resistant ovarian cancer with CA4P 63mg/m² given 18–20 hours prior to Paclitaxel™ 175 mg/m² and Carboplatin™ AUC 5 repeated 3-weekly. Hypertension, easily controlled by glyceryl trinitrate, a venodilator or amlodipine, was the most observed adverse-effect related to CA4P. Responses according to RECIST and GCIG criteria, have been seen in 6/44 and 15/44 patients respectively, demonstrating higher response rates than single agent chemotherapy [104, 105].

Table 3. Overview of the clinical trials involving VDAs and other chemotherapeutic agents

Name	Targeted therapy	Combination with	Development phase	Ref.
CA4P	Thyroid (ATC)	Taxol, carboplatin	II/III (2007-2010)	
	Ovarian, fallopian tube	Avastin	II (2011-2013)	[103]
	Ovarian	Taxol, carboplatin	II (2007-2009)	[104, 105]
	Naïve Lung cancer	Taxol, Avastin, carboplatin	II (2008-2010)	[106]
	Solid tumours	Taxol, carboplatin	II (2005-2008)	[107]
	Thyroid (ATC)	Doxorubicin, cisplatin, radiation	II (2003-2007)	
AVE-8062	NSCLC	Taxol, carboplatin	II (2011-2013)	
	Solid tumours	Taxol, carboplatin	I (2011-2013)	
	Neoplasm	Avastin	I (2010-2012)	
	Neoplasm	Taxotere, cisplatin	I (2010-2013)	
	Solid tumours	Taxotere	I (2008-2010)	[60]
BNC-105P	Renal cell	Everolimus	II (2010-2013)	
NPI-2358	NSCLC	Taxotere	I/II (2008-2011)	[63]
EPC-2407	Solid tumours	Cisplatin	I (2010-2012)	
ABT-751	NSCLC	Carboplatin	II (2004-2009)	
	Lung cancer	Taxotere	I/II (2007-2008)	
	NSCLC	Pemetrexed	I/II (2006-2009)	[108]
MPC-6827	Glioblastoma	Temozolomide/radiation	II (2010-2013)	
	Glioblastoma	Carboplatin	I (2008-2010)	
	Melanoma	Temozolomide	I (2008-2009)	
CYT-997	Glioblastoma	Carboplatin	I/II (2009-2011)	[109]
ZIO-301	Solid tumours	Capecitabine	I (2008-2009)	
	Solid tumours	Erlotinib	I (2008-2009)	
ZD-6126	Colorectal Cancer	5-FU, Leucovorin, Oxaliplatin	II (2003-2004)	

Preliminary results from a phase II study on stage IIIb/IV non-squamous NSCLC compared the efficacy of Carboplatin™, Paclitaxel™ and Avastin™ with or without CA4P given for 6 cycles of therapy showed 48% *vs.* 46% PR. However, patients in the CA4P group had a 52% improvement in overall survival compared with the control group with no additional toxicity observed [106].

Other VDAs in combination studies. Eight other VDAs underwent clinical trials in combination. So far, no results have been reported for BNC-105P, EPC-2407, MPC-6827 and ZIO-301 but trends indicate that VDA trials favour combinations with alkylating agents (Temozolomide™, Cisplatin™, Carboplatin™), antimetabolites (Pemetrexed™, Capecitabine™, 5-Fluorouracil™), antimitotic agents (Docetaxel™) or EGFR inhibitors (Erlotinib™).

Phase II trials on NSCLC of 75 mg/m^2 Docetaxel™ with or without sequentially 30 mg/m^2 or 20 mg/m^2 NPI-2358 showed improved PR in the combination arm (6/27 vs. 2/37). Interestingly, the incidence of the most frequent adverse-effect, neutropaenia, appeared lower in the combination arm (4% vs. 27%)[63]. Phase I/II trials on NSCLC combining the antimetabolite Pemetrexed™ (500 mg/m^2 day 1) with or without 200 mg oral ABT-751 days 1 to 14 on 21-day cycles showed improved overall survival rates in the squamous NSCLC subgroup[108].

Finally, patients with recurrent glioblastoma after radiation and Temozolomide™ received Carboplatin™ AUC 5 day 1 combined with CYT997 at 100-125 mg/m^2 by 24-hour continuous IV infusion beginning day 2 every 3 weeks. Imaging partial response consistent with vascular disruption, were seen in 3 out of 6 evaluable patients and cerebrovascular toxicity (DLT) was controlled by magnetic resonance angiography screening[109].

Combination studies therefore show promising additional activities consistent with preclinical studies.

Other potential combinations. Vascular shutdown offers attractive alternatives to treat cancer. An increasing number of *in vivo* studies are addressing combinations of several drugs, antibodies or treatments with VDAs.

A recent study[111] suggested that hypoxia generated by VDA treatment induced upregulation of genes associated with angiogenesis and drug resistance. Combination therapy of a VDA with a drug that is activated under hypoxic conditions such as Tirapazamine™ (SR4233)[112], AQ4N (Banoxantrone)[113, 114] or TH-302[115, 116] may therefore prove a promising strategy.

In vivo immunization therapy combined with low-dose CA4P significantly enhanced the therapeutic effect as compared with either treatment alone and results in an enhanced antitumor immune reactivity[117].

The last developments come from OXiGENE itself. They recently patented improved cancer treatments involving and a Chemokine Receptor Antagonist (e.g., a CXCR4 antagonist such as AMD-3100) or Chemokine Antagonist (e.g., a SDF-1 antagonist) sequentially or simultaneously in combination with combretastatins[118]. A similar patent described the combination of combretastatin VDAs with an $\alpha4\beta1$ integrin antagonist antibody[119].

Future Directions in the Development of VDAs

The crystal structure of alpha,beta-tubulin was first elucidated by Nogales in 1998 providing a much awaited insight into the protein's structure and function (PDB 1TUB).[120] In 2004, Ravelli solved the crystal structure of tubulin in complex with a colchicine analogue, showing that the CBS was located at the alpha,beta-interface of the homodimer (PDB 1SA0, Figure 2).[121] This achievement was soon followed by the co-crystallization of other CBS agents such as podophyllotoxin (PDB 1SA1),[121] CI-980 (PDB 3N2G & 3N2K),[122] ABT-751 (PDB 3HKC),[123] and T138067 (PDB 3HKE)[123] thus allowing for tubulin inhibitors optimization by molecular modeling.

Using the tubulin: colchicine crystal structure solved by Ravelli, several groups have carried out docking and virtual screening studies in order to identify new tubulin inhibitors. Nguyen identified several pharmacophoric points and developed binding models for 15 known CBS agents including colchicine, CA4, chalcone SD400, podophyllotoxin and ABT-751. These pharmacophore-based models led to the classification of CBS agents into seven groups according to their specific binding interactions with tubulin. [124] Ducki developed a predictive 5D-QSAR model for combretastatin-like analogues interacting at the CBS of tubulin and used it to predict the binding mode of some tubulin targeting chalcones. [125] The group later used the crystal structure of Ravelli and their QSAR model to design a family of alpha-arylchalcones which proved very efficient at inhibiting tubulin polymerization. [126, 127] These chalcones (SD400 and alpha-arylchalcone, Figure 6) are currently under preclinical evaluation. Seong docked MDL-27048 into the CBS of tubulin and proposed a binding model of this tubulin inhibitor. [128] The model was validated by designing new chalcones which proved very efficient as tubulin inhibitors (Chalcones MDL, Figure 6). Pelaez screened a library of 70,000 virtual molecules for their ability to interact at the CBS of tubulin and identified several molecule leads which are under evaluation.[129] Docking of a series of antimitotic pyrroles into the CBS of tubulin [130] allowed for drug optimization and led to the discovery of the tubulin inhibitors JG-03-14 (Figure 6). [131] The drug candidate is currently under preclinical evaluation. A pharmacophore model, built to include the antiproliferative features of a series of indole, was used by Wu to virtually screen a chemical database (130,000 compounds) and led to the discovery of a powerful antitubulin agent (Thiazole, Figure 6) which is currently being optimised. [132] Kim performed a pharmacophore-based virtual screening study of the CBS of tubulin with a focused library of molecules containing the 3,4,5-trimethophenyl moiety. [133] The study allowed the identification of several hits which were subsequently screened for their ability to inhibit cell proliferation, the cell cycle in G2/M phase and tubulin assembly. One molecule (Thiophene, Figure 6) emerged from this study and is currently under preclinical evaluation. More recently Bianucci [134] used tubulin polymerization inhibiting data of 550 compounds to develop a battery of 15 classification models capable of predicting the tubulin binding properties of small molecules. The predictivity of the model was close to 100% which should be useful in the design of novel tubulin inhibitors.

Alternative methods to discover novel tubulin inhibitors have relied on assay-based screening. Jordan [135] developed a series of assays to assess the binding stoichiometry and affinity of radiolabeled drugs for tubulin which could be used to identify novel tubulin inhibitors. Screpanti [136] used a fluorescence-based screen for kinetochore-microtubule interaction inhibitors and identified a novel antitubulin molecule (compound B, Figure 6).

In order for a tubulin inhibitor to enter clinical trials as a VDA, several *in vitro* and *in vivo* preclinical evaluations have to confirm the clinical potential of the new chemical entity (NCE). [13] The following tests are specific to VDAs but complementary to the preclinical assessments requested by the FDA [137] or EMEA [138] for the development of cancer drugs (safety pharmacology, pharmacokinetic, toxicology, maximum tolerated dose, in vivo antitumour efficacity, immunogenicity, genotoxicity, carcinogenicity, reproduction toxicity). Since there is a strong correlation between the antitubulin and the antivascular properties of VDAs, it is necessary to evaluate the ability of novel molecules to inhibit cell proliferation and microtubule formation in vitro.[139, 140] Tubulin inhibitors usually arrest cells in the G2/M phase of the cell cycle, which can be confirmed by flow cytometry [126, 139]. The HUVEC permeability assay is essential to evaluate the in vitro antivascular properties of the molecules and their ability to disrupt junction between cells in a monolayer [126, 141]. More importantly, this antivascular effect of these agents has to be confirmed in vivo. The use of intravital video microscopy (IVVM) allows real time monitoring and quantitative analysis (Vessel number, diameter, length, vascular density, permeability, blood flow velocity in living animals) [142-144]. The most commonly used technique remains dynamic contrast enhanced proton magnetic resonance imaging (DCE-MRI) which is used to assess the effects of VDA on tumour microcirculation (Tumour blood flow, permeability) and is now included in most clinical trials of VDAs.[55, 77, 145] The vascular extent and perfusion in tumours caused by VDAs can also be assessed using histological specimens with immunohistochemistry. [146, 147]

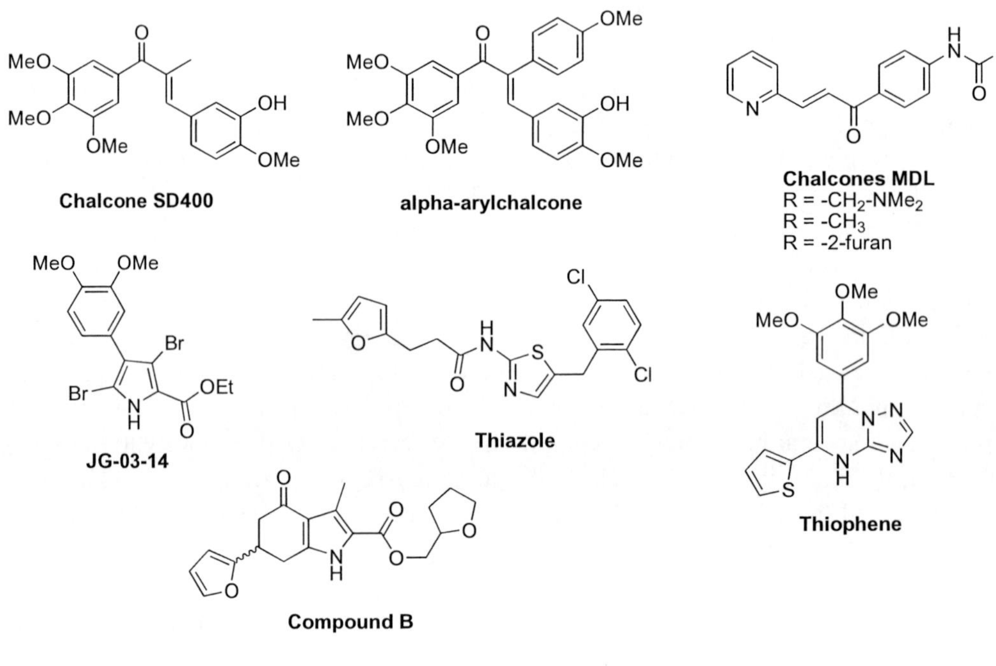

Figure 6. Tubulin inhibitors identified by molecular modeling studies.

Conclusion

Tubulin inhibitors interacting at the colchicine-binding site (CBS) represent the largest group of vascular-disrupting agents (VDAs). These VDAs cause a rapid and selective shutdown of tumour vasculature leading to extensive tumour necrosis. Although no VDAs has been formely approved for clinical use, this class of drugs represents an attractive alternative to conventional cytotoxic chemotherapy as it targets solid tumours which represents almost 90% of cancers.

Preclinical and clinical developments of VDA therapy have clearly established the concept of vascular disruption and the efficacy of such effect in cancer treatment. There are currently 16 VDAs in clinical trials for the treatment of various cancers with three drugs having recently reached phase III. Although patients experience tumour pain during VDA treatment, cardiotoxicity remains the most concerning side-effect. It has become clear that these agents will not be used as single agents mainly due to their inability to target the rim of the tumour which regrows. In this respect, combination chemotherapy, with other cancer drugs or radiotherapy, has overcome this resistance showing improved responses in many cases.

The development of future VDAs, being designed to be even more efficient and less toxic, holds promises. Further development will only be possible when the mechanism of action of tumour specificity is better understood and clinical data fully available to feed new designs. Once the clinical proof-of-concept of VDAs is established, they are expected to become powerful drugs for cancer teatment but also in other pathologies involving aberrant vascularisation: ocular diseases, primarily retinopathies and age related macular degeneration, pediatric hemangiomas, endometriosis, arthritis, psoriasis.

References

[1] Folkman, J., and George Klein, S. W. a. A. H. (1974) Tumor Angiogenesis, in *Advances in Cancer Research* pp 331-358, Academic Press.

[2] Makrilia, N., Lappa, T., Xyla, V., Nikolaidis, I., and Syrigos, K. (2009) The role of angiogenesis in solid tumours: An overview. *European Journal of Internal Medicine* 20, 663-671.

[3] Cristofaro, B., and Emanueli, C. (2009) Possible novel targets for therapeutic angiogenesis. *Current Opinion in Pharmacology* 9, 102-108.

[4] O'Reilly, M. S., Boehm, T., Shing, Y., Fukai, N., Vasios, G., Lane, W. S., Flynn, E., Birkhead, J. R., Olsen, B. R., and Folkman, J. (1997) Endostatin: An Endogenous Inhibitor of Angiogenesis and Tumor Growth. *Cell* 88, 277-285.

[5] O'Reilly, M. S., Holmgren, L., Shing, Y., Chen, C., Rosenthal, R. A., Moses, M., Lane, W. S., Cao, Y., Sage, E. H., and Folkman, J. (1994) Angiostatin: A novel angiogenesis inhibitor that mediates the suppression of metastases by a lewis lung carcinoma. *Cell* 79, 315-328.

[6] Less, J. R., Skalak, T. C., Sevick, E. M., and Jain, R. K. (1991) Microvascular architecture in a mammary carcinoma: Branching patterns and vessel dimensions. *Cancer Research* 51, 265-273.

[7] Dvorak, H. F., Nagy, J. A., Dvorak, J. T., and Dvorak, A. M. (1988) Identification and characterization of the blood vessels of solid tumors that are leaky to circulating macromolecules. *American Journal of Pathology* 133, 95-109.

[8] Patterson, D. M., and Rustin, G. J. S. (2007) Vascular Damaging Agents. *Clinical Oncology* 19, 443-456.

[9] Siemann, D. W. (2011) The unique characteristics of tumor vasculature and preclinical ecidence for its selective disruption by Tumor Disrupting Agents. *Cancer Treatment Reviews* 37, 63-74.

[10] Denekamp, J., Hill, S. A., and Hobson, B. (1983) Vascular occlusion and tumour cell death. *European Journal of Cancer and Clinical Oncology* 19, 271-275.

[11] Siemann, D. W., Bibby, M. C., Dark, G. G., Dicker, A. P., Eskens, F. A. L. M., Horsman, M. R., Marmé, D., and LoRusso, P. M. (2005) Differentiation and Definition of Vascular-Targeted Therapies. *Clinical Cancer Res.* 11, 416-420.

[12] Lippert, J. W. (2007) Vascular disrupting agents. *Bioorganic & Medicinal Chemistry* 15, 605-615.

[13] Mason, R. P., Zhao, D., Liu, L., Trawick, M. L., and Pinney, K. G. (2011) A perspective on vascular disrupting agents that interact with tubulin: preclinical tumor imaging and biological assessment. *Integrative Biology* 3, 375-387.

[14] Zask, A., Kaplan, J., Musto, S., and Loganzo, F. (2005) Hybrids of the Hemiasterlin Analogue Taltobulin and the Dolastatins Are Potent Antimicrotubule Agents. *Journal of the American Chemical Society* 127, 17667-17671.

[15] Smith, C. D., and Zhang, X. (1996) Mechanism of Action of Cryptophycin. *Journal of Biological Chemistry* 271, 6192-6198.

[16] Lee, R. M., and Gewirtz, D. A. (2008) Colchicine site inhibitors of microtubule integrity as vascular disrupting agents. *Drug Development Research* 69, 352-358.

[17] Tozer, G. M., Kanthou, C., Parkins, C. S., and Hill, S. A. (2002) The biology of the combretastatins as tumour vascular targeting agents. *International Journal of Experimental Pathology* 83, 21-38.

[18] Cai, S. X. (2007) Small Molecule Vascular Disrupting Agents: Potential New Drugs for Cancer Treatment. *Recent Patents on Anti-Cancer Drug Discovery* 2, 79-101.

[19] Pettit, G. R., Singh, S. B., Boyd, M. R., Hamel, E., Pettit, R. K., Schmidt, J. M., and Hogan, F. (1995) Antineoplastic Agents. 291. Isolation and Synthesis of Combretastatins A-4, A-5, and A-6. *Journal of Medicinal Chemistry* 38, 1666-1672.

[20] Chaplin, D., Prise, S., Tozer, G., Pettit, G., and Dark, G. (1997) Combretastatin A-4, an agent that displays selective toxicity towards tumour vasculature. *European Journal of Cancer* 33, S44-S44.

[21] Pelletier, P. S., and Caventon, J. (1820) *Ann. Chim. Phys.* 14, 69.

[22] Davis, P. D., Dougherty, G. J., Blakey, D. C., Galbraith, S. M., Tozer, G. M., Holder, A. L., Naylor, M. A., Nolan, J., Stratford, M. R. L., Chaplin, D. J., and Hill, S. A. (2002) ZD6126: A Novel Vascular-targeting Agent That Causes Selective Destruction of Tumor Vasculature *Cancer Research* 62, 7247-7253.

[23] Micheletti, G., Poli, M., Borsotti, P., Martinelli, M., Imberti, B., Taraboletti, G., and Giavazzi, R. (2003) Vascular-targeting Activity of ZD6126, a Novel Tubulin-binding Agent. *Cancer Research* 63, 1534-1537.

[24] Shi, W., and Siemann, D. W. (2005) Preclinical Studies of the Novel Vascular Disrupting Agent MN-029. *Anticancer Research* 25, 3899-3904.

[25] Kemnitzer, W., Kasibhatla, S., Jiang, S., Zhang, H., Zhao, J., Jia, S., Xu, L., Crogan-Grundy, C., Denis, R., Barriault, N., Vaillancourt, L., Charron, S., Dodd, J., Attardo, G., Labrecque, D., Lamothe, S., Gourdeau, H., Tseng, B., Drewe, J., and Cai, S. X. (2005) Discovery of 4-aryl-4H-chromenes as a new series of apoptosis inducers using a cell- and caspase-based high-throughput screening assay. 2. Structure-activity

relationships of the 7- and 5-, 6-, 8-positions. *Bioorganic & Medicinal Chemistry Letters* 15, 4745-4751.

[26] Gourdeau, H., Leblond, L., Hamelin, B., Desputeau, C., Dong, K., Kianicka, I., Custeau, D., Boudreau, C., Geerts, L., Cai, S.-X., Drewe, J., Labrecque, D., Kasibhatla, S., and Tseng, B. (2004) Antivascular and antitumor evaluation of 2-amino-4-(3-bromo-4,5-dimethoxy-phenyl)-3-cyano-4H-chromenes, a novel series of anticancer agents. *Molecular Cancer Therapeutics* 3, 1375-1384.

[27] Kasibhatla, S., Gourdeau, H., Meerovitch, K., Drewe, J., Reddy, S., Qiu, L., Zhang, H., Bergeron, F., Bouffard, D., Yang, Q., Herich, J., Lamothe, S., Cai, S. X., and Tseng, B. (2004) Discovery and mechanism of action of a novel series of apoptosis inducers with potential vascular targeting activity. *Molecular Cancer Therapeutics* 3, 1365-1374.

[28] Yoshimatsu, K., Yamaguchi, A., Yoshino, H., Koyanagi, N., and Kitoh, K. (1997) Mechanism of Action of E7010, an Orally Active Sulfonamide Antitumor Agent: Inhibition of Mitosis by Binding to the Colchicine Site of Tubulin. *Cancer Research* 57, 3208-3213.

[29] Kasibhatla, S., Baichwal, V., Cai, S. X., Roth, B., Skvortsova, I., Skvortsov, S., Lukas, P., English, N. M., Sirisoma, N., Drewe, J., Pervin, A., Tseng, B., Carlson, R. O., and Pleiman, C. M. (2007) MPC-6827: A Small-Molecule Inhibitor of Microtubule Formation That Is Not a Substrate for Multidrug Resistance Pumps. *Cancer Research* 67, 5865-5871.

[30] Burns, C. J., Harte, M. F., Bu, X., Fantino, E., Joffe, M., Sikanyika, H., Su, S., Tranberg, C. E., Wilson, N., Charman, S. A., Shackleford, D. M., and Wilks, A. F. (2009) Discovery of CYT997: a structurally novel orally active microtubule targeting agent. *Bioorganic & Medicinal Chemistry Letters* 19, 4639-4642.

[31] Bacher, G., Nickel, B., Emig, P., Vanhoefer, U., Seeber, S., Shandra, A., Klenner, T., and Beckers, T. (2001) D-24851, a Novel Synthetic Microtubule Inhibitor, Exerts Curative Antitumoral Activity in Vivo, Shows Efficacy toward Multidrug-resistant Tumor Cells, and Lacks Neurotoxicity. *Cancer Research* 61, 392-399.

[32] de Ines, C., Leynadier, D., Barasoain, I., Peyrot, V., Garcia, P., Briand, C., Rener, G. A., and Temple, C. (1994) Inhibition of Microtubules and Cell Cycle Arrest by a New 1-Deaza-7,8-dihydropteridine Antitumor Drug, CI 980, and by Its Chiral Isomer, NSC 613863. *Cancer Research* 54, 75-84.

[33] Bowdon, B. J., Waud, W. R., Wheeler, G. P., Hain, R., Dansby, L., and Temple, C. (1987) Comparison of 1,2-Dihydropyrido[3,4-b]pyrazines (1-Deaza-7,8-dihydropteridines) with Several Other Inhibitors of Mitosis. *Cancer Research* 47, 1621-1626.

[34] Shan, B., Medina, J. C., Santha, E., Frankmoelle, W. P., Chou, T. C., Learned, R. M., Narbut, M. R., Stott, D., Wu, P., Jaen, J. C., Rosen, T., Timmermans, P. B., and Beckman, H. (1999) Selective, covalent modification of beta-tubulin residue Cys-239 by T138067, an antitumor agent with in vivo efficacy against multi-drug resistant tumors. *Proceedings of the National Academy of Sciences USA* 96, 5686-5691.

[35] Lee, J., Bae, S., Lee, S.-h., Choi, H., Kim, Y. H., Kim, S. J., Park, G. T., Moon, S. K., Kim, D.-H., Lee, S., Ahn, S. K., Choi, N. S., and Lee, K. J. (2010) Discovery of a potent tubulin polymerization inhibitor: Synthesis and evaluation of water-soluble prodrugs of benzophenone analog. *Bioorganic & Medicinal Chemistry Letters* 20, 6327-6330.

[36] Lee, J., Kim, S. J., Choi, H., Kim, Y. H., Lim, I. T., Yang, H.-m., Lee, C. S., Kang, H. R., Ahn, S. K., Moon, S. K., Kim, D.-H., Lee, S., Choi, N. S., and Lee, K. J. (2010) Identification of CKD-516: A Potent Tubulin Polymerization Inhibitor with Marked Antitumor Activity against Murine and Human Solid Tumors. *Journal of Medicinal Chemistry* 53, 6337-6354.

[37] Engel, J. B. S., Tanja; Weidler, Claudia; Häusler, Sebastian; Krockenberger, Mathias; Rieger, Lorenz; Dietl, Johannes; Wischhusen, Jörg; Honig, Arnd (2009) Tubulin inhibitor AEZS 112 inhibits the growth of experimental human ovarian and endometrial cancers irrespective of caspase inhibition. *Oncology Reports* 22, 361-367.

[38] D'Amato, R. J., Lin, C. M., Flynn, E., Folkman, J., and Hamel, E. (1994) 2-Methoxyestradiol, an endogenous mammalian metabolite, inhibits tubulin polymerization by interacting at the colchicine site. *Proceedings of the National Academy of Sciences* 91, 3964-3968.

[39] Kamath, K., Okouneva, T., Larson, G., Panda, D., Wilson, L., and Jordan, M. A. (2006) 2-Methoxyestradiol suppresses microtubule dynamics and arrests mitosis without depolymerizing microtubules. *Molecular Cancer Therapeutics* 5, 2225-2233.

[40] Mabjeesh, N. J., Escuin, D., LaVallee, T. M., Pribluda, V. S., Swartz, G. M., Johnson, M. S., Willard, M. T., Zhong, H., Simons, J. W., and Giannakakou, P. (2003) 2ME2 inhibits tumor growth and angiogenesis by disrupting microtubules and dysregulating HIF. *Cancer Cell* 3, 363-375.

[41] [41]Moser, C., Lang, S., Mori, A., Hellerbrand, C., Schlitt, H., Geissler, E., Fogler, W., and Stoeltzing, O. (2008) ENMD-1198, a novel tubulin-binding agent reduces HIF-1alpha and STAT3 activity in human hepatocellular carcinoma(HCC) cells, and inhibits growth and vascularization in vivo. *BMC Cancer* 8, 206.

[42] Dittrich, C., Dumez, H., Calvert, H., Hanauske, A., Faber, M., Wanders, J., Yule, M., Ravic, M., and Fumoleau, P. (2003) Phase I and Pharmacokinetic Study of E7070, a Chloroindolyl-Sulfonamide Anticancer Agent, Administered on a Weekly Schedule to Patients with Solid Tumors. *Clinical Cancer Research* 9, 5195-5204.

[43] Haddad, R. I., Weinstein, L. J., Wieczorek, T. J., Bhattacharya, N., Raftopoulos, H., Oster, M. W., Zhang, X., Latham, V. M., Costello, R., Faucher, J., DeRosa, C., Yule, M., Miller, L. P., Loda, M., Posner, M. R., and Shapiro, G. I. (2004) A Phase II Clinical and Pharmacodynamic Study of E7070 in Patients with Metastatic, Recurrent, or Refractory Squamous Cell Carcinoma of the Head and Neck. *Clinical Cancer Research* 10, 4680-4687.

[44] Raymond, E., ten Bokkel Huinink, W. W., TaÃ¯eb, J., Beijnen, J. H., Faivre, S., Wanders, J., Ravic, M., Fumoleau, P., Armand, J. P., and Schellens, J. H. M. (2002) Phase I and Pharmacokinetic Study of E7070, a Novel Chloroindolyl Sulfonamide Cell-Cycle Inhibitor, Administered as a One-Hour Infusion Every Three Weeks in Patients With Advanced Cancer. *Journal of Clinical Oncology* 20, 3508-3521.

[45] Smyth, J. F., Aamdal, S., Awada, A., Dittrich, C., Caponigro, F., SchÃ¶ffski, P., Gore, M., Lesimple, T., Djurasinovic, N., Baron, B., Ravic, M., Fumoleau, P., and Punt, C. J. A. (2005) Phase II study of E7070 in patients with metastatic melanoma. *Annals of Oncology* 16, 158-161.

[46] Yokoi, A., Kuromitsu, J., Kawai, T., Nagasu, T., Hata Sugi, N., Yoshimatsu, K., Yoshino, H., and Owa, T. (2002) Profiling Novel Sulfonamide Antitumor Agents with

Cell-based Phenotypic Screens and Array-based Gene Expression Analysis. *Molecular Cancer Therapeutics* 1, 275-286.

[47] Chen, C.-P., Hu, C.-B., Yeh, K.-C., Song, J.-S., Yeh, T.-K., Tung, F.-F., Hwang, L.-L., Tseng, H.-Y., Huang, Y.-C., Shy, H.-S., Hsieh, S.-H., Shen, C.-C., Wang, H.-S., Hsieh, H.-P., Liou, J.-P., Chao, Y.-S., and Chen, C.-T. (2010) Antiangiogenic Activities and Cisplatin-combined Antitumor Activities of BPR0L075. *Anticancer Research* 30, 2813-2822.

[48] Cheung, C. H. A., Chen, H.-H., Kuo, C.-C., Chang, C.-Y., Coumar, M., Hsieh, H.-P., and Chang, J.-Y. (2009) Survivin counteracts the therapeutic effect of microtubule de-stabilizers by stabilizing tubulin polymers. *Molecular Cancer* 8, 43.

[49] Desai, A. R., M. J. Moshier, K. Tipton, M. Dooley, W. Hocknell, K. Dahl, T. Sherman, M., Limentani S. (2006) A phase 1, dose-escalation trial of STA-5312, a microtubule inhibitor with a novel binding site, in advanced or metastatic solid malignancies. *Journal of Clinical Oncology, 2006 ASCO Annual Meeting Proceedings Part I.* 24, 13040.

[50] Koya, K., Sun, L., Li, H., James, D., Tatsuta, N., Korbut, T., Zhou, D., Liang, G., Wu, Y., Du, Z., Chen, S., Barsoum, J., Dahl, T. A., Ono, M., and Chen, L. B. (2004) STA-5312, a novel tubulin inhibitor, demonstrates the anti-cancer activity against chemotherapy-resistant cancers. *AACR Meeting Abstracts* 2004, 1250a.

[51] Li, H., Xia, Z., Chen, S., Koya, K., Ono, M., and Sun, L. (2007) Development of a Practical Synthesis of STA-5312, a Novel Indolizine Oxalylamide Microtubule Inhibitor. *Organic Process Research & Development* 11, 246-250.

[52] Dowlati, A., Robertson, K., Cooney, M., Petros, W. P., Stratford, M., Jesberger, J., Rafie, N., Overmoyer, B., Makkar, V., Stambler, B., Taylor, A., Waas, J., Lewin, J. S., McCrae, K. R., and Remick, S. C. (2002) A Phase I Pharmacokinetic and Translational Study of the Novel Vascular Targeting Agent Combretastatin A-4 Phosphate on a Single-Dose Intravenous Schedule in Patients with Advanced Cancer. *Cancer Research* 62, 3408-3416.

[53] Cooney, M. M. S., P. Agarwala, S. Wang, D. Flick, S. Bergant, S. Bhakta, S. Lavertu, P. Ortiz, J. and Remick, S. . (2006) Phase II study of combretastatin A4 phosphate (CA4P) in patients with advanced anaplastic thyroid carcinoma (ATC) *Journal of Clinical Oncology, ASCO Annual Meeting Proceedings* 24, 5580.

[54] Rustin, G. J. S., Galbraith, S. M., Anderson, H., Stratford, M., Folkes, L. K., Sena, L., Gumbrell, L., and Price, P. M. (2003) Phase I Clinical Trial of Weekly Combretastatin A4 Phosphate: Clinical and Pharmacokinetic Results. *Journal of Clinical Oncology* 21, 2815-2822.

[55] Stevenson, J. P., Rosen, M., Sun, W., Gallagher, M., Haller, D. G., Vaughn, D., Giantonio, B., Zimmer, R., Petros, W. P., Stratford, M., Chaplin, D., Young, S. L., Schnall, M., and OÂ'Dwyer, P. J. (2003) Phase I Trial of the Antivascular Agent Combretastatin A4 Phosphate on a 5-Day Schedule to Patients With Cancer: Magnetic Resonance Imaging Evidence for Altered Tumor Blood Flow. *Journal of Clinical Oncology* 21, 4428-4438.

[56] Cooney, M. M., Radivoyevitch, T., Dowlati, A., Overmoyer, B., Levitan, N., Robertson, K., Levine, S. L., DeCaro, K., Buchter, C., Taylor, A., Stambler, B. S., and Remick, S. C. (2004) Cardiovascular Safety Profile of Combretastatin A4 Phosphate in

a Single-Dose Phase I Study in Patients with Advanced Cancer. *Clinical Cancer Research* 10, 96-100.

[57] Patterson, D. M. C., N.Saleem, A . Stratford, M. Dickinson, C. Asselin, M. Ross, P. Walicke, P. Price, P. and Rustin, G. J. . (2008) Phase I evaluation of vascular disrupting agent OXi4503 *Journal of Clinical Oncology, ASCO Annual Meeting Proceedings* 26, 3551.

[58] Zweifel, M., Patterson, D. M., Padhani, A. R., Middleton, M. R., Price, P., Stratford, M. R., Ross, P., Halford, S., Z., G., and J., R. G. (2010) Phase I pharmacokinetic and pharmacodynamic evaluation of the vascular disrupting agent OXi4503 in patients with advanced solid tumors. . *Journal of Clinical Oncology, ASCO Annual Meeting Proceedings* 28, 2594.

[59] Patterson, D. M. R., P. Koetz, B. Saleem, A. Stratford, M. Stirling, J. Padhani, A. Asselin, M. Price, P. and Rustin, G. J. (2007) Phase I evaluation of OXi4503, a vascular disrupting agent, in patients with advanced solid tumours *Journal of Clinical Oncology, ASCO Annual Meeting Proceedings* 25, 14146.

[60] Tresca, P. T., D. van Doorn, L. Fontaine, H. Gaast, A. V. Veyrat-Follet, C. Oprea, C. Dieras, V. and Eskens, F. (2010) Phase I and pharmacologic study of the vascular disrupting agent ombrabulin (Ob) combined with docetaxel (D) in patients (pts) with advanced solid tumors. *Journal of Clinical Oncology, ASCO Annual Meeting Proceedings* 28, 3023.

[61] Leske, A. F., Bibby, D. C., Lavranos, T. C., Beaumont, D., Gasic, J., Rischin, D., Desai, J., Chong, G., Wong, S. S., Flynn, B., and Kremmidiotis, G. (2011) in *23rd Lorne Cancer Conference*.

[62] Mita, A. C., Yee, L. K., Papadopoulos, K. P., Heath, E. I., Romero, O., Lloyd, G. K., Cropp, G., Spear, M. A., M., M. M., and LoRusso, P. M. (2008) Phase I study of NPI-2358 (a novel vascular disrupting agent) in patients with solid tumors and lymphomas *Journal of Clinical Oncology, ASCO Annual Meeting Proceedings* 26, 3525.

[63] Mita, A. C., Heist, R. S., Aren, O., Mainwaring, P. N., Bazhenova, L., Gadgeel, S. M., Blum, R. H., Polikoff, J., Biswas, J., and Spear, M. A. (2010) Phase II study of docetaxel with or without plinabulin (NPI-2358) in patients with non-small cell lung cancer (NSCLC). . *J Clin Oncol* 28, 7592.

[64] Traynor, A., Gordon, M., Alberti, D., Mendelson, D., Munsey, M., Wilding, G., Gammans, R., and Read, W. (2010) A dose escalation, safety, and tolerability study of MN-029 in patients with advanced solid tumors. *Investigational New Drugs* 28, 509-515.

[65] Ricart, A., Ashton, E., Cooney, M., Sarantopoulos, J., Brell, J., Feldman, M., Ruby, K., Matsuda, K., Munsey, M., Medina, G., Zambito, A., Tolcher, A., and Remick, S. (2011) A phase I study of MN-029 (denibulin), a novel vascular-disrupting agent, in patients with advanced solid tumors. *Cancer Chemotherapy and Pharmacology*, 1-12.

[66] Anthony, S. P., Read, W., Rosen, P. J., Tibes, R., Park, D., Everton, D., Tseng, B., J., W., and Von Hoff, D. D. (2008) Initial results of a first-in-man phase I study of EPC2407, a novel small molecule microtubule inhibitor anticancer agent with tumor vascular endothelial disrupting activity *Journal of Clinical Oncology, ASCO Annual Meeting Proceedings* 26, 2531.

[67] Read, W. L., Rosen, P., Lee, P., Anthony, S., Korn, R., Raghunand, N., Tseng, B., Whisnant, J., Von Hoff, D., and Tibes, R. (2009) Pharmacokinetic and

pharmacodynamic results of a 4-hr IV administration phase I study with EPC2407, a novel vascular disrupting agent *Journal of Clinical Oncology, ASCO Annual Meeting Proceedings* 27, 3569.

[68] Fox, E., Maris, J., Cohn, S., Goodspeed, W., Goodwin, A., Kromplewski, M., Medina, D., Xiong, H., Krivoshik, A., Widemann, B., Adamson, P., and Balis, F. (2010) Pharmacokinetics of orally administered ABT-751 in children with neuroblastoma and other solid tumors. *Cancer Chemotherapy and Pharmacology* 66, 737-743.

[69] Benson, A. B., Kindler, H. L., Jodrell, D., Hagey, A., Coates, A. I., Meek, K. A., Cernohous, P., and B., G. G. (2005) Phase 2 Study of ABT-751 in Patients with Refractory Metastatic Colorectal Carcinoma (CRC). *Journal of Clinical Oncology, 2005 ASCO Annual Meeting Proceedings.* 23, 3537.

[70] Hagey, A. E., Figlin, R. A., Moldawer, N., Sosman, J. A., Chi, K. N., Medina, D. M., Meek, K. A., Cernohous, P., and B., G. G. (2005) Preliminary Phase 2 Results of ABT-751 in Subjects with Advanced Renal Cell Carcinoma (RCC). *Journal of Clinical Oncology, 2005 ASCO Annual Meeting Proceedings.* 23, 4603.

[71] Washington, D. K., Storniolo, A. V., Saleh, M., Tan-Chiu, E., Hagey, A., Medina, D. M., Meek, K. A., Cernohous, P., and B., G. G. (2005) Phase 2 Results of ABT-751 in Subjects with Taxane-Refractory Breast Cancer: Interim Analysis. *Journal of Clinical Oncology, 2005 ASCO Annual Meeting Proceedings.* 23, 724.

[72] Mauer, A. M., Cohen, E. E. W., Ma, P. C., Kozloff, M. F., Schwartzberg, L., Coates, A. I., Qian, J., Hagey, A. E., and Gordon, G. B. (2008) A Phase II Study of ABT-751 in Patients with Advanced Non-small Cell Lung Cancer. *Journal of Thoracic Oncology* 3, 631-636

[73] Fox, E., Maris, J. M., Widemann, B. C., Goodspeed, W., Goodwin, A., Kromplewski, M., Fouts, M. E., Medina, D., Cohn, S. L., Krivoshik, A., Hagey, A. E., Adamson, P. C., and Balis, F. M. (2008) A Phase I Study of ABT-751, an Orally Bioavailable Tubulin Inhibitor, Administered Daily for 21 Days Every 28 Days in Pediatric Patients with Solid Tumors. *Clinical Cancer Research* 14, 1111-1115.

[74] Hande, K. R., Hagey, A., Berlin, J., Cai, Y., Meek, K., Kobayashi, H., Lockhart, A. C., Medina, D., Sosman, J., Gordon, G. B., and Rothenberg, M. L. (2006) The Pharmacokinetics and Safety of ABT-751, a Novel, Orally Bioavailable Sulfonamide Antimitotic Agent: Results of a Phase 1 Study. *Clinical Cancer Research* 12, 2834-2840.

[75] Tsimberidou, A.-M., Akerley, W., Schabel, M. C., Hong, D. S., Uehara, C., Chhabra, A., Warren, T., Mather, G. G., Evans, B. A., Woodland, D. P., Swabb, E. A., and Kurzrock, R. (2010) Phase I Clinical Trial of MPC-6827 (Azixa), a Microtubule Destabilizing Agent, in Patients with Advanced Cancer. *Molecular Cancer Therapeutics* 9, 3410-3419.

[76] Lickliter, J., Francesconi, A., Smith, G., Burge, M., Coulthard, A., Rose, S., Griffin, M., Wilks, A., Wyld, D., and Vasey, P. (2008) Phase I evaluation of CYT997, a novel cytotoxic and vascular-disrupting agent, in patients with advanced cancer *Journal of Clinical Oncology, ASCO Annual Meeting Proceedings* 26, 3504.

[77] Lickliter, J. D., Francesconi, A. B., Smith, G., Burge, M., Coulthard, A., Rose, S., Griffin, M., Milne, R., McCarron, J., Yeadon, T., Wilks, A., Cubitt, A., Wyld, D. K., and Vasey, P. A. (2010) Phase I trial of CYT997, a novel cytotoxic and vascular-disrupting agent. *British Journal of Cancer* 103, 597-606.

[78] Francesconi, A., Kotasek, D., Burge, M., Smith, G., and Lickliter, J. (2009) Phase I evaluation of orally-administered CYT997, a novel cytotoxic vascular-disrupting agent, in patients with advanced cancer *Journal of Clinical Oncology, ASCO Annual Meeting Proceedings* 27, 3568.

[79] Northfelt DW, G. P., Block H, Sindermann H, Teifel M, Von Hoff D, Mendelson D. (2009) in *AACR* pp Poster #5567.

[80] Kuppens, I., Witteveen, P., Schot, M., Schuessler, V., Daehling, A., Beijnen, J., Voest, E., and Schellens, J. (2007) Phase I dose-finding and pharmacokinetic trial of orally administered indibulin (D-24851) to patients with solid tumors. *Investigational New Drugs* 25, 227-235.

[81] Beerepoot, L. V., Radema, S. A., Witteveen, E. O., Thomas, T., Wheeler, C., Kempin, S., and Voest, E. E. (2006) Phase I clinical evaluation of weekly administration of the novel vascular-targeting agent, ZD6126, in patients with solid tumors. *Journal of Clinical Oncology* 24, 1491-1498

[82] LoRusso, P., Gadgeel, S., Wozniak, A., Barge, A., Jones, H., DelProposto, Z., DeLuca, P., Evelhoch, J., Boerner, S., and Wheeler, C. (2008) Phase I clinical evaluation of ZD6126, a novel vascular-targeting agent, in patients with solid tumors. *Investigational New Drugs* 26, 159-167.

[83] Pazdur, R., Meyers, C., Diaz-Canton, E., Abbruzzese, J. L., Patt, Y., Grove, W., and Ajani, J. (1997) Phase II Trial of Intravenous CI-980 (NSC 370147) in Patients with Metastatic Colorectal Carcinoma. *American Journal of Clinical Oncology* 20, 573-576.

[84] Whitehead, R. P., Unger, J. M., Flaherty, L. E., Eckardt, J. R., Taylor, S. A., Didolkar, M. S., Samlowski, W., and Sondak, V. K. (2001) Phase II Trial of CI-980 in Patients with Disseminated Malignant Melanoma and no Prior Chemotherapy – A Southwest Oncology Group Study. *Investigational New Drugs* 19, 239-243.

[85] Sklarin, N. T., Benson, L., Roca, J., Einzig, A. I., Wiernik, P. H., Whitfield, L. R., Lathia, C., Kowal, C., and Grove, W. (1995) Phase I study of CI-980: 24-hr infusion schedule. *ASCO Annual Meeting*, 1559.

[86] Meyers, C. A., Kudelka, A. P., Conrad, C. A., Gelke, C. K., Grove, W., and Pazdur, R. (1997) Neurotoxicity of CI-980, a novel mitotic inhibitor. *Clinical Cancer Research* 3, 419-422.

[87] Berlin, J., Venook, A., Bergsland, E., Rothenberg, M., Lockhart, A. C., and Rosen, L. (2008) Phase II Trial of T138067, a Novel Microtubule Inhibitor, in Patients with Metastatic, Refractory Colorectal Carcinoma. *Clinical Colorectal Cancer* 7, 44-47.

[88] Posey, J., Johnson, P., Mok, T., Hirmand, M., Dahlberg, S., L., K., and Leung, T. (2005) Results of a phase 2/3 open-label, randomized trial of T138067 versus doxorubicin (DOX) in chemotherapy-naïve, unresectable hepatocellular carcinoma (HCC) *Journal of Clinical Oncology, 2005 ASCO Annual Meeting Proceedings.* 23, 4035.

[89] Leung, T. W., Feun, L., Posey, J., Stagg, R. J., Levy, M. D., and Venook, A. P. (2002) A phase II study of T138067-sodium in patients (pts) with unresectable hepatocellular carcinoma (HCC). *Proceedings of the American Society of Clinical Oncology; ASCO Annual Meeting* 21, 572.

[90] Kirby, S., Gertler, S. Z., Mason, W., Watling, C., Forsyth, P., Aniagolu, J., Stagg, R., Wright, M., Powers, J., and Eisenhauer, E. A. (2005) Phase 2 study of T138067-sodium in patients with malignant glioma: Trial of the National Cancer Institute of Canada Clinical Trials Group. *Neuro-Oncology* 7, 183-188.

[91] Venook, A. P., Rowinsky, E. K., Donehower, R. C., Leung, T., Floren, L. C., and Levy, M. (2004) Safety and pharmacokinetics (PK) of T138067 (T67) administered as a weekly 3-hour infusion in subjects with hepatocellular carcinoma (HCC) in a phase 1 study. *Journal of Clinical Oncology, 2004 ASCO Annual Meeting Proceedings* 22, 4087.

[92] Honess DJ, H., F, Chaplin, DJ, Tozer, GM. (2002) Comparison of strategies to overcome the hypertensive effect of combretastatin-A-4-phosphate in a rat model. *British Journal of Cancer* 86, S118.

[93] Fox, E., Maris, J. M., Widemann, B. C., Meek, K., Goodwin, A., Goodspeed, W., Kromplewski, M., Fouts, M. E., Medina, D., Cho, S. Y., Cohn, S. L., Krivoshik, A., Hagey, A. E., Adamson, P. C., and Balis, F. M. (2006) A Phase 1 Study of ABT-751, an Orally Bioavailable Tubulin Inhibitor, Administered Daily for 7 Days Every 21 Days in Pediatric Patients with Solid Tumors. *Clinical Cancer Research* 12, 4882-4887.

[94] Raab G, B. G., Komarnitsky PB, Wallner BP and Kutscher B. *Ziopharm Oncology*, New York.

[95] Wienecke, A., and Bacher, G. (2009) Indibulin, a Novel Microtubule Inhibitor, Discriminates between Mature Neuronal and Nonneuronal Tubulin. *Cancer Research* 69, 171-177.

[96] Grosios, K., Holwell, S. E., McGown, A. T., Pettit, G. R., and Bibby, M. C. (1999) In vivo and in vitro evaluation of combretastatin A-4 and its sodium phosphate prodrug. *Br J Cancer* 81, 1318-1327.

[97] Anderson, H. L., Yap, J. T., Miller, M. P., Robbins, A., Jones, T., and Price, P. M. (2003) Assessment of Pharmacodynamic Vascular Response in a Phase I Trial of Combretastatin A4 Phosphate. *Journal of Clinical Oncology* 21, 2823-2830.

[98] Horsman, M. R., and Siemann, D. W. (2006) Pathophysiologic Effects of Vascular-Targeting Agents and the Implications for Combination with Conventional Therapies. *Cancer Research* 66, 11520-11539.

[99] Li, L., Rojiani, A., and Siemann, D. W. (1998) Targeting the tumor vasculature with combretastatin A-4 disodium phosphate: effects on radiation therapy. *International Journal of Radiation Oncology*Biology*Physics* 42, 899-903.

[100] Murata, R., Siemann, D. W., Overgaard, J., and Horsman, M. R. (2001) Interaction between combretastatin A-4 disodium phosphate and radiation in murine tumors. *Radiotherapy and Oncology* 60, 155-161.

[101] Horsman, M. R., Murata, R., and Overgaard, J. (2002) 152. Oral Combination studies with combretastatin and radiation: effects in early and late responding normal tissues. *Radiotherapy and Oncology* 64, S50-S50.

[102] Siemann, D. W., and Shi, W. (2008) Dual Targeting of Tumor Vasculature: Combining Avastin and Vascular Disrupting Agents (CA4P or OXi4503). *Anticancer Research* 28, 2027-2031.

[103] Nathan, P. D., Judson, I., Padhani, A., Harris, A., Carden, C. P., Smythe, J., Collins, D., Leach, M., Walicke, P., and Rustin, G. J. (2008) A phase I study of combretastatin A4 phosphate (CA4P) and bevacizumab in subjects with advanced solid tumors *Journal of Clinical Oncology, ASCO Annual Meeting Proceedings* 26, 3550.

[104] Zweifel, M., Jayson, G. C., Reed, N. S., Osborne, R., Hassan, B., Ledermann, J., Shreeves, G., Poupard, L., Lu, S. P., Balkissoon, J., Chaplin, D. J., and Rustin, G. J. S. (2011) Phase II trial of combretastatin A4 phosphate, carboplatin, and paclitaxel in

patients with platinum-resistant ovarian cancer. *Annals of Oncology*, 10.1093/annonc/mdq708.

[105] Zweifel, M., Jayson, G., Reed, N., Osborne, R., Hassan, B., Shreeves, G., Poupard, L., Walicke, P. A., Balkissoon, J., Chaplin, D., and Rustin, G. (2009) Combretastatin A-4 phosphate (CA4P) carboplatin and paclitaxel in patients with platinum-resistant ovarian cancer: Final phase II trial results *Journal of Clinical Oncology, ASCO Annual Meeting Proceedings* 27, 5502.

[106] Garon, E. B., Kabbinavar, F. F., Neidhart, J. A., Neidhart, J. D., Gabrail, N. Y., Oliveira, M. R., Lu, S. P., and Balkissoon, J. (2010) Randomized phase II trial of a tumor vascular disrupting agent fosbretabulin tromethamine (CA4P) with carboplatin (C), paclitaxel (P), and bevacizumab (B) in stage IIIb/IV nonsquamous non-small cell lung cancer (NSCLC): The FALCON trial. *Journal of Clinical Oncology, ASCO Annual Meeting Proceedings* 28, 7587.

[107] Akerley, W. L., Schabel, M., Morrell, G., Horvath, E., Yu, M., Johnsson, B., and Arbogast, K. (2007) A randomized phase II trial of combretatstatin A4 phosphate (CA4P) in combination with paclitaxel and carboplatin to evaluate safety and efficacy in subjects with advanced imageable malignancies *Journal of Clinical Oncology, ASCO Annual Meeting Proceedings* 25, 14060.

[108] Rudin, C. M., Mauer, A., Smakal, M., Juergens, R., Spelda, S., Wertheim, M., Coates, A., McKeegan, E., Ansell, P., Zhou, X., Qian, J., Pradhan, R., Dowell, B., Krivoshik, A., and Gordon, G. (2011) Phase I/II Study of Pemetrexed With or Without ABT-751 in Advanced or Metastatic Non-Small-Cell Lung Cancer. *Journal of Clinical Oncology* 29, 1075-1082.

[109] Lickliter, J., Fida, R., Wheeler, H., Kichenadasse, G., Yang, Q., Ganju, V., and Briggs, P. (2010) Carboplatin combined with the vascular-disrupting agent CYT997 for recurrent glioblastoma multiforme. *Journal of Clinical Oncology, ASCO Annual Meeting Proceedings* 28, e13591.

[110] Rustin, G. J., Shreeves, G., Nathan, P. D., Gaya, A., Ganesan, T. S., Wang, D., Boxall, J., Poupard, L., Chaplin, D. J., Stratford, M. R. L., Balkissoon, J., and Zweifel, M. (2010) A Phase Ib trial of CA4P (combretastatin A-4 phosphate), carboplatin, and paclitaxel in patients with advanced cancer. *British Journal of Cancer* 102, 1355-1360.

[111] Tozer, G. M., Kanthou, C., Lewis, G., Prise, V. E., Vojnovic, B., and Hill, S. A. (2008) Tumour vascular disrupting agents: combating treatment resistance. *British Journal of Radiology* 81, S12-20.

[112] Brown, J. M. (1993) SR 4233 (Tirapazamine): a new anticancer drug exploiting hypoxia in solid tumours. *British Journal of Cancer* 67, 1163-1170.

[113] Albertella, M. R., Loadman, P. M., Jones, P. H., Phillips, R. M., Rampling, R., Burnet, N., Alcock, C., Anthoney, A., Vjaters, E., Dunk, C. R., Harris, P. A., Wong, A., Lalani, A. S., and Twelves, C. J. (2008) Hypoxia-Selective Targeting by the Bioreductive Prodrug AQ4N in Patients with Solid Tumors: Results of a Phase I Study. *Clinical Cancer Research* 14, 1096-1104.

[114] Patterson, L. H. (2002) Bioreductively Activated Antitumor N-Oxides: The Case Of AQ4N, A Unique Approach To Hypoxia-Activated Cancer Chemotherapy. *Drug Metabolism Reviews* 34, 581-592.

[115] Weiss, G. J., Infante, J. R., Chiorean, E. G., Borad, M. J., Bendell, J. C., Molina, J. R., Tibes, R., Ramanathan, R. K., Lewandowski, K., Jones, S. F., Lacouture, M. E.,

Langmuir, V. K., Lee, H., Kroll, S., and Burris, H. A. (2011) Phase 1 Study of the Safety, Tolerability, and Pharmacokinetics of TH-302, a Hypoxia-Activated Prodrug, in Patients with Advanced Solid Malignancies. *Clinical Cancer Research* 17, 2997-3004.

[116] Duan, J.-X., Jiao, H., Kaizerman, J., Stanton, T., Evans, J. W., Lan, L., Lorente, G., Banica, M., Jung, D., Wang, J., Ma, H., Li, X., Yang, Z., Hoffman, R. M., Ammons, W. S., Hart, C. P., and Matteucci, M. (2008) Potent and Highly Selective Hypoxia-Activated Achiral Phosphoramidate Mustards as Anticancer Drugs. *Journal of Medicinal Chemistry* 51, 2412-2420.

[117] Badn, W., Kalliomaki, S., Widegren, B., and Sjogren, H. O. (2006) Low-Dose Combretastatin A4 Phosphate Enhances the Immune Response of Tumor Hosts to Experimental Colon Carcinoma. *Clinical Cancer Research* 12, 4714-4719.

[118] Kerbel, R., Shaked, Y. Methods for enhancing the efficacy of vascular disrupting agents. OXiGENE, Inc., USA. WO2008109076A1. pp 82. (issued 12/09/2008).

[119] Kerbel, R., Shaked, Y. Methods for enhancing the efficacy of vascular disrupting agents. OXiGENE, Inc., USA. US2011/0110940A1. pp 24. (issued 12/05/2011).

[120] Nogales, E., Wolf, S. G., and Downing, K. H. (1998) Structure of the [alpha][beta] tubulin dimer by electron crystallography. *Nature* 391, 199-203.

[121] Ravelli, R. B. G., Gigant, B., Curmi, P. A., Jourdain, I., Lachkar, S., Sobel, A., and Knossow, M. (2004) Insight into tubulin regulation from a complex with colchicine and a stathmin-like domain. *Nature* 428, 198-202.

[122] Barbier, P., Dorleans, A., Devred, F., Sanz, L., Allegro, D., Alfonso, C., Knossow, M., Peyrot, V., and Andreu, J. M. (2010) Stathmin and Interfacial Microtubule Inhibitors Recognize a Naturally Curved Conformation of Tubulin Dimers. *Journal of Biological Chemistry* 285, 31672-31681.

[123] Dorleans, A., Gigant, B. t., Ravelli, R. B. G., Mailliet, P., Mikol, V., and Knossow, M. (2009) Variations in the colchicine-binding domain provide insight into the structural switch of tubulin. *Proceedings of the National Academy of Sciences* 106, 13775-13779.

[124] Nguyen, T. L., McGrath, C., Hermone, A. R., Burnett, J. C., Zaharevitz, D. W., Day, B. W., Wipf, P., Hamel, E., and Gussio, R. (2005) A Common Pharmacophore for a Diverse Set of Colchicine Site Inhibitors Using a Structure-Based Approach. *Journal of Medicinal Chemistry* 48, 6107-6116.

[125] Ducki, S., Mackenzie, G., Lawrence, N. J., and Snyder, J. P. (2005) Quantitative Structure-Activity Relationship (5D-QSAR) Study of Combretastatin-like Analogues as Inhibitors of Tubulin Assembly. *Journal of Medicinal Chemistry* 48, 457-465.

[126] Ducki, S., Rennison, D., Woo, M., Kendall, A., Chabert, J. F. D., McGown, A. T., and Lawrence, N. J. (2009) Combretastatin-like chalcones as inhibitors of microtubule polymerization. Part 1: Synthesis and biological evaluation of antivascular activity. *Bioorganic & Medicinal Chemistry* 17, 7698-7710.

[127] Ducki, S., Mackenzie, G., Greedy, B., Armitage, S., Chabert, J. F. D., Bennett, E., Nettles, J., Snyder, J. P., and Lawrence, N. J. (2009) Combretastatin-like chalcones as inhibitors of microtubule polymerisation. Part 2: Structure-based discovery of alpha-aryl chalcones. *Bioorganic & Medicinal Chemistry* 17, 7711-7722.

[128] Kim, D. Y., Kim, K.-H., Kim, N. D., Lee, K. Y., Han, C. K., Yoon, J. H., Moon, S. K., Lee, S. S., and Seong, B. L. (2006) Design and Biological Evaluation of Novel Tubulin Inhibitors as Antimitotic Agents Using a Pharmacophore Binding Model with Tubulin. *Journal of Medicinal Chemistry* 49, 5664-5670.

[129] Peláez, R., López, J. L., and Medarde, M. (2007) Application of Chemoinformatic Tools for the Analysis of Virtual Screening Studies of Tubulin Inhibitors. . *Advances in Soft Computing* 44, 411-417.

[130] Tripathi, A., Fornabaio, M., Kellogg, G. E., Gupton, J. T., Gewirtz, D. A., Yeudall, W. A., Vega, N. E., and Mooberry, S. L. (2008) Docking and hydropathic scoring of polysubstituted pyrrole compounds with antitubulin activity. *Bioorganic & Medicinal Chemistry* 16, 2235-2242.

[131] Mooberry, S. L., Weiderhold, K. N., Dakshanamurthy, S., Hamel, E., Banner, E. J., Kharlamova, A., Hempel, J., Gupton, J. T., and Brown, M. L. (2007) Identification and Characterization of a New Tubulin-Binding Tetrasubstituted Brominated Pyrrole. *Molecular Pharmacology* 72, 132-140.

[132] Chiang, Y.-K., Kuo, C.-C., Wu, Y.-S., Chen, C.-T., Coumar, M. S., Wu, J.-S., Hsieh, H.-P., Chang, C.-Y., Jseng, H.-Y., Wu, M.-H., Leou, J.-S., Song, J.-S., Chang, J.-Y., Lyu, P.-C., Chao, Y.-S., and Wu, S.-Y. (2009) Generation of Ligand-Based Pharmacophore Model and Virtual Screening for Identification of Novel Tubulin Inhibitors with Potent Anticancer Activity. *Journal of Medicinal Chemistry* 52, 4221-4233.

[133] Kim, N. D., Park, E.-S., Kim, Y. H., Moon, S. K., Lee, S. S., Ahn, S. K., Yu, D.-Y., No, K. T., and Kim, K.-H. Structure-based virtual screening of novel tubulin inhibitors and their characterization as anti-mitotic agents. *Bioorganic & Medicinal Chemistry* 18, 7092-7100.

[134] Massarelli, I., Imbriani, M., James, T. L., Mundula, T., and Bianucci, A. M. (2011) Development of classification model batteries for predicting inhibition of tubulin polymerization by small molecules. *Chemometrics and Intelligent Laboratory Systems* 107, 206-214.

[135] Smith, J. A., and Jordan, M. A. (2010) Determination of Drug Binding to Microtubules In Vitro. *Methods in Cell Biology Chapter* 16, 289-299.

[136] Screpanti, E., Santaguida, S., Nguyen, T., Silvestri, R., Gussio, R., Musacchio, A., Hamel, E., and De Wulf, P. (2010) A Screen for Kinetochore-Microtubule Interaction Inhibitors Identifies Novel Antitubulin Compounds. *PLoS ONE* 5, e11603.

[137] DeGeorge, J. J., Ahn, C.-H., Andrews, P. A., Brower, M. E., Giorgio, D. W., Goheer, M. A., Lee-Ham, D. Y., McGuinn, W. D., Schmidt, W., Sun, C. J., and Tripathi, S. C. (1998) Regulatory considerations for preclinical development of anticancer drugs. *Cancer Chemotherapy and Pharmacology* 41, 173-185.

[138] EuropeanMedicinesAgency. (January 1999) Note for *Guidance on the Pre-Clinical Evaluation of Anticancer Medicinal Products*.

[139] Lawrence, N. J., McGown, A. T., Ducki, S., and Hadfield, J. A. (2000) The interaction of chalcones with tubulin. *Anti-Cancer Drug Design* 15, 135-141.

[140] Woods, J. A., Hadfield, J. A., Pettit, G. R., Fox, B. W., and McGown, A. T. (1995) The interaction with tubulin of a series of stilbenes based on combretastatin A4. *British Journal of Cancer* 72, 705.

[141] Galbraith, S. M., Chaplin, D. J., Lee, F., Stratford, M. R. L., Locke, R. J., Vojnovic, B., and Tozer, G. M. (2001) Effects of combretastatin A4 phosphate on endothelial cell morphology in vitro and relationship to tumour vascular targeting activity in vivo. *Anticancer Research* 21, 93-102.

[142] Iga, A. M., Sarkar, S., Sales, K. M., Winslet, M. C., and Seifalian, A. M. (2006) Quantitating Therapeutic Disruption of Tumor Blood Flow with Intravital Video Microscopy. *Cancer Research* 66, 11517-11519.

[143] Tozer, G. M., Prise, V. E., and Wilson, J. (2001) Mechanisms associated with tumor vascular shut-down induced by combretastatin A-4 phosphate: intravital microscopy and measurement of vascular permeability. *Cancer Research* 61, 6413-6422.

[144] Hori, K., and Saito, S. (2003) Microvascular mechanisms by which the combretastatin A-4 derivative AC7700 (AVE8062) induces tumour blood flow stasis. *British Journal of Cancer* 89, 1334-1344.

[145] Luo, Y., Hradil, V. P., Frost, D. J., Rosenberg, S. H., Gordon, G. B., Morgan, S. J., Gagne, G. D., Cox, B. F., Tahir, S. K., and Fox, G. B. (2009) ABT-751, a novel tubulin-binding agent, decreases tumor perfusion and disrupts tumor vasculature. *Anti-Cancer Drugs* 20, 483-492.

[146] Zhao, D., Jiang, L., Hahn, E. W., and Mason, R. P. (2005) Tumor physiologic response to combretastatin A4 phosphate assessed by MRI. *International Journal of Radiation Oncology*Biology*Physics* 62, 872-880.

[147] Zhao, D., Richer, E., Antich, P. P., and Mason, R. P. (2008) Antivascular effects of combretastatin A4 phosphate in breast cancer xenograft assessed using dynamic bioluminescence imaging and confirmed by MRI. *The FASEB Journal* 22, 2445-2451.

In: Tubulin: Structure, Functions and Roles in Disease
Editors: Wei Yamauchi and Antonio Sokić ISBN: 978-1-62100-141-6

Chapter VII

Evolution and Coevolution of Tubulin's Carboxy-Terminal Tails and Mitochondria*

Dan L. Sackett

Laboratory of Integrative and Medical Biophysics, Program in Physical Biology, Eunice Kennedy Shriver National Institute of Child Health and Human Development, National Institutes of Health, Bethesda, MD, US

Introduction

Microtubules and mitochondria are two of the defining features of a eukaryotic cell. Tubulin, the subunit protein of microtubules (MT) is found in all eukaryotic species and in essentially all eukaryotic cells, while mitochondria are found in most eukaryotic species though there are a significant number that lack mitochondria sensu stricto. Both tubulin and mitochondria have been part of the eukaryotic spectrum of characteristics for the vast majority of the ~1-2 billion years that eukaryotes have existed, though the details at early time, including when the eukaryotic line began, remain points of vigorous discussion [1-9].

The interaction of tubulin and mitochondria will be the subject of the sections that follow. This interaction occurs both at the level of MT and mitochondria as well as at the level of the unpolymerized tubulin dimer and mitochondria. This latter interaction is the one featured in the previous two chapters by Saks, et al., and by Rostovtseva, and is the focus here as well. In the following sections, relevant features of MT and tubulin structure and evolution will be discussed, followed by mitochondrial structure and variation. Finally the coevolution of tubulin and mitochondria will be considered in light of the binding of tubulin's tail peptides to the mitochondrial outer membrane (MOM) through the voltage dependent anion channel (VDAC), also known as mitochondrial porin.

* A version of this chapter also appears in Mitochondria: Structure, Functions and Dysfunctions, edited by Oliver L. Svensson, published by Nova Science Publishers, Inc. It was submitted for appropriate modifications in an effort to encourage wider dissemination of research.

Microtubules, Tubulin Structure and Evolution

Tubulin, the subunit protein of microtubules (MT), is a heterodimer composed of two similar but nonidentical subunits named α- and β-tubulin. These represent the vast majority of tubulin in cells, but other members of the tubulin family are known as well [10]. These forms are much less abundant than the αβ heterodimer, and though important biologically [11,12] are not known to be important to mitochondrial function and will not be further considered here. Here "tubulin" will mean the αβ-tubulin dimer. Both α- and β-tubulin are acidic proteins of molecular weight ~50 kDa that bind guanine nucleotides and polymerize into MT in a GTP-dependent manner [13], and both are highly conserved in sequence throughout evolution. While many species have only one α- and one β-tubulin gene, many others have multiple genes for one or for both subunits [14]. These isoforms may differ from each other at only a few, or at multiple residues. These differences are found largely, but not only, in the carboxyl terminus of the subunits (see below).

FtsZ is a monomeric, ~40 kDa, guanine nucleotide binding protein found in all bacteria and archaea (and in some eukaryotes), which polymerizes in a GTP-dependent manner [15-19]. FtsZ polymers are single protein thick filaments whose collective action is essential for cell division in bactereia and archaea and to organelle division (chloroplasts and mitochondria) in eukaryotes that contain it. Despite limited sequence homology, the 3D structure of FtsZ [20] has high structural similarity to tubulin that, combined with the properties just listed, strongly indicates that FtsZ is a prokaryotic antecedent of tubulin [21].

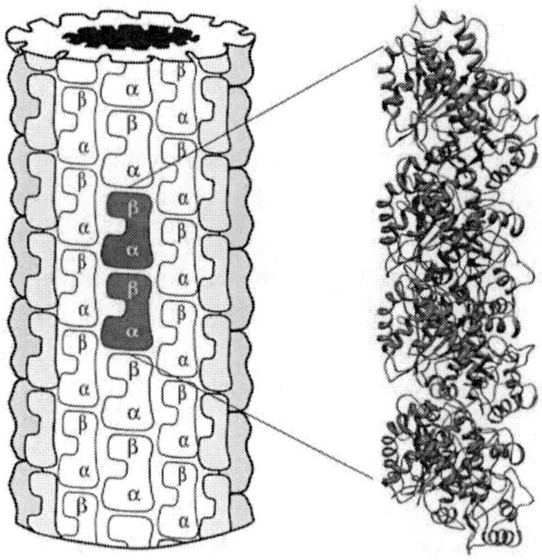

Figure 1. Microtubule and tubulin dimer structure.

Microtubules are cylindrical polymers of tubulin. The tubulin αβ-heterodimers polymerize in a head-to-tail manner, forming 'protofilaments', which associate laterally to form a closed cylinder. In cells, microtubules normally contain 13 protofilaments. Note that α- and β-tubulins are essentially the same size and that their folding patterns are extremely similar. The expanded view of one protofilament shows two dimers from the same perspective as the microtubule, so the outside surface is viewed. Figure is modified from reference [145].

The native structures of α- and β-tubulin are very similar to each other as well as to γ-tubulin and FtsZ [20-25]. All consist of three domains – 2 globular domains comprising 90+% of the protein, and an unstructured C-terminal tail. The globular domains consist of an N-terminal domain that binds GTP or GDP and represents about ½ of the protein, and a slightly smaller C-terminal globular domain to which the unstructured tail is grafted. All four proteins have binding regions that facilitate longitudinal assembly of the protein into filaments consisting of head-to-tail, single file alignments of the subunit protein. Tubulin has additional binding sites that allow the filaments to join together laterally to form sheets of filaments (under some conditions) or (usually) closed cylinders of filaments now called "protofilaments" (see Figure 1). These cylindrical polymers are of course microtubules, and usually consist of 13 protofilaments in cells. FtsZ lacks these lateral binding sites and hence forms only single protein thick filaments [17].

The sequence of the globular domains of tubulin are very conserved evolutionarily [23], as are the intron positions in tubulin genes [27], and the calculated physical properties of the protein [28]. A clear example of the conservation of function is the demonstration that a chimeric β-tubulin consisting of half chicken and half yeast sequences was expressed and incorporated normally into the microtubules of mouse cells in culture [29]. The N-terminal GTP/GDP binding domain has most sequence identity through the tubulins and FtsZ [19,21,26]. Though the C-terminal globular domain retains high structural similarity between the tubulins and FtsZ, there is much lower sequence similarity in this region [21,26].

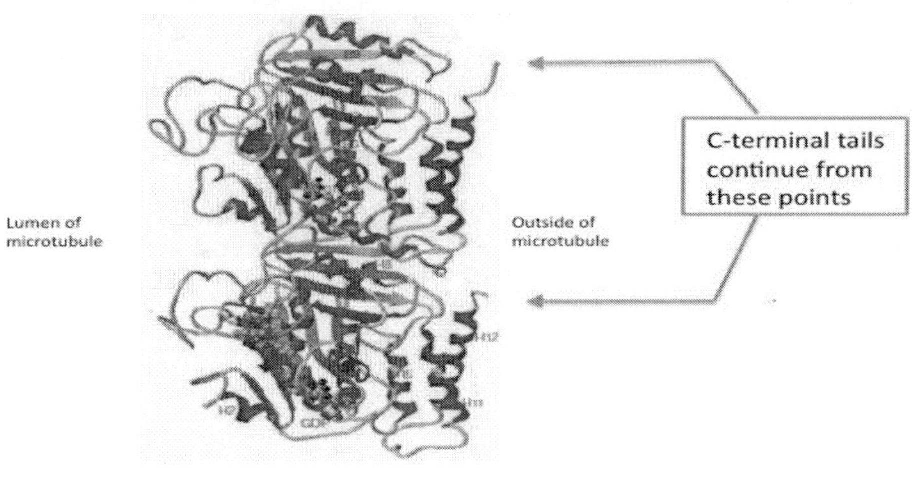

Figure 2. Tubulin crystallographic structure.

Ribbon diagrams of the structure of the tubulin dimer are shown based on electron crystallography of tubulin sheet polymers (2-D crystals). The view is rotated 90^{0} around the long protofilament axis relative to Figure 1, so the outside surface of the microtubule now faces to the right. The top subunit is α. The C-terminal tails (CTT) are not resolved, but extend from the points indicated, namely the last residues that were resolved in the structure. These residues also serve as the points of alignment of the sequences in Figure 4. The figure is modified from reference 22.

The sequence conservation of α- and β-tubulins has made them an often used probe of evolutionary relationships between organisms [30]. Tubulin sequences have been used in identifying particular species, e.g. as discriminating markers of Leishmania parasite species [31], as well as revealing relationships between groups of organisms, such as providing evidence for the fungal origin of microsporidia [32-35]. Comparisons of tubulin sequences have also revealed functional constraints on particular isotypes, as in a slowly changing testis-specific β-tubulin of flies [36,37]. Genetic mechanisms in evolution have also been probed by tubulin sequence comparisons, including eukaryote-to-eukaryote lateral gene transfer [38], and restructuring of gene arrays in Leishmania [39,40]. Tubulin sequence has also provided a window into basal organisms such as Trichomonas [41], Jakobid flagellates [42], and the early stages of eukaryote evolution [43]. Quite interesting in this regard is the discovery of αβtubulin in the bacterial genus Prosthecobacter [44], which will be discussed further in a section below.

Tubulin's Carboxy-Terminal Tails

Both α- and β-tubulin are globular proteins of ~4.5 nm diameter, with an unfolded, highly anionic, C-terminal "tail" (CTT) of 10-15 residues. The CTT are not seen in electron or X-ray structures or in reconstructions from cryoelectron microscopy (see Figure 2), presumably due to their sequence heterogeneity and (especially) their flexibility. Nonetheless, these tails on tubulin were shown to be unfolded and exposed on the MT surface by sensitivity to proteases, especially to subtilisin [45-47] Tubulin with the CTT removed by subtilisin is hence called tubulin-S [48,49]. Computational studies have examined and confirmed the flexibility of the tails [50]. This is illustrated in Figure 3, which also shows the extent of the tails. If the tails were fully extended, they would protrude ~ 5.5 nm (for a tail of 15 residues) from the globular part of the protein, or equivalently, from the microtubule surface [51]. Of course, peptides are rarely fully extended in physiological ionic strength and the CTT show considerable flexibility while still extending from the body of the protein (Figure 3).

Table 1. Comparison of the formal charge on the Carboxy-terminal 15 residues of α-, β-, and γ-tubulin, and FtsZ

Protein	Mean Formal charge1	s.d.2	|s.d. / mean|
α-tubulin3	-7.3	1.1	0.15
β-tubulin3	-9.0	0.9	0.1
γ-tubulin3	-1.4	1.2	0.9
FtsZ with C-terminal motif 4	+0.3	2.9	9.7
FtsZ without C-terminal motif 4	+1.1	2.8	2.5

1. In all cases n=9
2. Standard deviation for n=9
3. All were from the same organisms: Homo (vertebrate), Drosophila (insect), Arabidopsis (plant), Zea (plant), Leishmania (trypanosome), Tetrahymena (ciliate), Schizosaccharomyces (yeast), Dictyostelium (amoeba), and Giardia (diplomonad).
4. Sequences from Figure 7 in [19].

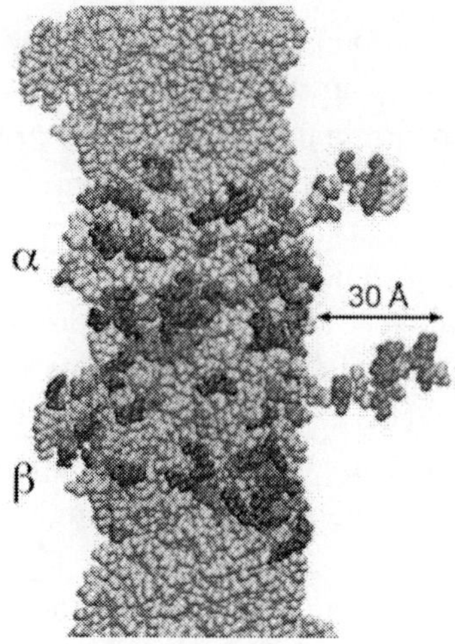

Figure 3. Tubulin dimer with C-terminal tails.

A tubulin protofilament is shown, with the dimers in the same orientation as in Figure 2. A single dimer is highlighted, with non-charged residues colored in light green for α, and steel blue for β. Anionic (acidic) residues (Glu, Asp) are colored red and cationic (basic) residues (Lys, Arg) are colored blue. The C-terminal tails are shown on both subunits. The configuration of the tails is a single snapshot from a molecular dynamics simulation; many other configurations are also found. Notice that the tails are not fully extended but still protrude considerably from the outside surface of the protein. Figure was provided by Dr. David Sept, University of Michigan.

The tails are strongly anionic. While the CTT represent only about 3% of the total mass of the protein, they contain about 40% of the total charge [45,51]. The anionic CTT is not found in the other (less abundant) members of the tubulin family, nor in the likely bacterial ancestral protein FtsZ, suggesting a function limited to the abundant αβ-dimer. Other tubulins (e.g. γ-tubulin, which is involved in MT nucleation) and FtsZ often have unfolded tails, but they differ from the CTT of αβ-tubulin. Tubulin tails are conserved in length (~ 15 residues average; α = 10, β = 20) and charge (vide infra), while the tails of other tubulins, such as γ-tubulin, are not highly charged, and the tails of FtsZ are quite variable in length and charge, but usually longer than tubulin (e.g. FtsZ of E. coli has a tail of 39 residues [19]. The charge differences are illustrated in Table 1, which shows the average charge for the terminal 15 residues of α-, β-, and γ-tubulin, as well as for FtsZ from bacteria and archaea. The FtsZ sequences are separated into those with an FtsZ C-terminal motif and those lacking it [19]. While the terminal 15 residues is a somewhat arbitrary number, this is chosen in order to facilitate comparison with the CTT of αβ-tubulin. It is clear that the CTT of α- or β-tubulin are more anionic than the others, and that they are more homogeneous, since the standard deviation is much lower than the mean for these, but equal to the mean or much greater for γ-tubulin and FtsZ, respectively.

The tails are not directly involved in the formation of interdimer surfaces during polymerization, consistent with their exposure on the surface of the MT. In fact, tubulin-S polymerizes more readily (with a lower critical concentration) than does intact tubulin [45,46,48]. The tail of FtsZ also is not required for polymerization, though it appears to be important for proper cellular localization [52]. Unlike the case of tubulin, the FtsZ tail is quite variable in length, and is not anionic in character. In fact the FtsZ tail in some species is cationic, and often contains a near-terminal basic / amide region [19].

CTT function is essential, since genetic deletion of the tails is lethal [53]. While the tails may be interchangeable in some circumstances, other examples are known in which particular tails are required: particular CTT have been shown to be required for axoneme assembly and function [54-56]. These experiments also reveal important intra-molecular interactions between CTT and the body of the protein, since a chimera of the appropriate tail grafted onto the globular body from a different isotype did not substitute for the natural form [54]. Among the known functions of the CTT relevant to axoneme function is the regulation of binding of motor proteins such as kinesin and dynein [57-59], though it is not clear if intra-molecular interactions alter this.

While always highly acidic, the exact sequence of the CTT varies between isotypes of tubulin and between species more than any other part of the protein. Isotypes of α- and β-tubulin are expressed in multicellular organisms in a tissue- and developmental stage-specific manner [14]. Clear examples of tubulin isotypes whose expression is tailored to one or a few cell types are presented by testis-specific β-tubulin in flies [36,37] and platelet-specific β-tubulin in mammals [60]. In single cell organisms, isotypes may be expressed in a life-cycle stage-specific manner [40, 61].

A result of tissue-specific expression of different isotypes is the expression of different arrays of tubulin tail sequences in biologically distinct cells. This might support tissue- and developmental- differences based on different groups of motor proteins being recruited to the MT arrays, though this has not been shown. A counterexample to this suggestion is found in the moss Physcomitrella, whose multiple β-isotypes are differentially expressed in different cell types, yet the isotypes show minimal differences in the CTT [62].

CTT structure can vary due to posttranslational modification (PTM) as well as to sequence differences. There are many known PTMs of tubulin, including acetylation, detyrosylation, phosphorylation, polyglutamylation, and polyglycylation [63,64]. All are reversible, all but acetylation and detyrosylation (which occur on α-tubulin) occur on both subunits, and all but acetylation occur on the CTT. It is also notable that all occur on the MT, and are reversed on the dimer: they are not only post-translational modifications but also post-assembly modifications. All but acetylation are exposed on the outer surface of the MT since they occur on the CTT. PTMs have been shown to regulate binding of many proteins to the MT surface, including the motor protein kinesin [65]. The complete list of functions of tubulin PTMs is not clear, but these functions seem likely to be significant, since the PTMs are conserved throughout the eukaryotes and many are unique to tubulin or nearly so. Aspects of the CTT evolution will be considered further in the section on coevolution of tubulin and mitochondria.

Microtubule- and Tubulin-Binding Proteins – Role of the CTT

Many proteins interact with the tubulin dimer and / or MT surfaces. MT-binding proteins are known collectively as Microtubule Associated Proteins, or MAPs [13, 65], often divided into "structural MAPs" and motor proteins. The structural MAPs, which bind to the MT surface and regulate MT stability, include such proteins as MAP-2 and tau [66]. The motor proteins [67,68] include kinesin [69] and dynein [70], which use ATP hydrolysis to move cargo along the MT surface. This cargo includes mitochondria, as discussed further below. The binding of both structural and motor MAPs to MT is either dependent on or regulated by tubulin's CTT [57,58, 71-73]. Some MT-destabilizing proteins also use the CTT as a binding site, such as the MT-cleaving protein spastin [74,75] and the depolymerizing kinesin, MCAK [76].

Other proteins do not so clearly fit this dichotomy, either because they bind to tubulin dimers as well as or instead of MT (e.g. the MT-destabilizing protein stathmin [77, 78]), or because their binding to MT is weaker than structural MAPs under dilute conditions of biochemical experiments. The latter group includes glycolytic enzymes and may be relevant to tubulin / MT-mitochondrial involvement in the energy metabolism of the cell.

A number of glycolytic enzymes have been found to bind to tubulin dimer and to MT [79]. These include aldolase, pyruvate kinase, lactate dehydrogenase(M), glyceraldehyde-3-phosphate dehydrogenase, and phosphofructokinase [80-83]. Enzyme binding is with tubulin CTT, on tubulin dimers or MT [85,86]. Enzyme binding may alter MT assembly and MT-MT association, as well as altering enzyme activity [81-83, 87,88]. Particularly relevant to tubulin-mitochondrial interaction, hexokinase may bind to tubulin and/or MT [89].

Table 2. Variation in Mitochondrial Properties

Mitochondria / cell a,b	
Human sperm	few
Hepatocyte	~1 k
Oocyte	~10 k
Yeast	~10
Amoeba (Chaos)	~500 k
Trypanosome (Plasmodium)	1 (but unusual c)
Mitochondrial genomes / cell a	
Hepatocyte	~10 k
Oocyte	10-100 k
Yeast	~50
Genome size, kbpa	
Human	~16.5 (other animals similar)
Plant (Arabidopsis)	367 (other plants 200-2400)
Trypanosome (Plasmodium)	6
Genes a	Proteins
Human	13
Plant (Arabidopsis)	31
Trypanosome (Plasmodium)	3

a. All data from refs [96-98]
b. Based on microscopic counts
b. See refs [99-101]
Membrane Tubulin.

Tubulin occurs as soluble dimers and as polymerized MT, but has also been long reported to occur in tight association with a number of cell membranes including plasma membranes, endocytic vesicles (i.e. clathrin coated vesicles) and other vesicles, and mitochondrial outer membranes [90,91]. Some of these associations appear to be directly with lipid, while others are mediated by a variety of membrane proteins, mostly unknown. The mitochondrial outer membrane (MOM) was among the first inner cell membranes reported to feature tightly associated tubulin. This binding was shown to be saturable, high affinity, reversible, and mediated by a protein on the MOM [92]. Subsequently tubulin has been shown to be in a tight complex with VDAC [93], the MOM voltage-dependent anion channel through which ADP and ATP move from the cytoplasm into or out of the mitochondria. We recently showed that this interaction was between the CTT of tubulin dimers and the channel of VDAC [94]. The binding is functionally significant since it causes reduction in channel conductance and reduces oxidative phosphorylation in functioning mitochondria i.e. raises the Km for ADP [94,95]. Biologically, the result is that binding of tubulin dimer to the MOM will regulate mitochondrial activity by limiting VDAC-mediated mitochondrial ADP-ATP exchange due to tubulin CTT blocking the VDAC channel. The CTT were shown to be the active part of tubulin since tubulin-S did not cause channel closure. The effects of this binding on channel function and mitochondrial metabolism are further developed in the chapters by Rostovtseva and by Saks, et al. We will return to this interaction below.

Mitochondrial Structure, Variation

Mitochondria originated by endosymbiosis of formerly free-living α-proteobacteria 1-2 x 109 years ago, followed by transformation from an endosymbiont to an organelle by massive gene transfer to the nucleus of the host cell and specialization of the two membranes surrounding the symbiont / organelle [1-9]. There remain much active research and vigorous discussion about the details of this, and the early events that followed, but it is clear that the process gave rise to the mitochondrial forms that we find throughout the eukaryotes today.

Mitochondria are quite variable in properties (see Table 2), and this merits a small discussion, since it contrasts markedly with the properties of tubulin that are relevant to the interaction with mitochondria. Our view of mitochondria (as seen in e.g. biochemistry or bioenergetics textbooks) is substantially driven by the extensive work on animal and yeast mitochondria [e.g. reference 102], and hence for many, "mitochondria" and "oxidative phosphorylation" are synonymous or nearly so. Yet many eukaryotes with mitochondria live in hypoxic or anoxic environments or in environments that alternate between aerobic and anaerobic [103-105]. These include animals such as some marine invertebrates, as well as unicellular inhabitants of marine or freshwater sediments, and symbionts or parasites living in the tissues or guts of other organisms. Anaerobic mitochondria produce ATP, but using a terminal electron acceptor (endogenous or environmental) other than oxygen [103]. In addition to classical aerobic mitochondria, anaerobic, and facultative anerobic mitochondria, other mitochondria-related organelles are known. Relict mitochondria, hydrogenosomes, and mitosomes are double membrane-bound organelles found in a number of anaerobic lineages from many different protistan groups [104-107].

The vast majority of eukaryotes, unicellular and multicellular are aerobic and contain "standard" mitochondria that perform oxidative phosphorylation. But these also exhibit a large range of properties (Table 2). Genome size ranges from 6 kb in trypanosomes to 2400 kb in some plants. The number of proteins coded ranges from 3 in trypanosomes to 97 in the freshwater protozoan Reclinomonas [108]. Morphologies also are variable. For example mitochondria may appear as multiple bacteria-size entities or one single reticulum extending throughout the cell, and may contain inner membrane cristae or not [96, 109, 110]. Thus mitochondrial structure, function, and genetics shows a huge range of different current end points deriving from the original symbiont.

Despite all the variation inside the mitochondria, a common feature of all mitochondrial forms is the double membrane that surrounds them, and the need to import and export materials through it. The inner membrane contains the machinery of ATP production, but most components of that system and all of its substrates and products only arrive or leave there by passing through the mitochondrial outer membrane (MOM).

The outer membrane is the only part of the mitochondria that the rest of the cell contacts, through proteins in the MOM. These notably include linker proteins that allow the mitochondria to anchor to and move on the MT (discussed in the next section), and channel proteins that allow proteins and metabolites to move into and out of the mitochondria. Thus these proteins render the MOM not merely a barrier but a site of cytoplasmic regulation of mitochondrial function [111, 112].

The main channel proteins of the MOM are TOM40, which allows movement of proteins, and VDAC, which allows movement of small molecules [113-115]. Both proteins are β-barrel proteins and likely evolved from the β-barrel proteins of the outer membrane of bacteria [116]. Both MOM proteins are found throughout eukaryotes, as expected for proteins mediating functions essential to mitochondrial activity. Since VDAC is the gateway for metabolite movement into and out of the mitochondria it is a major site of regulation of mitochondrial metabolism [111, 112]. The particular β-barrel structure of VDAC (discussed in greater detail in the chapter by Rostovetseva) is more conserved than the sequence; that is, the secondary and tertiary structure is more conserved than the primary sequence [115, 117]. This makes it difficult to trace the protein phylogenetically as is possible with tubulin, but nonetheless it is clear that VDAC is present in animals, plants, and fungi [86f]. The original isolation and description of VDAC was from Paramecium [118] and that plus the recent demonstration of VDAC in Trypanosoma brucei [119] extends VDAC very likely throughout all eukaryotes.

Mitochondria Interactions with Microtubules and Tubulin

Classical, oxidative mitochondria interact with both tubulin dimers and with MT, but in different ways. MT are important for mitochondrial localization, movement, fusion, and fission, while tubulin dimer interactions are with the MOM and regulate VDAC activity and hence oxidative ADP phosphorylation.

Mitochondria are dynamic both in position, moving through the cell, and in shape, undergoing fission and fusion whose balance maintains or alters the shape. Both of these dynamic aspects of mitochondria were described nearly a century ago [120,121], but has recently been the focus of renewed interest as movements have been tied to the cytoskeleton [122,123] and proteins involved in fission and fusion of the outer and the inner membranes have been identified [124, 125]. Mitochondria localize and move on the cytoskeleton. MT are the main cytoskeletal fiber in most (animal) cells studied [126, 127], but actin microfilaments play this role in Saccharomyces cerevisiae [128, 129], and in plant cells [123]. In the fission yeast Schizosaccharomyces pombe, MT appear to be the principal cytoskeletal anchor as in animal cells [130, 131]. Microtubules also appear to anchor mitochondria in the giant amoeba Reticulomyxa [122]. Mitochondrial anchoring and movement on MT or microfilaments is undescribed in most other eukaryote groups.

Movement along the cytoskeleton through mammalian cells depends on both MT (for longer movement) and actin microfilaments (which may assist MT over shorter distances). Movement in cells such as neurons or leukocytes serves to direct mitochondria to regions of high ATP demand [132, 133], though in cells such as adult cardiomyocytes or skeletal muscle cells, mitochondria are essentially immobile [132, 134]. Movement depends on linker or adapter proteins on the mitochondrial outer surface, and on motor proteins such as kinesin and dynein [135, 136].

While moving, mitochondria undergo extensive fission, driven by dynamin related proteins and others [137], and fusion driven by mitofusin proteins and others [125]. These processes allow mixture and exchange of mitochondrial matrix components [110], and also comprise a quality control mechanism that allows damaged mitochondria to be removed by selective autophagy [138, 139].

The net effect of mitochondrial anchoring and movement on MT is that MT interact with the outer surface of the mitochondria, but through adaptor proteins attached to motor proteins [136]. The MOM-MT interaction is not direct but involves a spacing due to these intermediate proteins. This means that it is unlikely that the CTT of tubulin, exposed on the outer surface of the MT, can reach or bind to the MOM. This is very different from the interaction of tubulin dimer with the MOM, which involves direct contact, and actual insertion of the tubulin CTT into the channel of the MOM protein VDAC.

As already discussed, tubulin dimers bind to the outer surface of the MOM by binding to VDAC. This was initially demonstrated by a direct binding study [92], which was motivated by the finding of specific colchicine binding to purified mitochondrial fractions, as well as to other membrane fractions [91]. Since colchicine binding is quite specific for tubulin, these earlier results indicated that tubulin was present in, and presumably could bind to, the MOM. This was confirmed by demonstration of simple saturable, high affinity, reversible binding [92]. Coimmunoprecipitation later demonstrated that the tubulin complex was with VDAC [93]. But these results did not indicate the biological result of tubulin binding to VDAC.

The functional result of tubulin-VDAC binding has recently been demonstrated by electrophysiological recordings of current through the reconstituted channel, as well as by ADP-dependent oxygen consumption by intact mitochondria [94, 95, 140]. Both methods demonstrate that tubulin binding restricts passage through VDAC, though some small ions are still able to pass. This blockade requires the tubulin CTT, which appears to insert into the channel. Similar results were found with VDAC from mammalian and fungal sources, and with tubulin from mammalian or protozoal (Leishmania) sources. This indicates that the tubulin-VDAC interaction is very widely found in nature, and suggests why certain features of the tubulin CTT appear conserved in evolution. It also suggests that tubulin is the highly conserved protein regulator of VDAC reported some time ago [141].

Table 3. Conservation of Tubulin CTT length and chargea

	Alpha (n = 86)	Beta (n = 132)
Charge	-7.2 1.4	-10.3 ± 1.5
Length (residues)	9.9 ± 1.6	19.2 ±2.8

a. Data from [94].

A. Alpha sequences

Species	Accession	Sequence
Eukaryotes with mitochondria		
Homo sapiens (Kα-1)	NM_006082	VGVDSV EGEGEEEGEEY
Gallus gallus	CAA30852	VGLDSY EDEEEGEE
" (a2)	XM_419249	VGTDSM DGEDEGEEY
" (a8)	NM_205444	VATDLF EDENEAGDS
Danio rerio	AAB84143	VGAESV EGEDEGEEY
Notophthalmus viridescens	Q91060	VGLDSY EGEEDEGEEY
Xenopus laevis	P08537	VGADSA DAEDEGEEY
Brugia malyi	XP_001899189	VGIDSL EGEGEGEEY
Drosophila melanogaster (a85E)	NM_079573	VGIDST TELGEDEEY
" (a67C)	NM_079285	VGLDNA EEGGDEDFDEF
Homarus americanus	Q94570	VGIDTA DGEDDEEANDY
Aplysia californican	Q8T6A5	VGVDSV DAEGEDEGEEY
Arabidopsis thaliana	NM_121983	VGAEGG DDEEDEGEDY
Triticum aestivum	P14640	VGAEFD EGEDGDEGDEY
Zea mays	P22275	VAAEGG SDDGDEEEEY
Leishmania tarentolae	ABC40566	VGAESA DDMGEEDVEEY
Toxoplasma gondii	P10873	VGIETA EGEGEEEGYGDEY
Blepharisma japonicum	Q08629	VGIETA EAEGEEEGYGEEL
Tetrahymena thermonphila	AAA21350	VGIETA EGEGEEEGY
Saccharomyces cerevisiae	NP_013625	VGADSY AEEEEF
Schizosaccharomyces pombe	CAA16866	VGGDSM DNEMYEADEEY
Neurospora crassa	P38668	VAGDYN DVDVDAEY
Schizophyllum commune	CAA60035	VGTDSA DAEEEGEY
Dictyostelium discoidium	P32255	VSASTE GEEQEEEY
Pelvetia fastigiata	Q40831	VGAETA DGDGEEEEFGEEY
Euglena gracilis	P33625	VGAESA DVEGEEDVEEY
Naeglaria gruberi	P11237	VGTESQ EGDGEEGEDGGDQ

A. (Continued)

Species	Accession	Sequence
Eukaryotes lacking mitochondria – near normal CTT		
Giardia lamblia	XP_001706843	IGAETL GDGEGEDMEEDDAY
Trichomonaas vaginalis	XP_001330666	VAAESV EGGDEEEDGGEM
Encephalitozoon intestinalis	AAN78301	IGAETL GDGEGEDMEEDDAY
Eukaryotes lacking mitochondria – truncated or decharged CTT		
Guillardia theta	Q9SCC8	VGSESQ DLISNSFF
Hemiselmis andersenii	XP_001712353	VGNENN EEDILEY
Encephalitozoon cuniculi	NP_586048	ISSNAE PVDEY
Entamoeba dispar	XP_001739532	LAIDNT VEGESMTAQ
Bacteria		
Prosthecobacter dejongeii (BTubA)	AAO12155	YQVAEE SGAKAKVQDSAGDTGMRAAAAGVSDDAR- -GSMSLRDLVDRRR
" (BTubB)	AAO12159	SYRDAS
Prosthecobacter vanneervenii (BTubA)	CAJ14012	YQVAEE SGAKAKVQDSSADYPRSSASSDDSR- -SGMSLRDLVDRRRA
" (BTubB)	CAJ14013	SYRDAS
Prosthecobacter debontii (BTubA)	AM041149	YQVAEE SGAKAKIQDVSGETVSRSSSMDDPR- -STMSLRDLVERRR
" (BTubB)	AM041149	SYRDAS

B. Beta sequences

Species	Accession	Sequence
Eukaryotes with mitochondria		
Homo sapiens	AAB59507	QQYQD ATADEQGEFEEEGEDEA
Gallus gallus (b2b)	NM_001004400	QQYQD ATADEQGEFEEEFEEDEA
" (b6)	NM_001031012	QQYQE ATANDGEEAFEDDEEEINE
Danio rerio	NP_001070241	QQYQE ATADDEADFGEEGE
Xenopus laevis	P13602	QQYQD ATADEQGEFEEEEDEA
Brugia malyi	XP_001896615	QQYQD ATADEEQDLQEGESEYIEQEE
Drosophila melanogaster (56D)	NP_523795	QQYQE ATADEDAEFEEEQEAEVDEN

Species	Accession	Sequence
" (97EF)	NP_65166	QQYQE ATADDEVEFDDEQAEQ
Homarus americanus	Q25009	QQYQE ATADDEAEFEEEGEVEGEYA
Aplysia californican	AAP13560	QQYQD ATAEDEGEFDEEEGDEGGEEYA
Arabidopsis thaliana	BAB10059	QQYQD ATADEEEGYEYEEDEVEVQEEQ
Triticum aestivum	AAD10492	QQYQD ATADEEEELYEDEDDADLQE
Zea mays	Q43695	QQYQD ATAEEYDEEEQDGEEEHD
Leishmania tarentolae	ABC40567	QQYQD ATVEEEGEYDEEEEPT
Toxoplasma gondii	P10878	QQYQD ATAEEEGEFDEEEGEMGAEEGA
Tetrahymena thermonphila	P41352	QQYQD ATAEEEGEFEEEEGEN
Saccharomyces cerevisiae	P02557	QQYQE ATVEDDEEVDENGDFGAPQNQDEPITENFE
Schizosaccharomyces pombe	P05219	QQYQE AGIDEGDEDYEIEEEKEPLEY
Neurospora crassa	XP_957669	QQYQD AQVDEEEEEYEEEAPLEGEE
Schizophyllum commune	CAA44972	QQYQD ATVEEEGEYEEEVIEDGE
Dictyostelium discoidium	P32256	QQYSN QETEEDGGEYQEEHEEHEEQAEN
Ectocarpus variabilis	AAA33284	QQYQD ATAEEEGEFDEDEELDDAMG
Euglena gracilis	P12457	QQYQD ATVEEEGEFDEEEDVEQY
Naeglaria gruberi	P34108	QQYQD ATAEEEGEFDENEGAEQEEQPADY
Eukaryotes lacking mitochondria – near normal CTT		
Entamoeba dispar	EDR29554	QQYYN TPLILDEGFEPEGELKTEVQQEETLPTNIESV
Giardia lamblia	XP_001707388	QQYQE AGVDEGEEFEEEEDFGDEYA
Trichomonaas vaginalis	L05468	EMYET AGVEEQGEGEEDEEGEAAA
Eukaryotes lacking mitochondria – truncated or decharged CTT		
Guillardia theta	XP_001713469	QQYQD AKMDNDAFEDQDLY
Hemiselmis andersenii	XP_001712325	QQYQE AKADELGYEDDDLY
Enterocytozoon bieneusi	AAB72136	QQYRD SQVGGY
Encephalitozoon cuniculi	CAD26226	QQYQD ATIEDAEEFLVN
Encephalitozoon intestinalis	AAN78301	QQYQD ATVEDAEEFLVN
Bacteria		

B. (Continued)

Species	Accession	Sequence
Prosthecobacter dejongeii (BTubA)	AAO12155	YQVAEE SGAKAKVQDSAGDTGMRAAAAGVSDDAR - -GSMSLRDLVDRRR
" (BTubB)	AAO12159	SYRDAS
Prosthecobacter vanneervenii (BTubA)	CAJ14012	YQVAEE SGAKAKVQDSSADYPRSSASSDDSR- -SGMSLRDLVDRRRA
" (BTubB)	CAJ14013	SYRDAS
Prosthecobacter debontii (BTubA)	AM041149	YQVAEE SGAKAKIQDVSGETVSRSSSMDDPR- -STMSLRDLVERRR
" (BTubB)	AM041149	SYRDAS

Figure 4. Sequence alignments of C-terminal tails.

Tails sequences from a number of organisms are aligned using the last resolved residue in Figure 2 as a reference point. Sequences for α-tubulin are shown in (A), and sequences for β-tubulin are shown in (B). Anionic residues (Glu, Asp) are highlighted in red and cationic residues (Lys, Arg) are highlighted in blue. Organisms are grouped by characteristics of mitochondria and CTT character: Eukaryotes with mitochondria, all of which have normal CTT (defined in Table 3), Eukaryotes lacking mitochondria, which still have normal CTT, Eukaryotes lacking mitochondria, which have truncated or decharged CTT, and bacteria (Prosthecobacter), which have highly variable CTT. Data are from [94].

Coevolution of Tubulin CTT and Mitochondria

The sequences of the tubulin CTT vary considerably between species, much more than the rest of the protein, which is highly conserved. However, the number of residues and the number of charges in each tail are highly conserved in all species for which sequences are available. This can be seen by examining the alignments of CTT sequences in Figure 4, and the summary statistics in Table 3. Alignment of the tail sequences shows that the exact position of the charges can vary, while the total charge is almost unchanged. Similarly the extent of conservation of length is remarkable. This is particularly true if compared to FtsZ. The tails of FtsZ not only vary significantly in charge (Table 1), but also in length. The sequences aligned in reference 19 exhibit tail lengths from ~25 to >200 residues in length. Tubulin CTT by contrast are about 10 residues for α- and 20 residues for β- with a fairly narrow distribution (standard deviation / mean of ~15%). There are exceptions to this, however.

Species that have α- or β-tubulin sequences with shortened or less charged tails fall into two categories. One small group is species with multiple tubulin genes, one isotype of which has a shortened or decharged tail. In this case the other tubulins can presumably perform the functions requiring a charged CTT, such as regulating mitochondrial function through control of VDAC permeability. An example of this may be the human β-VII tubulin (TUBB4Q) [142]. This is the only one of the 9 human β-tubulins that differs from the results in Table 3 [143]. When aligned as in Figure 4, the CTT of TUBB4Q is only 8 residues long and has only one negative charge; its biology is unknown.

The other, and perhaps more interesting group with altered tails consists of species that lack mitochondria. In these cases, VDAC regulation is not required since normal mitochondrial function is not extant. It is important to note that not all eukaryotes that lack mitochondria have shortened or decharged CTT. However, in organisms that lack multiple isotypes of tubulin, tubulins with shortened or decharged CTT occur only in organisms that lack mitochondria. A final case is provided by bacteria of the genus Prosthecobacter. Some species contain genes for α- and β-tubuiln [36]. These organisms likely acquired tubulin genes through horizontal gene transfer from an unknown eukaryote [144]. The structure of the globular part of the proteins is very similar to that of mammalian tubulin [144] but the CTT is quite different, showing great variation in length and charge, including addition of many basic residues and acquisition of overall cationic character (see Figure 4). These bacterial species of course have no mitochondria and the CTT of these tubulins appear unrestrained in length or charge.

The picture that emerges is of a long evolutionary linkage of tubulin tail sequences and mitochondrial outer membranes. The interaction between the tubulin CTT and VDAC in the MOM may constitute the function for which selection has maintained the CTT characteristics of length and charge. Much of the early history is no longer available, but development of the tubulin CTT may have been part of the "domestication" of the endosymbiont that became the mitochondria. Tubulin, and its charged tails, continues to do so today by regulation of the flow through the VDAC gateway into the mitochondria.

Conclusion

The evolutionary transition from prokaryotes to eukaryotes involved (among many other things) transition from a free-living α-proteobacteria to an endosymbiont to a mitochondrial organelle, integrated into the energy metabolism of the newly constituted eukaryotic cell. Concurrently, tubulin evolved from an FtsZ-like precursor, which involved developing lateral binding sites and anionic CTTs. The lateral binding sites were required to allow tubulin to form MT rather than the single protomer-wide filaments formed by FtsZ. The anionic CTT was required to allow (among other things) tubulin dimer to bind to VDAC and thereby regulate the metabolite flux across the mitochondrial outer membrane. It may be that this function has served as a selective constraint on tubulin's CTT in the long period since the evolution of mitochondria-containing eukaryotic cells.

Acknowledgments

The author wishes to thank Dr. David Sept, University of Michigan, for Figure 3 and for many stimulating discussions. This work was supported by Intramural Research Funds of the

Eunice Kennedy Shriver National Institute of Child Health and Human Development, National Institutes of Health, Bethesda, MD, 20892, USA.

References

[1] Margulis, L., Dolan, M. F. & Guerrero, R. (2000). The chimeric eukaryote: origin of the nucleus from the karyomastigont in amitochondriate protists. *Proc Natl Acad Sci, U.S.A.* 97(13), 6954-9.

[2] Baldauf, S. L. (2003). The deep roots of eukaryotes. *Science.*, 300(5626), 1703-6.

[3] Embley, T. M. & Martin, W. (2006). Eukaryotic evolution, changes and challenges. *Nature.*, 440(7084), 623-30.

[4] Margulis, L., Chapman, M., Guerrero, R. & Hall, J. (2006). The last eukaryotic common ancestor (LECA), acquisition of cytoskeletal motility from aerotolerant spirochetes in the Proterozoic Eon. *Proc Natl Acad Sci, U.S.A.*, 103(35), 13080-5.

[5] de Duve, C. (2007). The origin of eukaryotes: a reappraisal. *Nat Rev Genet.*, 8(5), 395-403.

[6] Cox, C. J., Foster, P. G., Hirt, R. P., Harris, S. R. & Embley, T. M. (2008). The archaebacterial origin of eukaryotes. *Proc Natl Acad Sci, U.S.A.* 105(51), 20356-61.

[7] Cavalier-Smith, T. (2009). Predation and eukaryote cell origins: a coevolutionary perspective. *Int J Biochem Cell Biol.*, 41(2), 307-22.

[8] Davidov, Y. & Jurkevitch, E. (2009). Predation between prokaryotes and the origin of eukaryotes. *Bioessays.*, 31(7), 748-57.

[9] Gabaldón, T. & Huynen, M. A. (2007). From endosymbiont to host-controlled organelle: the hijacking of mitochondrial protein synthesis and metabolism. *PLoS Comput Biol.*, 3(11), e219.

[10] Dutcher, S. K. (2001). The tubulin fraternity: alpha to eta. Curr Opin Cell Biol., 13(1), 49-54.

[11] Gull, K. (2001). Protist tubulins: new arrivals, evolutionary relationships and insights to cytoskeletal function. *Curr Opin Microbiol.*, 4(4), 427-32.

[12] Dutcher, S. K. (2003). Long-lost relatives reappear: identification of new members of the tubulin superfamily. *Curr Opin Microbiol.*, 6(6), 634-40.

[13] Wade, R. H. (2009). On and around microtubules: an overview. Mol Biotechnol., 43(2), 177-91.

[14] Ludueña, R. F. (1998). Multiple forms of tubulin: different gene products and covalent modifications. *Int Rev Cytol.*, 178, 207-75.

[15] Erickson, H. P. (1995). FtsZ, a prokaryotic homolog of tubulin? *Cell.*, 80(3), 367-70.

[16] Mukherjee, A. & Lutkenhaus, J. (1994). Guanine nucleotide-dependent assembly of FtsZ into filaments. *J Bacteriol.*, 176(9), 2754-8.

[17] Erickson, H. P. (2007). Evolution of the cytoskeleton. *Bioessays.*, 29(7), 668-77.

[18] Löwe, J. & Amos, L. A. (2009). Evolution of cytomotive filaments: the cytoskeleton from prokaryotes to eukaryotes. *Int J Biochem Cell Biol.*, 41(2), 323-9.

[19] Vaughan, S., Wickstead, B., Gull, K. & Addinall, S. G. (2004). Molecular evolution of FtsZ protein sequences encoded within the genomes of archaea, bacteria, and eukaryota. *J Mol Evol.*, 58(1), 19-29.

[20] Löwe, J. & Amos, L. A. (1998). Crystal structure of the bacterial cell-division protein FtsZ. *Nature.*, 391(6663), 203-6.

[21] Nogales, E., Downing, K. H., Amos, L. A. & Löwe, J. (1998). Tubulin and FtsZ form a distinct family of GTPases. *Nat Struct Biol.*, 5(6), 451-8.

[22] Nogales, E., Wolf, S. G. & Downing, K. H. (1998). Structure of the alpha beta tubulin dimer by electron crystallography. *Nature.*, 391(6663), 199-203.

[23] Löwe, J., Li, H., Downing, K. H. & Nogales, E. (2001). Refined structure of alpha beta-tubulin at 3.5 A resolution. *J Mol Biol.*, 313(5), 1045-57.

[24] Gigant, B., Curmi, P. A., Martin-Barbey, C., Charbaut, E., Lachkar, S., Lebeau, L., Siavoshian, S., Sobel, A. & Knossow, M. (2000). The 4 A X-ray structure of a tubulin:stathmin-like domain complex. *Cell.*, 102(6), 809-16.

[25] Aldaz, H., Rice, L. M., Stearns, T. & Agard, D. A. (2005). Insights into microtubule nucleation from the crystal structure of human gamma-tubulin. *Nature.*, 435(7041), 523-7.

[26] Wade, R. H., Garcia-Saez, I. & Kozielski, F. (2009). Structural variations in protein superfamilies: actin and tubulin. *Mol Biotechnol.*, 42(1), 49-60.

[27] Perumal, B. S., Sakharkar, K. R., Chow, V. T., Pandjassarame, K. & Sakharkar, M. K. (2005). Intron position conservation across eukaryotic lineages in tubulin genes. *Front Biosci.*, 10, 2412-9.

[28] Tuszynski. J. A., Carpenter, E. J., Huzil, J. T., Malinski, W., Luchko, T. & Luduena, R. F. (2006). The evolution of the structure of tubulin and its potential consequences for the role and function of microtubules in cells and embryos. *Int J Dev Biol.*, 50(2-3), 341-58.

[29] Bond, J. F., Fridovich-Keil, J. L., Pillus, L., Mulligan, R. C. & Solomon, F. (1986). A chicken-yeast chimeric beta-tubulin protein is incorporated into mouse microtubules in vivo. *Cell.*, 44(3), 461-8.

[30] Huang, C. H., Lee, F. L. & Tai, C. J. (2009). The beta-tubulin gene as a molecular phylogenetic marker for classification and discrimination of the Saccharomyces sensu stricto complex. *Antonie Van Leeuwenhoek.*, 95(2), 135-42.

[31] Mendoza-León, A., Luis, L. & Martínez, C. (2002). The beta-tubulin gene region as a molecular marker to distinguish Leishmania parasites. *Methods Mol Biol.*, 179, 61-83.

[32] Schardl, C. L., Craven, K. D., Speakman, S., Stromberg, A., Lindstrom, A. & Yoshida, R. (2008). A novel test for host-symbiont codivergence indicates ancient origin of fungal endophytes in grasses. *Syst Biol.*, 57(3), 483-98.

[33] Keeling, P. J. (2003). Congruent evidence from alpha-tubulin and beta-tubulin gene phylogenies for a zygomycete origin of microsporidia. *Fungal Genet Biol.*, 38(3), 298-309.

[34] Lee, R. C., Williams, B. A., Brown, A. M., Adamson, M. L., Keeling, P. J. & Fast, N. M. (2008). Alpha- and beta-tubulin phylogenies support a close relationship between the microsporidia Brachiola algerae and Antonospora locustae. *J Eukaryot Microbiol.*, 55(5), 388-92.

[35] Keeling, P. J., Luker, M. A. & Palmer, J. D. (2000). Evidence from beta-tubulin phylogeny that microsporidia evolved from within the fungi. *Mol Biol Evol.*, 17(1), 23-31.

[36] Nielsen, M. G., Caserta, J. M., Kidd, S. J. & Phillips, C. M. (2006). Functional constraint underlies 60 million year stasis of Dipteran testis-specific beta-tubulin. *Evol Dev.*, 8(1), 23-9.

[37] Nielsen, M. G. & Raff, E. C. (2002). The best of all worlds or the best possible world? Developmental constraint in the evolution of beta-tubulin and the sperm tail axoneme. *Evol Dev.*, 4(4), 303-15.

[38] Simpson, A. G., Perley, T. A. & Lara, E. (2008). Lateral transfer of the gene for a widely used marker, alpha-tubulin, indicated by a multi-protein study of the phylogenetic position of Andalucia (Excavata). *Mol Phylogenet Evol.*, 47(1), 366-77.

[39] Jackson, A. P., Vaughan, S. & Gull, K. (2006). Evolution of tubulin gene arrays in Trypanosomatid parasites: genomic restructuring in Leishmania. *BMC Genomics.*, 7, 261.

[40] Jackson, A. P., Vaughan, S. & Gull, K. (2006). Comparative genomics and concerted evolution of beta-tubulin paralogs in Leishmania spp. *BMC Genomics.,* 7, 137.

[41] Noël, C., Gerbod, D., Fast, N. M., Wintjens, R., Delgado-Viscogliosi, P., Doolittle, W. F. & Viscogliosi, E. (2001). Tubulins in Trichomonas vaginalis: molecular characterization of alpha-tubulin genes, posttranslational modifications, and homology modeling of the tubulin dimer. *J Eukaryot Microbiol.*, 48(6), 647-54.

[42] Edgcomb, V. P., Roger, A. J., Simpson, A. G., Kysela, D. T. & Sogin, M. L. (2001). Evolutionary relationships among "jakobid" flagellates as indicated by alpha- and beta-tubulin phylogenies. *Mol Biol Evol.*, 18(4), 514-22.

[43] Moriya, S., Tanaka, K., Ohkuma, M., Sugano, S. & Kudo, T. (2001). Diversification of the microtubule system in the early stage of eukaryote evolution: elongation factor 1 alpha and alpha-tubulin protein phylogeny of termite symbiotic oxymonad and hypermastigote protists. *J Mol Evol.*, 52(1), 6-16.

[44] Jenkins, C., Samudrala, R., Anderson, I., Hedlund, B. P., Petroni, G., Michailova, N., Pinel, N., Overbeek, R., Rosati, G. & Staley, J. T. (2002). Genes for the cytoskeletal protein tubulin in the bacterial genus Prosthecobacter. *Proc Natl Acad Sci, U.S.A.*, 99(26), 17049-54.

[45] Sackett, D. L., Bhattacharyya, B. & Wolff, J. (1985). Tubulin subunit carboxyl termini determine polymerization efficiency. *J Biol Chem.*, 260(1), 43-5.

[46] Serrano, L., de la Torre, J., Maccioni, R. B. & Avila, J. (1984). Involvement of the carboxyl-terminal domain of tubulin in the regulation of its assembly. *Proc Natl Acad Sci, U.S.A.*, 81(19), 5989-93.

[47] Lobert, S. & Correia, J. J. (1992). Subtilisin cleavage of tubulin heterodimers and polymers. *Arch Biochem Biophys.*, 296(1), 152-60.

[48] Bhattacharyya, B., Sackett, D. L. & Wolff, J. (1985). Tubulin, hybrid dimers, and tubulin S. Stepwise charge reduction and polymerization. *J Biol Chem.*, 260(18), 10208-16.

[49] Knipling, L., Hwang, J. & Wolff, J. (1999). Preparation and properties of pure tubulin S. *Cell Motil Cytoskeleton.*, 43(1), 63-71. PubMed PMID: 10340704.

[50] Luchko, T., Huzil, J. T., Stepanova, M. & Tuszynski, J. (2008). Conformational analysis of the carboxy-terminal tails of human beta-tubulin isotypes. *Biophys J.*, 94(6), 1971-82.

[51] Sackett, D. L. (1995). Structure and function in the tubulin dimer and the role of the acidic carboxyl terminus. *Subcell Biochem.*, 24, 255-302.

[52] Shen, B. & Lutkenhaus, J. (2009). The conserved C-terminal tail of FtsZ is required for the septal localization and division inhibitory activity of MinC(C)/MinD. *Mol Microbiol.*, 72(2), 410-24.

[53] Duan, J. & Gorovsky, M. A. (2002). Both carboxy-terminal tails of alpha- and beta-tubulin are essential, but either one will suffice. *Curr Biol.*, 12(4), 313-6.

[54] Popodi, E. M., Hoyle, H. D., Turner, F. R. & Raff, E. C. (2008). Cooperativity between the beta-tubulin carboxy tail and the body of the molecule is required for microtubule function. *Cell Motil Cytoskeleton.*, 65(12), 955-63.

[55] Popodi, E. M., Hoyle, H. D., Turner, F. R. & Raff, E. C. (2005). The proximal region of the beta-tubulin C-terminal tail is sufficient for axoneme assembly. *Cell Motil Cytoskeleton.*, 62(1), 48-64.

[56] Nielsen, M. G., Turner, F. R., Hutchens, J. A. & Raff, E. C. (2001). Axoneme-specific beta-tubulin specialization: a conserved C-terminal motif specifies the central pair. *Curr Biol.,* 11(7), 529-33.

[57] Wang, Z. & Sheetz, M. P. (2000). The C-terminus of tubulin increases cytoplasmic dynein and kinesin processivity. *Biophys J.*, 78(4), 1955-64.

[58] Lakämper, S. & Meyhöfer, E. (2006). Back on track - on the role of the microtubule for kinesin motility and cellular function. *J Muscle Res Cell Motil.,* 27(2), 161-71.

[59] Skiniotis, G., Cochran, J. C., Müller, J., Mandelkow, E., Gilbert, S. P. & Hoenger, A. (2004). Modulation of kinesin binding by the C-termini of tubulin. *EMBO J.*, 23(5), 989-99.

[60] Schwer, H. D., Lecine, P., Tiwari, S., Italiano, J. E., Jr, Hartwig, J. H. & Shivdasani, R. A. (2001). A lineage-restricted and divergent beta-tubulin isoform is essential for the biogenesis, structure and function of blood platelets. *Curr Biol.,* 11(8), 579-86.

[61] Bellatin, J. A., Murray, A. S., Zhao, M. & McMaster, W. R. (2002). Leishmania mexicana: identification of genes that are preferentially expressed in amastigotes. *Exp Parasitol.*, 100(1), 44-53.

[62] Jost, W., Baur, A., Nick, P., Reski, R. & Gorr, G. (2004). A large plant beta-tubulin family with minimal C-terminal variation but differences in expression. *Gene*, 340(1), 151-60.

[63] Hammond, J. W., Cai, D. & Verhey, K. J. (2008). Tubulin modifications and their cellular functions. *Curr Opin Cell Biol.*, 20(1), 71-6.

[64] Westermann, S. & Weber, K. (2003). Post-translational modifications regulate microtubule function. *Nat Rev Mol Cell Biol.*, 4(12), 938-47.

[65] Amos, L. A. & Schlieper, D. (2005). Microtubules and maps. *Adv Protein Chem.*, 71, 257-98.

[66] Dehmelt, L. & Halpain, S. (2005). The MAP2/Tau family of microtubule-associated proteins. *Genome Biol.*, 6(1), 204.

[67] Marx, A., Müller, J. & Mandelkow, E. (2005). The structure of microtubule motor proteins. *Adv Protein Chem.*, 71, 299-344.

[68] Mallik, R. & Gross, S. P. (2004). Molecular motors: strategies to get along. *Curr Biol.*, 14(22), R971-82.

[69] Hirokawa, N., Noda, Y., Tanaka, Y. & Niwa, S. (2009). Kinesin superfamily motor proteins and intracellular transport. *Nat Rev Mol Cell Biol.*, 10(10), 682-96.

[70] Wickstead, B. & Gull, K. (2007). Dyneins across eukaryotes: a comparative genomic analysis. *Traffic.*, 8(12), 1708-21.

[71] Serrano, L., Avila, J. & Maccioni, R. B. (1984). Controlled proteolysis of tubulin by subtilisin: localization of the site for MAP2 interaction. *Biochemistry.,* 23(20), 4675-81.

[72] Serrano, L., Montejo de Garcini, E., Hernández, M. A. & Avila, J. (1985). Localization of the tubulin binding site for tau protein. *Eur J Biochem.*, 153(3), 595-600.

[73] Littauer, U. Z., Giveon, D., Thierauf, M., Ginzburg, I. & Ponstingl, H. (1998). Common and distinct tubulin binding sites for microtubule-associated proteins. *Proc Natl Acad Sci, U.S.A.*, 83(19), 7162-6.

[74] Roll-Mecak, A. & Vale, R. D. (2008). Structural basis of microtubule severing by the hereditary spastic paraplegia protein spastin. *Nature.*, 451(7176), 363-7.

[75] White, S. R., Evans, K. J., Lary, J., Cole, J. L. & Lauring, B. (2007). Recognition of C-terminal amino acids in tubulin by pore loops in Spastin is important for microtubule severing. *J Cell Biol.*, 176(7), 995-1005. 5/16/125/16/12.

[76] Hertzer, K. M. & Walczak, C. E. (2008). The C-termini of tubulin and the specific geometry of tubulin substrates influence the depolymerization activity of MCAK. *Cell Cycle.*, 7(17), 2727-37.

[77] Steinmetz, M. O. (2006). Structure and thermodynamics of the tubulin-stathmin interaction. *J Struct Biol.,* 158(2), 137-47.

[78] Curmi, P. A., Gavet, O., Charbaut, E., Ozon, S., Lachkar-Colmerauer, S., Manceau, V., Siavoshian, S., Maucuer, A. & Sobel, A. (1999). Stathmin and its phosphoprotein family: general properties, biochemical and functional interaction with tubulin. *Cell Struct Funct.*, 24(5), 345-57.

[79] Knull, H. R. & Walsh, J. L. (1992). Association of glycolytic enzymes with the cytoskeleton. *Curr Top Cell Regul.*, 33, 15-30.

[80] Huitorel, P. & Pantaloni, D. (1985). Bundling of microtubules by glyceraldehyde-3-phosphate dehydrogenase and its modulation by ATP. *Eur J Biochem.*, 150(2), 265-9.

[81] Somers, M., Engelborghs, Y. & Baert, J. (1990). Analysis of the binding of glyceraldehyde-3-phosphate dehydrogenase to microtubules, the mechanism of bundle formation and the linkage effect. *Eur J Biochem.*, 193(2), 437-44.

[82] Lehotzky, A., Telegdi, M., Liliom, K. & Ovádi, J. (1993). Interaction of phosphofructokinase with tubulin and microtubules. Quantitative evaluation of the mutual effects. *J Biol Chem.*, 268(15), 10888-94.

[83] Durrieu, C., Bernier-Valentin, F. & Rousset, B. (1987). Microtubules bind glyceraldehyde 3-phosphate dehydrogenase and modulate its enzyme activity and quaternary structure. *Arch Biochem Biophys.*, 252(1), 32-40.

[84] Itin, C., Burki, Y., Certa, U. & Döbeli, H. (1993). Selective inhibition of Plasmodium falciparum aldolase by a tubulin derived peptide and identification of the binding site. *Mol Biochem Parasitol.,* 58(1), 135-43.

[85] Volker, K. W. & Knull, H. R. (1993). Glycolytic enzyme-tubulin interactions: role of tubulin carboxy terminals. *J Mol Recognit.*, 6(4), 167-77.

[86] Volker, K. W. & Knull, H. (1997). A glycolytic enzyme binding domain on tubulin. *Arch Biochem Biophys., 338*(2), 237-43.

[87] Volker, K. W., Reinitz, C. A. & Knull, H. R. (1995). Glycolytic enzymes and assembly of microtubule networks. *Comp Biochem Physiol B Biochem Mol Biol.,* 112(3), 503-14.

[88] Knull, H. & Minton, A. P. (1996). Structure within eukaryotic cytoplasm and its relationship to glycolytic metabolism. *Cell Biochem Funct.,* 14(4), 237-48.

[89] Wágner, G., Kovács, J., Löw, P., Orosz, F. & Ovádi, J. (2001). Tubulin and microtubule are potential targets for brain hexokinase binding. *FEBS Lett.* 509(1), 81-4.

[90] Stephens, R.E. (1986). Membrane tubulin. *Biol Cell.*, 57(2), 95-109.

[91] Wolff, J. (2009). Plasma membrane tubulin. *Biochim Biophys Acta.*, 1788(7), 1415-33.

[92] Bernier-Valentin, F. & Rousset, B. (1982). Interaction of tubulin with rat liver mitochondria. *J Biol Chem.,* 257(12), 7092-9.

[93] Carré, M., André, N., Carles, G., Borghi, H., Brichese, L., Briand, C. & Braguer, D. (2002). Tubulin is an inherent component of mitochondrial membranes that interacts with the voltage-dependent anion channel. *J Biol Chem.*, 277(37), 33664-9.

[94] Rostovtseva, T. K., Sheldon, K. L., Hassanzadeh, E., Monge, C., Saks, V., Bezrukov, S. M. & Sackett, D. L. (2008). Tubulin binding blocks mitochondrial voltage-dependent anion channel and regulates respiration. *Proc Natl Acad Sci, U.S.A.,* 105(48), 18746-51.

[95] Monge, C., Beraud, N., Kuznetsov, A. V., Rostovtseva, T., Sackett, D., Schlattner, U., Vendelin, M. & Saks, V. A. (2008). Regulation of respiration in brain mitochondria and synaptosomes: restrictions of ADP diffusion in situ, roles of tubulin, and mitochondrial creatine kinase. *Mol Cell Biochem.*, 318(1-2), 147-65.

[96] Scheffler, I. E. (2008). *Mitochondria* (2nd Ed.). Hoboken, John Wiley & Sons.

[97] Garcia-Rodriguez, L. J. (2007). Basic properties of mitochondria. *Meth. Cell. Biol.,* 80, 809-812.

[98] Burger, G. & Lang, B. F. (2003). Parallels in genome evolution in mitochondria and bacterial symbionts. *IUBMB Life.*, 55(4-5), 205-12.

[99] de Souza, W., Attias, M. & Rodrigues, J. C. (2009). Particularities of mitochondrial structure in parasitic protists (Apicomplexa and Kinetoplastida). *Int J Biochem Cell Biol.*, 41(10), 2069-80.

[100] Mather, M. W. & Vaidya, A. B. (2008). Mitochondria in malaria and related parasites: ancient, diverse and streamlined. *J Bioenerg Biomembr.,* 40(5), 425-33.

[101] Vaidya, A. B. & Mather, M. W. (2009). Mitochondrial evolution and functions in malaria parasites. *Annu Rev Microbiol.*, 63, 249-67.

[102] Lehninger, A. L. (1964). The Mitochondrion. NewYork. W.A. Benjamin, Inc.

[103] Tielens, A. G., Rotte, C., van Hellemond, J. J. & Martin, W. (2002). Mitochondria as we don't know them. *Trends Biochem Sci.*, 27(11), 564-72.

[104] Barbera, M. J., Ruiz-Trillo, I., Leigh, J., Hug, L. A. & Roger, A. J. (2007). The diversity of mitochondrion-related organelles amongst eukaryotic microbes. In: Origin of Mitochondria and Hydrogenosomes (eds. W.F. Martin and . Muller). Berlin, Springer-Verlag.

[105] Bryant, C. (ed.). (1991). Metazoan Life without Oxygen. New York. Chapman and Hall.

[106] van der Giezen, M. & Tovar, J. (2005). Degenerate mitochondria. *EMBO Rep.,* 6(6), 525-30.

[107] Howe, C. J. (2008). Cellular evolution: what's in a mitochondrion? *Curr Biol.*, 18(10), R429-31.

[108] Lang, B. F., Burger, G., O'Kelly, C. J., Cedergren, R., Golding, G. B., Lemieux, C., Sankoff, D., Turmel, M. & Gray, M. W. (1997). An ancestral mitochondrial DNA resembling a eubacterial genome in miniature. *Nature.*, 387(6632), 493-7.

[109] Mitra, K., Wunder, C., Roysam, B., Lin, G. & Lippincott-Schwartz, J. (2009). A hyperfused mitochondrial state achieved at G1-S regulates cyclin E buildup and entry into S phase. *Proc Natl Acad Sci, U.S.A.* Jul 21, 106(29)., 11960-5.

[110] Nunnari, J., Marshall, W. F., Straight, A., Murray, A., Sedat, J. W. & Walter, P. (1997). Mitochondrial transmission during mating in Saccharomyces cerevisiae is determined by mitochondrial fusion and fission and the intramitochondrial segregation of mitochondrial DNA. *Mol Biol Cell.*, 8(7), 1233-42.

[111] Colombini, M. (2004). VDAC: the channel at the interface between mitochondria and the cytosol. *Mol Cell Biochem.*, 256-257(1-2), 107-15.

[112] Lemasters, J. J. & Holmuhamedov, E. (2006). Voltage-dependent anion channel (VDAC) as mitochondrial governator--thinking outside the box. *Biochim Biophys Acta.*, 1762(2), 181-90.

[113] Blachly-Dyson, E. & Forte, M. (2001). VDAC channels. *IUBMB Life.*, 52(3-5), 113-8.

[114] Rostovtseva, T. K. & Bezrukov, S. M. (2008). VDAC regulation: role of cytosolic proteins and mitochondrial lipids. *J Bioenerg Biomembr.*, 40(3), 163-70.

[115] Mannella, C. A. & Kinnally, K. W. (2008). Reflections on VDAC as a voltage-gated channel and a mitochondrial regulator. *J Bioenerg Biomembr.*, 40(3), 149-55.

[116] Paschen, S. A., Waizenegger, T., Stan, T., Preuss, M., Cyrklaff, M., Hell, K., Rapaport, D. & Neupert, W. (2003). Evolutionary conservation of biogenesis of beta-barrel membrane proteins. *Nature.*, 426(6968), 862-6.

[117] Young, M. J., Bay, D. C., Hausner, G. & Court, D. A. (2007). The evolutionary history of mitochondrial porins. *BMC Evol Biol.*, 7, 31.

[118] Schein, S. J., Colombini, M. & Finkelstein, A. (1976). Reconstitution in planar lipid bilayers of a voltage-dependent anion-selective channel obtained from paramecium mitochondria. *J Membr Biol.*, 30(2), 99-120.

[119] Pusnik, M., Charrière, F., Mäser, P., Waller, R. F., Dagley, M. J., Lithgow, T. & Schneider. A. (2009). The single mitochondrial porin of Trypanosoma brucei is the main metabolite transporter in the outer mitochondrial membrane. *Mol Biol Evol., Mar*; 26(3), 671-80.

[120] Lewis, M.R., Lewis, W.H. (1914). Mitochondria (and other cytoplasmic structures) in tissue cultures. *Amer. J. Anatomy* 17: 339-401.

[121] Lewis, M.R., Lewis, W.H. (1914). Mitochondria in tissue culture. *Science* 39: 330-333.

[122] Bereiter-Hahn, J. & Vöth, M. (1994). Dynamics of mitochondria in living cells: shape changes, dislocations, fusion, and fission of mitochondria. *Microsc Res Tech.*, 27(3), 198-219.

[123] von Braun, S. S. & Schleiff, E. (2007). Movement of endosymbiotic organelles. *Curr Protein Pept Sci.*, 8(5), 426-38.

[124] Detmer, S. A. & Chan, D. C. (2007). Functions and dysfunctions of mitochondrial dynamics. *Nat Rev Mol Cell Biol.*, 8(11), 870-9.

[125] Okamoto, K. & Shaw, J. M. (2005). Mitochondrial morphology and dynamics in yeast and multicellular eukaryotes. *Annu Rev Genet.*, 39, 503-36.

[126] Heggeness, M. H., Simon, M. & Singer, S. J. (1978). Association of mitochondria with microtubules in cultured cells. *Proc Natl Acad Sci, U.S.A.*, 75(8), 3863-6.

[127] Chan, K. Y. & Bunt, A. H. (1978). An association between mitochondria and microtubules in synaptosomes and axon terminals of cerebral cortex. *J Neurocytol.*, 7(2), 137-43.

[128] Yang, H. C., Simon, V., Swayne, T. C. & Pon, L. (2001). Visualization of mitochondrial movement in yeast. *Methods Cell Biol.*, 65, 333-51.

[129] Drubin, D. G., Jones, H. D. & Wertman, K. F. (1993). Actin structure and function: roles in mitochondrial organization and morphogenesis in budding yeast and identification of the phalloidin-binding site. *Mol Biol Cell.*, 4(12), 1277-94.

[130] Yaffe, M. P., Harata, D., Verde, F., Eddison, M., Toda, T. & Nurse, P. (1996). Microtubules mediate mitochondrial distribution in fission yeast. *Proc Natl Acad Sci, U.S.A.*, 93(21), 11664-8.

[131] Yaffe, M. P., Stuurman, N. & Vale, R. D. (2003). Mitochondrial positioning in fission yeast is driven by association with dynamic microtubules and mitotic spindle poles. *Proc Natl Acad Sci U.S.A.*, 100(20), 11424-8.

[132] Kuznetsov, A. V., Hermann, M., Saks, V., Hengster, P. & Margreiter, R. (2009). The cell-type specificity of mitochondrial dynamics. *Int J Biochem Cell Biol.*, 10, 1928-39.

[133] Sánchez-Madrid, F. & Serrador, J. M. (2007). Mitochondrial redistribution: adding new players to the chemotaxis game. *Trends Immunol.,* 28(5), 193-6.

[134] Anesti, V. & Scorrano, L. (2006). The relationship between mitochondrial shape and function and the cytoskeleton. *Biochim Biophys Acta.,* 1757(5-6), 692-9.

[135] Hollenbeck, P. J. & Saxton, W. M. (2005). The axonal transport of mitochondria. *J Cell Sci.*, 118(Pt 23), 5411-9.

[136] Hollenbeck, P. J. (2005). Mitochondria and neurotransmission: evacuating the synapse. *Neuron.*, 4, 47(3), 331-3.

[137] Lackner, L. L. & Nunnari, J. M. (2009). The molecular mechanism and cellular functions of mitochondrial division. *Biochim Biophys Acta.*, 1792(12), 1138-44.

[138] Twig, G., Hyde, B. & Shirihai, O. S. (2008). Mitochondrial fusion, fission and autophagy as a quality control axis: the bioenergetic view. *Biochim Biophys Acta.,* 1777(9), 1092-7.

[139] Benard, G., Bellance, N., James, D., Parrone, P., Fernandez, H., Letellier, T. & Rossignol, R. (2007). Mitochondrial bioenergetics and structural network organization. *J Cell Sci.*, 120(Pt 5), 838-48.

[140] Saks, V. A., Kuznetsov, A. V., Khuchua, Z. A., Vasilyeva, E. V., Belikova, J. O., Kesvatera, T. & Tiivel, T. (1995). Control of cellular respiration in vivo by mitochondrial outer membrane and by creatine kinase. A new speculative hypothesis: possible involvement of mitochondrial-cytoskeleton interactions. *J Mol Cell Cardiol.*, 27(1), 625-45.

[141] Liu, M. Y. & Colombini, M. (1991). Voltage gating of the mitochondrial outer membrane channel VDAC is regulated by a very conserved protein. *Am J Physiol.*, 260(2 Pt 1), C371-4.

[142] van Geel, M., van Deutekom, J. C., van Staalduinen, A., Lemmers, R. J., Dickson, M. C., Hofker, M. H., Padberg, G. W., Hewitt, J. E., de Jong, P. J. & Frants, R. R. (2000). Identification of a novel beta-tubulin subfamily with one member (TUBB4Q) located near the telomere of chromosome region 4q35. *Cytogenet Cell Genet.*, 88(3-4), 316-21.

[143] Luchko, T., Huzil, J. T., Stepanova, M. & Tuszynski, J. (2008). Conformational analysis of the carboxy-terminal tails of human beta-tubulin isotypes. *Biophys J.*, 94(6), 1971-82.

[144] Schlieper, D., Oliva, M. A., Andreu, J. M. & Löwe, J. (2005). Structure of bacterial tubulin BtubA/B: evidence for horizontal gene transfer. *Proc Natl Acad Sci, U.S.A*. 102(26), 9170-5.

[145] Sackett, D. L. & Sept, D. (2009). Protein-protein interactions: Making drug design second nature. *Nature Chem*, 1(8), 596-597.

In: Tubulin: Structure, Functions and Roles in Disease
Editors: Wei Yamauchi and Antonio Sokić ISBN: 978-1-62100-141-6

Chapter VIII

Myelin Basic Protein Interactions with Actin and Tubulin *in vitro*: Binding, Assembly, and Regulation*

Joan M. Boggs†

Department of Molecular Structure and Function, Research Institute,
Hospital for Sick Children, 555 University Ave., Toronto, Ontario, Canada
and Department of Laboratory Medicine and Pathobiology,
University of Toronto, Toronto, Ontario, Canada

Abstract

The sole function of MBP has long been thought to be adhesion of the cytosolic surfaces of the oligodendrocyte membranes of central nervous system (CNS) myelin. However, MBP is known to bind to a number of proteins, including actin and tubulin, and can cause their polymerization and bundling *in vitro*. It can also tether actin filaments and bundles to a membrane surface. These interactions can be regulated by physiological post-translational modifications of MBP (primarily Ser/Thr phosphorylation and Arg deimination to citrulline), by Ca^{2+}-calmodulin binding to MBP, and by physiological changes in membrane surface potential. *In vivo*, both actin and tubulin are isolated together with MBP in a low density, detergent-insoluble, glycosphingolipid-enriched fraction of myelin, which also contains caveolin and kinases, and thus may be a membrane signaling domain. Some MBP is colocalized with actin *in vivo* at the membrane edges of cultured oligodendrocytes and with cytoskeletal veins of microfilaments and microtubules. In OLs from the shiverer mutant mouse, which lacks MBP, actin microfilaments do not form bundles and are not colocalized with

---

* A version of this chapter also appears in Myelin Basic Protein, edited by Joan M. Boggs, published by Nova Science Publishers, Inc. It was submitted for appropriate modifications in an effort to encourage wider dissemination of research.

† E-mail address: jmboggs@sickkids.ca; Tel: 1-416-813-5919; Fax: 1-416-813-5022; Address: Department of Molecular Structure and Function, Research Institute, Hospital for Sick Children, 555 University Ave., Toronto, Ontario, Canada M5G 1X8 and Department of Laboratory Medicine and Pathobiology, University of Toronto, Toronto, Ontario, Canada M5G 1L5.

microtubules. Moreover, microtubular structures are abnormal in size, and cell processes are usually smaller than normal with a larger cell body. In addition, extracellular signaling via glycosphingolipids to the cytoskeleton does not occur. These observations and changes in shiverer OLs suggest that interactions of MBP with actin microfilaments and microtubules occur also *in vivo* in OLs, and that these interactions are important for cell function, membrane process formation, and myelination.

Introduction

Myelin basic protein (MBP) was first shown to bind to other proteins such as actin, tubulin, calmodulin, and tropomyosin in the 1980s (Barylko and Dobrowolski, 1984; Dobrowolski et al., 1986a; Dobrowolski et al., 1986b; Modesti and Barra, 1986; Grand and Perry, 1980; Chan et al., 1990). MBP not only bound to actin and tubulin, but polymerized them into actin filaments and microtubules, and bundled the filaments and microtubules together into larger assemblies. The binding of Ca^{2+}-calmodulin to MBP dissociated it from actin and tubulin, and resulted in depolymerization of the filaments and microtubules. MBP was found to be one of two proteins in rat brain which could stabilize microtubules from depolymerizing in the cold, called STOP (Stable Tubule Only Polypeptide) activity (Pirollet et al., 1992) (see Chapter VII). These early intriguing observations were largely ignored until recently, because these proteins are negatively-charged, whereas MBP is highly positively-charged. It thus appeared at first glance that the interaction was merely electrostatic, and therefore non-specific and not of physiological consequence.

It is now known that polycationic proteins can interact with a number of polyanionic ligands, proteins and surfaces in functional ways (Jones et al., 2004). Some proteins, such as MARCKS (myristoylated alanine-rich C kinase substrate), gravin, K-Ras4B, and Src have an unstructured basic cluster which binds polyanionic ligands (Murray et al., 1997; McLaughlin et al., 2005), but also have other domains which are acidic. This clustered arrangement of amino acid types is different from MBP, whose basic and acidic residues are distributed throughout its sequence. The high net charge and low hydrophobicity of basic proteins such as MBP, microtubule-associated proteins (MAP) such as tau and MAP2, α-synuclein (Tompa, 2005), or proteins with a basic effector domain such as MARCKS (Harauz et al., 2000), maximize intramolecular electrostatic repulsion leading to an extended disordered structure (Hill et al., 2002; Harauz et al., 2004). Such proteins have sufficient flexibility to bind to various charged surfaces and ligands, and to acquire whatever local conformation is necessary to optimize binding to several different targets (Dyson and Wright, 2002; Fuxreiter et al., 2004). Moreover, they are often multifunctional regulatory proteins associated with signal transduction, adhesion, cell-cycle regulation, gene expression, alternative splicing of pre-mRNAs, or chaperone action (Benmerah et al., 2003; Tompa et al., 2005; Tompa, 2005; Pandey et al., 2004; Haynes and Iakoucheva, 2006). Disordered proteins with high surface charge which are involved in multiple interactions may integrate hubs of various biological activities (Uversky et al., 2005; Patil and Nakamura, 2006).

In this regard, MBP also interacts with several other proteins in addition to binding to negatively-charged proteins such as actin and tubulin. Its binding to calmodulin has been studied extensively, and three different domains of MBP may be involved (Libich et al., 2003a; Libich et al., 2003b; Polverini et al., 2004) (see Chapter IX). In addition, MBP has a domain that has been predicted to be a PXXP SH3-target consensus sequence (Moscarello, 1997), *viz.*, TPRTPPP (residues 92-98) (murine sequence). This domain has recently been shown to form a poly-proline helix and to bind SH3 domains of several proteins (Polverini et al., 2008; see Chapter IX). MBP has also been shown to bind to clathrin and catalyze its polymerization into artificial baskets that appear structurally similar to the baskets assembled by other assembly proteins (Prasad et al., 1995). Thus, MBP has also been suggested to be a multifunctional protein (Boggs, 2006; Dyer, 1997; Martenson, 1980; Pedraza et al., 1997). In this Chapter, we will focus on its interactions with actin and tubulin.

MBP-Mediated Interactions with Actin and Tubulin, and Effect of Post-Translational Modifications

MBP binds to G-actin in solution at an MBP/actin mole ratio of 1:2 and causes its polymerization into filaments under otherwise non-polymerizing, low ionic strength conditions (Dobrowolski et al., 1986a; Roth et al., 1993) (Figure 1A).

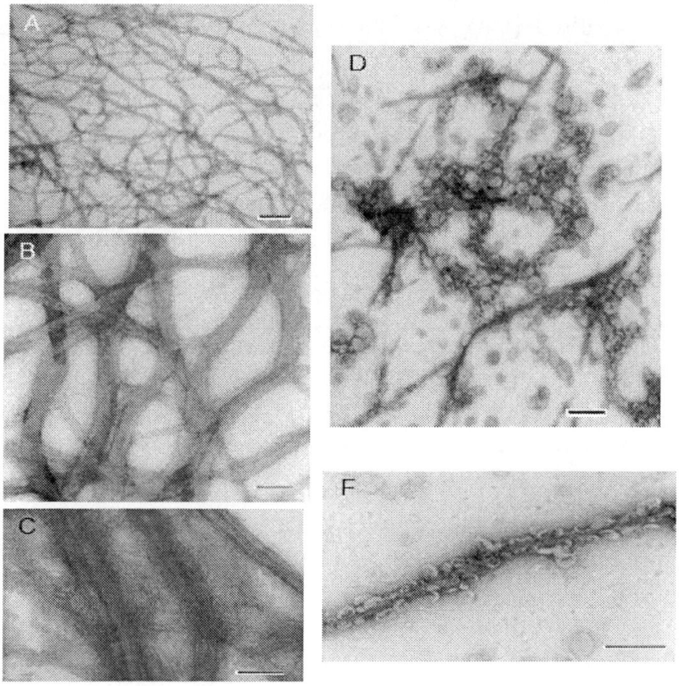

Figure 1. Electron micrographs of (A) F-actin, bar = 100 nm; (B) F-actin in the presence of equimolar MBP/C1 in F-buffer, bar =100 nm; (C) F-actin in the presence of equimolar P-MBP/C1 in F-buffer, bar = 100 nm. Strands of unbundled filaments next to bundled filaments can be seen in panel C, in contrast to panel B where virtually all actin filaments are bundled; (D) F-actin in the presence of equimolar MBP/C1 and negatively-charged phosphatidylcholine/phosphatidylglycerol 4:1 (m/m) large unilamellar lipid vesicles, bar = 500 nm; (F) as in (D) but higher magnification, bar = 500 nm. The negatively-charged lipid vesicles are bound to F-actin bundles, via MBP.

It also binds to F-actin in a 1:1 mole ratio and induces the formation of ordered bundles of F-actin filaments (Barylko and Dobrowolski, 1984) (Figure 1B). Calcium-activated calmodulin (Ca^{2+}CaM), dissociates MBP from actin bundles (Barylko and Dobrowolski, 1984) and from actin filaments resulting in their depolymerization (Dobrowolski et al., 1986a). MBP may interact with the cytoskeleton in OLs, in cytosolic inclusions in myelin, and even in compact myelin. Actin and tubulin are present in compact myelin where they may be associated with MBP and the radial component, a series of tight junctions, containing tight junction proteins claudin-11 (Gow et al., 1999) and zonula occludens-1 ZO-1 (Li et al., 2004), that pass through many layers of myelin (Karthigasan et al., 1994; Yoshikawa, 2001). Colocalization studies by immunohistochemistry in immature cultured OLs indicated that MBP was closely associated with microtubules and actin microfilaments (Wilson and Brophy, 1989; LoPresti et al., 1995; Muller et al., 1997) (see Chapter VII). MBP has been co-immunoprecipitated with tubulin from brain tissue (Taketomi et al., 2002).

MBP interacts with G-actin and F-actin through electrostatic interactions and the interaction is inhibited by an increase in salt concentration (Boggs et al., 2005; Boggs et al., 2006; Hill and Harauz, 2005). Thus, the interaction of MBP with actin might be expected to be affected by post-translational modifications which decrease the net positive charge of MBP. It is modified by a variety of such post-translational modifications *in vivo*, including deamidation, Ser/Thr phosphorylation (Chou et al., 1976; Zand et al., 1998), and deimination of Arginine (Arg) to Citrulline (Cit) (Wood and Moscarello, 1989), also termed citrullination (see Chapters II and III). It can be isolated from myelin as a series of charge components (or isomers), termed C1, C2, C3, C4, C5, C6, and C8 due to various combinations of these modifications. Component C1 is the least modified, most positively-charged form of the protein, and C8 comprises the least positively-charged isomers due to citrullination. Recombinant murine forms of C1 (rmC1) and C8 (rmC8) have been prepared; the latter contains Gln substituted for Cit at 6 sites of deiminated Arg (Bates et al., 2002). The dissociation constant of the less charged isomer rmC8 for actin was a little greater than that of rmC1, and rmC8 had somewhat less ability to polymerize actin and bundle F-actin filaments than rmC1 (Boggs et al., 2005). Moreover, rmC8 was more readily dissociated from actin by Ca^{2+}-calmodulin than rmC1. In contrast to the effect of deimination, phosphorylation of C1 *in vitro* at Thr 94 and Thr 97 (bovine sequence numbering) using MAPK (Erickson et al., 1990; Hirschberg et al., 2003) significantly decreased the ability of MBP to polymerize actin and to bundle actin filaments (Figure 1C), despite having no effect on binding to F-actin or on the ability of Ca^{2+}-CaM to dissociate the complex (Boggs et al., 2006). The rate and extent of actin polymerization induced by naturally occurring MBP charge isomers C1-C6 decreased with reduction in positive charge (Hill and Harauz, 2005).

The greater effect of deimination of MBP on its binding to actin compared to that of phosphorylation may be due to the fact that the 6 deiminated Arg residues were distributed throughout the sequence, whereas the 2 phosphorylated Thr residues were in a localized region. Earlier studies indicated that MBP has at least two actin-binding sites (Roth et al., 1993). Phosphorylation at specific sites may interfere with only one of these regions, whereas deimination, which is spread over the entire sequence, may interfere with both but in a less severe way. The comparable rates of polymerization induced by rmC8 and rmC1, despite the reduced net charge and hence electrostatic interactions of rmC8, suggest that hydrophobic and/or hydrogen-bonding interactions may also contribute to actin binding by MBP. Indeed, the polymerizing activities of the different size 14-21.5 kDa recombinant isoforms also did not depend on their net charges (Hill and Harauz, 2005). Phenylalanyl residues in the MARCKS effector domain were found to contribute to its actin-binding affinity and rate of F-actin nucleation (Wohnsland et al., 2000). A common actin-binding motif is an amphipathic helix with hydrophobic side chains on one side, which binds to a hydrophobic pocket in actin between actin subdomains 1 and 3 (Dominguez, 2004). The two sets of Phe-Phe pairs at positions 42-43 and 86-87 in MBP (murine 18.5-kDa sequence numbering) may contribute to its actin binding.

MBP was found to be one of only two proteins in rat brains to be able to protect microtubules from cold-induced depolymerization (Pirollet et al., 1992; Galliano et al., 2006). It inhibits tubulin carboxypeptidase activity (Modesti and Barra, 1986) and it polymerizes tubulin and bundles microtubules in a calmodulin-dependent way (Hill et al., 2005) (Figure 2).

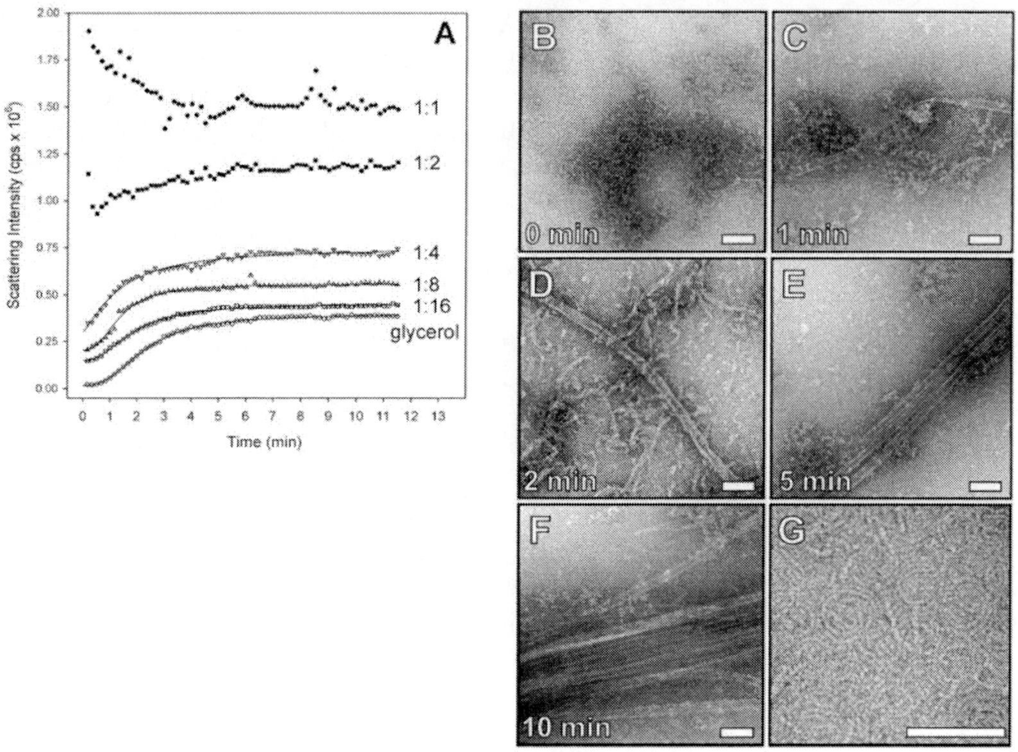

Figure 2. Assembly of tubulin by unmodified 18.5 kDa bovine MBP. (A) Time-dependence of increase in light scattering due to assembly of tubulin promoted by MBP at the molar ratios of MBP:tubulin indicated, or by 10% glycerol. (B-F) Electron micrographs of negatively-stained samples of tubulin after addition of MBP at a mole ratio to tubulin of 1:8, taken at the times indicated. Scale bars are 100 nm. A microtubule can be seen in panel D, and bundles of microtubules can be seen in panels E and F. (G) Tubulin ring structures observed 1 min after addition of MBP to tubulin at a 1:1 mole ratio. The scale bar is 50 nm. Reproduced from Hill et al. (2005) with permission from the American Chemical Society.

A reduction in positive charge due to deamidation in C2 decreased the ability of MBP to assemble tubulin, but further reductions in net positive charge in naturally occurring charge components C2-C6, enhanced the ability of MBP to assemble tubulin, in contrast to their effects on actin polymerization (Hill et al., 2005). The reduced-charge variant rmC8 had a similar ability to assemble tubulin as rmC1, also in contrast to its decreased interaction with actin. Thus, the interaction between MBP and tubulin seems less dependent on electrostatic effects than its interactions with actin. Site-specific phosphorylation of MBP may be the modification in these charge isomers which enhances the interaction with tubulin, as is the case with tau phosphorylated at certain sites (Feijoo et al., 2005). The rm21.5-, rm17.22- and rm18.5-kDa C1 had similar microtubule bundling and F-actin bundling activity, but that of rm14-kDa C1 was less (Hill and Harauz, 2005; Hill et al., 2005).

Cultured oligodendrocytes (OLs) produce large membrane sheets with large cytoskeletal veins and a lacy network of smaller veins, containing actin and tubulin (Dyer and Benjamins, 1989; Wilson and Brophy, 1989; Boggs and Wang, 2004). MBP is colocalized with these cytoskeletal veins. The cytoskeletal appearance in shiverer mouse OLs, which lack MBP, suggests that MBP interacts with actin and tubulin *in vivo*. In these mutant OLs, actin microfilaments did not form bundles (Dyer et al., 1995). The microtubular structures were also abnormal in size, and cell processes were usually smaller than normal with a larger cell body. In shiverer OLs transfected with the MBP gene, in which the average MBP expression was less than normal, a normal microtubule appearance and actin bundles were seen only in those membrane sheets where the MBP expression appeared normal. They were not seen in membrane sheets where the MBP expression was abnormally low (Dyer et al., 1997).

MBP-mediated Binding of Actin to Membranes and Effect of Post-Translational Modifications

MBP can tether actin filaments to the surface of negatively-charged lipid vesicles (Figure 1 D,F), suggesting that it may be able to act as a membrane actin-binding protein (Boggs and Rangaraj, 2000). In the case of the MARCKS effector domain, theoretical calculations showed that the electrostatic potential was quite positive above the peptide, even when associated with acidic lipids (Murray et al., 1999). A similar effect for MBP would allow it to interact with actin while simultaneously binding to the lipid negative surface charge. MBP was observed to be partially colocalized with actin at the edge of membrane sheets of OLs where it might link actin to the membrane (Boggs et al., 2006) (Figure 3).

Many actin-binding proteins insert into the lipid bilayer, allowing them to anchor actin filaments to the membrane and possibly to also sense signals transmitted through the membrane. This insertion into the lipid bilayer has been detected by labeling with hydrophobic photolabels such as 3-(trifluoromethyl)-3-(*m*-[^{125}I]iodophenyl)diazirine (TID) (Isenberg and Goldmann, 1995), which react only with groups in the acyl chain region of the bilayer. The spin label studies, described in Chapter VI, indicate that hydrophobic side chains of MBP insert into the lipid bilayer; MBP can also be labeled by TID when bound to lipid (Jo and Boggs, 1995; Boggs et al., 1999). Like many actin-binding proteins which are myristoylated at the N-terminus (Behrisch et al., 1995; Isenberg and Goldmann, 1992), MBP is also acylated at the N-terminus, although in MBP the chain length is heterogeneous and no longer than 10 carbons (Moscarello et al., 1992). Other similarities, including analogous structural motifs, between MBP and the actin-binding MARCKS protein have been pointed out by Harauz et al. (2000).

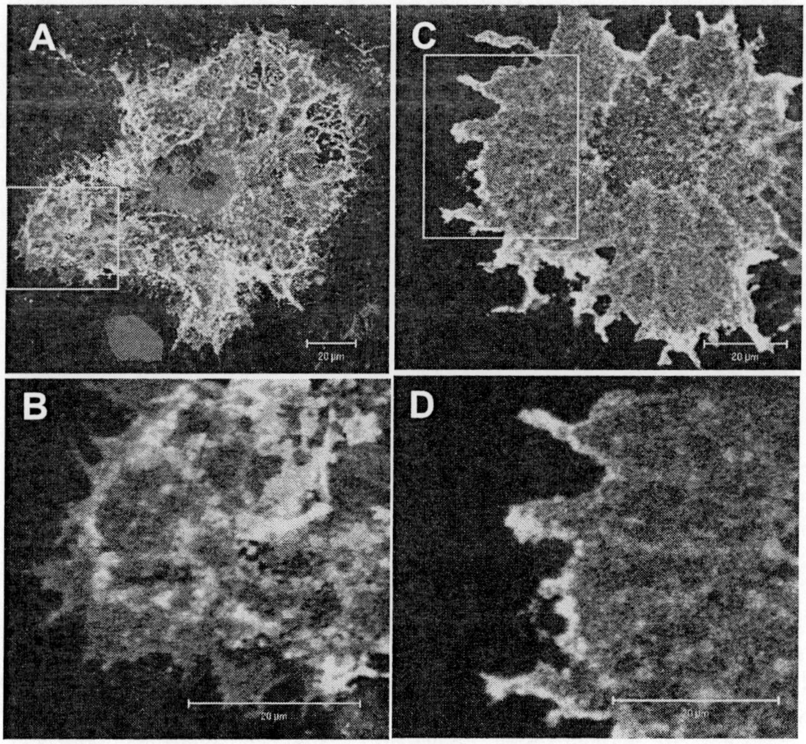

Figure 3. Confocal microscope images of cultured oligodendrocytes that have been fixed, permeabilized, and costained with rat anti-MBP Ab and Cy^{TM}2-conjugated-anti-rat IgG (green) and rhodamine-conjugated phalloidin (binds to F-actin; red). Merged images are shown. Panels B and D show enlarged images of the boxed sections marked in panels A and C, respectively. Colocalization of MBP and actin is seen at the edges of the membrane sheets as well as on cytoskeletal veins. In panel A, actin precedes MBP at the membrane edge, and in panel B, actin and MBP are both present together at the membrane edges and leading processes. The cell shown in panels C and D may be more mature than that in panels A and B. Scale bar = 20 μm in each case.

The ability of MBP to bind actin to the myelin or oligodendrocyte membrane may allow it to participate in signaling (Dyer et al., 1994). Insertion of hydrophobic side chains or its N-terminal acyl group into the bilayer would allow it to sense mechanical signals transmitted through the membrane, *e.g.*, disordering effects on the lipid bilayer or bilayer curvature strain (McMahon and Gallop, 2005), and respond through a structural change, which could then impact on the cytoskeleton. Actin binding to MBP decreased the labeling of MBP by the hydrophobic photolabel TID, indicating that it decreased the hydrophobic interactions of MBP with the bilayer (Boggs and Rangaraj, 2000). This may be due to transfer of hydrophobic residues of MBP, such as the Phe-Phe pairs, from the lipid bilayer to the hydrophobic pocket in actin. This change in interaction of MBP with the bilayer could then create a cytosol-to-membrane signal caused by changes in interaction of the cytoskeleton with the membrane.

The amount of actin bound to the MBP-lipid vesicles decreased with increasing net negative surface charge of the lipid vesicles (Boggs and Rangaraj, 2000) (Figure 4).

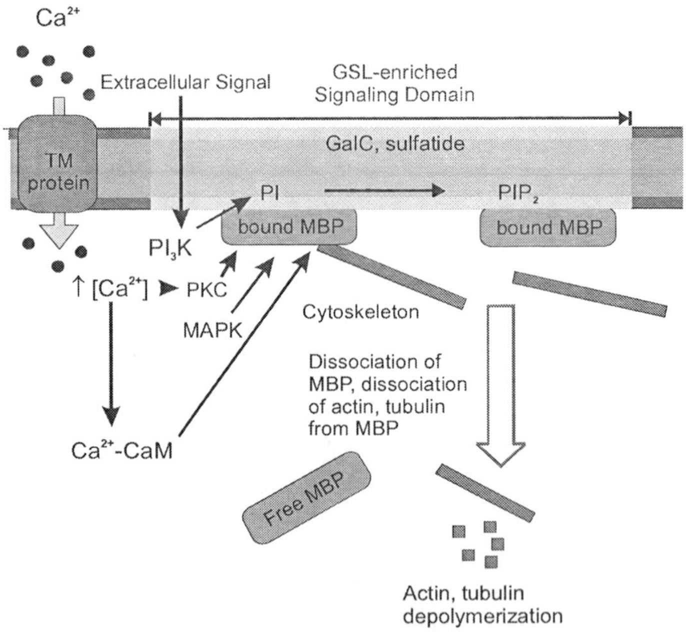

Figure 4. Schematic of mechanisms of regulation of binding of actin filaments and microtubules to MBP in oligodendrocytes and of transmission of extracellular signals to the cytoskeleton, via MBP. MBP bound to negatively charged lipids, phosphatidylserine and phosphatidylinositol (PI), on the cytosolic side of the membrane tethers actin filaments to the membrane. Extracellular signals impinging on GSL-enriched lipid rafts or membrane signaling domains in the OL membrane, such as those imparted by anti-GalC/sulfatide antibodies or by GalC/sulfatide-containing liposomes, as discussed in the text, cause Ca^{2+}-entry and/or activate kinases such as PI_3K and result in depolymerization of the cytoskeleton. The binding of Ca^{2+} to calmodulin causes dissociation of the actin from MBP, and of MBP from the membrane. It also causes dissociation of microtubules from MBP. Ca^{2+} activation of PKC causes phosphorylation of MBP which may decrease interactions of MBP with actin filaments and microtubules, as has been shown for phosphorylation of MBP with MAPK. PI_3K causes phosphorylation of PI and increased amounts of PIP_2, which are clustered by MBP, a "pipmodulin". The increased PIP_2 increases the membrane negative surface potential causing dissociation of negatively charged actin filaments from the membrane-bound MBP. The free actin filaments may then depolymerize, also causing depolymerization of microtubules, to which they may be bound, mediated by a number of cytoskeleton-binding proteins, and possibly also by MBP.

The binding of Ca^{2+}-CaM also caused dissociation of MBP and the MBP-actin complex from lipid vesicles (Boggs and Rangaraj, 2000; Boggs et al., 2005; Boggs et al., 2006) as it does for the basic domains of other proteins such as MARCKS (McLaughlin et al., 2005). Although basic residues of MBP are distributed throughout its sequence, in contrast to MARCKS where they are clustered, MBP also has several potential CaM binding sites (Polverini et al., 2004) (see Chapter IX), which may allow different regions of MBP to be dissociated from the bilayer by CaM. The deiminated protein component, C8, purified from human brain, and *in vitro*-phosphorylated component C1 (on Thr 94 and Thr 97 using MAPK) had significantly less ability to bind actin to lipid bilayers (Boggs et al., 2005; Boggs et al., 2006). The effect of phosphorylation was much greater than that of citrullination at six sites. Although average electrostatic forces are the primary determinant of the interaction of MBP with actin, phosphorylation may have an additional effect due to a site-specific electrostatic effect or to a conformational change. The 21.5 kDa isoform containing the exon II (now called exon 6, see Chapter I)-encoded sequence bound actin to the lipid bilayer less well than the 18.5 kDa form lacking this sequence (Boggs et al., 2005). The ability of other size isoforms to bind actin to the lipid bilayer has not been investigated.

Thus, deimination, a posttranslational modification of MBP, which normally occurs early in life but which is increased in adults with MS, and phosphorylation of MBP, which occurs in response to various extracellular signals in both myelin and OLs, both attenuate the ability of MBP to polymerize and bundle actin, and to bind actin to a membrane surface (Figure 4). Phosphorylation of a number of membrane actin-binding proteins such as MARCKS, regulates their interaction with the cytoskeleton and the membrane (Bubb et al., 1999).

An increase in negative surface potential of the lipid bilayer also decreased the binding of actin filaments to the lipid-bound MBP due to repulsion of the negatively-charged actin filaments. The increase in negative surface charge was produced by increasing the mole ratio of negatively-charged lipid to neutral lipid, phosphatidylcholine (PC) (Boggs and Rangaraj, 2000), or by incorporating PI, PIP, or PIP_2, at identical mole ratios of inositol lipid to PC (Boggs and Rangaraj, 2004) (Figure 4). We have shown that MBP can behave as a "PIP_2-modulin" similar to MARCKS, GAP43, and CAP23 (Laux *et al.* 2000) and cluster PIP_2 into localized domains (Harauz et al., 2008). Thus, changes in lipid composition resulting from signal transduction events could also regulate MBP-mediated binding of actin to the membrane.

MBP-cytoskeleton Interactions *in Vivo*

During the period of rapid myelination, in the first year of life in humans, the OL must produce large amounts of membrane (Simons and Trotter, 2007). Many events involved in myelination such as OL precursor migration in the CNS, differentiation, cell process extension, and membrane production may depend in part on the OL cytoskeleton (Kachar et al., 1986; Richter-Landsberg, 2007). Its interactions with the cytosolic side of the plasma membrane would allow it to participate in transmission of signals between the extracellular environment and the cytosol (Lunn et al., 1997). Such signals are received from extracellular matrix, growth hormones, and the axon. Continued interaction and communication between the axon and the mature compact myelin sheath occurs throughout life (Chakraborty et al., 1999; Witt and Brady, 2000; Simons and Trajkovic, 2006). Interactions between the adhered membranes of myelin in the multilayered myelin sheath may also provide signals which are transmitted across the membrane and across many layers of myelin.

Interactions of MBP with the cytoskeleton may allow it to play a role in signaling in OLs. Anti-galactosylceramide (GalC) antibody added to OLs mediates signals which cause Ca^{2+} entry, GalC and sulfatide redistribution in the membrane, phosphorylation changes of MBP, depolymerization of microtubules and other signal transduction events in cultured OLs (Dyer, 1993; Dyer and Benjamins, 1989) (Figure 4). GalC and sulfatide may be ligands for each other, which cause similar signaling effects (Boggs et al., 2004; Boggs et al., 2008a) as found for GSLs in other cells (Hakomori, 2002). The divalent cation Ca^{2+} can bridge sugars by binding sugar oxygens to its coordination sphere (Cook and Bugg, 1975), and it causes adhesion of liposomes containing GalC to liposomes containing sulfatide (Stewart and Boggs, 1993). Addition of GalC/sulfatide-containing liposomes to cultured OLs to interact with GalC and sulfatide in the OL membrane by these Ca^{2+}-mediated *trans* carbohydrate-carbohydrate interactions across apposed membranes had similar effects on lipid distribution and depolymerization of the cytoskeleton as anti-GalC antibody and also caused changes in protein distribution, including that of MBP (Boggs and Wang, 2001).

This extracellular signal impinging on GalC/sulfatide may be a signal for compaction of the MBP-containing domains (Dyer, 2002; Boggs et al., 2008a). MBP is required for these effects, since they do not occur in shiverer OLs, and are decreased in OLs in which MBP expression is inhibited with MBP siRNA, suggesting that MBP is involved in transduction of the signal (Dyer et al., 1994; Boggs et al., 2008b). Interactions of OLs with neurons also caused redistribution of OL lipids, in particular clustering of GalC (Fitzner et al., 2006). This neuron-induced GalC clustering also depended on the presence of MBP.

Stabilization of the actin cytoskeleton with jasplakinolide prevented all effects of the GalC/sulfatide-containing liposomes, including GalC and MBP redistribution and microtubule depolymerization, suggesting that the stability of microtubules in OLs depends on the integrity of the actin cytoskeleton (Boggs and Wang, 2004). The MBP redistribution caused by depolymerization of the actin cytoskeleton suggests that actin is linked directly or indirectly to MBP in OLs. In support of this idea, MBP was present in a Triton X-100 insoluble fraction from OLs, which resembled intact cytoskeleton and which contained actin and tubulin (Wilson and Brophy, 1989).

MBP might also bind actin to microtubules. In shiverer mouse OLs, actin filaments were not colocalized with microtubular structures as they were in wild type OLs (Dyer et al., 1995). MBP may also interact with actin and tubulin in the myelin sheath, in addition to OLs. Detergent extraction of myelin indicates that MBP is located in several different membrane domains or associated with different proteins. Some MBP distributes to two low density glycosphingolipid/cholesterol-enriched Triton-X-100 insoluble fractions, one of which is associated with actin, tubulin, caveolin, and kinases (Karthigasan et al., 1994; Arvanitis et al., 2005; Gillespie et al., 1989; Pereyra et al., 1988). Some MBP is also found in the Triton-X-100 supernatant associated with MAG, PLP, MAPK, and some phospholipid (Arvanitis et al., 2002). Although these isolated fractions do not necessarily come from different membrane domains in myelin *in situ* (Lichtenberg et al., 2005), the presence of MBP in several different fractions suggests that it can interact with different constituents or complexes in myelin, some of which interact with kinases. In this way, MBP may also participate in signaling in myelin. The p42/p44 MAPK in myelin is active and phosphorylates exogenous MBP (Arvanitis et al., 2005) and endogenous MBP, CNP, and tubulin (Y. Gong, W. Min and J.M. Boggs, unpublished). Different charge isomers and size isoforms of bovine MBP also are differentially associated with several detergent-insoluble fractions of myelin. A CHAPS-resistant low density fraction contained most of the phospho-Thr97-MBP whereas most of the citrullinated and methylated MBP were in a heavier CHAPS-insoluble fraction (DeBruin et al., 2005). The 18.5 kDa MBP was in the low density fraction and 21.5 kDa MBP was in the heavier CHAPS-insoluble fraction. This observation suggests that these different species of MBP associate preferentially with different myelin constituents and may perform different functions in these complexes.

In addition to kinases and phosphatases which can mediate rapid turnover of phosphate groups, myelin also contains mechanisms which control kinase activity, *e.g.*, phospholipase C to generate DAG, adenyl cyclase to generate cAMP, and MEK to activate p42/p44 MAPK (reviewed in Ledeen, 1992; Arvanitis et al., 2002). Myelin also contains relatively high amounts of the signal transduction lipids phosphatidylinositol (PI), phosphatidic acid, phosphatidylinositol 4-phosphate (PIP), and phosphatidylinositol 4,5-bisphosphate (PIP_2) (Deshmukh et al., 1980). The lightest myelin fraction, which comes from compact myelin, was even more enriched in these signal transduction lipids. PIP_2 is 1.5% of the total phospholipid in myelin and 1.7% in the lightest fraction, compared to 1% reported for the plasma membrane of other cells (McLaughlin and Murray, 2005). MBP stimulates a PI-specific phospholipase C purified from human myelin, but deimination of MBP prevents this effect (Tompkins and Moscarello, 1993). Thus, the more highly charged isomers of MBP may be directly involved in a signaling system involving phosphoinositides in myelin. There is considerable turnover of phosphoinositides in myelin *in vivo* (Kahn and Morell, 1988), which may be regulated by communication with the axon (Witt and Brady, 2000). In addition to being a source of second messengers, PIP_2 is a potent regulator of many actin-binding proteins (Janmey et al., 1999). Thus, mechanisms are present in both OLs and myelin to regulate the interaction of MBP with the membrane and with the cytoskeleton by phosphorylation of MBP and by changes in lipid composition.

Trans interactions between GSLs in membrane signaling domains in apposed cells trigger signaling and have been proposed to result in formation of a glycosynapse between the cells (Hakomori, 2002). Glycosynapses could also form between the apposed extracellular surfaces in compact myelin through contact between GalC and sulfatide in signaling domains (Boggs et al., 2004; Boggs et al., 2008a). Ions and water released into the periaxonal space following the axonal action potential may be able to traverse throughout the extracellular space in compact myelin by passing through the tight junction pores of the radial component (Dyer, 2002) and could regulate this interaction. The signals may in turn be transmitted across the membrane to MBP and the cytoskeletal elements in myelin, as found in cultured OLs for anti-GalC antibody and GalC/sulfatide-containing liposomes, possibly by causing Ca^{2+} entry into the cytosolic domains. The accumulation of Ca^{2+} into the cytosolic domains of compact myelin, mediated by NMDA receptors, has recently been detected within compact myelin (Micu et al., 2005). Subsequent effects on the cytoskeleton may regulate opening and closing of the tight junction pores between the adjacent myelin layers (Dyer, 2002; Boggs et al., 2008a) as occurs for intestinal epithelial cells (van Itallie and Anderson, 2004). This process may allow transmission of signals from the axon throughout compact myelin.

Conclusion

In order to produce myelin, the OL must extend membranous processes and produce large membrane sheets which then must interact with and wrap around the axon. These events require the cytoskeleton for membrane extension, transport of membrane constituents, organization of membrane domains, and localization of signaling molecules. Movement of the leading edge of membrane processes is driven by actin polymerization, with the microtubules located proximal to the actin network. The actin filaments provide tracks for the microtubules to invade regions of new growth (Song et al., 2001). Many cytosolic proteins which bind cytoskeletal proteins to each other, and membrane peripheral proteins which connect the membrane to the cytoskeleton have been identified in various cells (Sonnenberg and Liem, 2007; Niggli, 1995); some are ubiquitous and others are specific to certain cells (Greengard et al., 1994; Worton, 1995). Since MBP has the ability to assemble actin and tubulin and to bind actin filaments to a membrane surface *in vitro,* since it is such an abundant protein in OLs and myelin (30% of the myelin protein and 10% of myelin by weight), and since it is the only myelin-specific structural protein shown so far to be essential for myelination, it is reasonable to suggest that it is a multifunctional protein, and that one of its functions *in vivo*, is to bind cytoskeletal proteins to the cytosolic membrane surface, and perhaps also to each other (Boggs, 2006). The OL receives extracellular signals from growth factors, extracellular matrix, and the axon, which regulate myelination and influence the mature myelin sheath throughout life. MBP is necessary for transmission of some of these signals to the cytoskeleton. Its interactions with the cytoskeleton and regulation of these interactions by Ca^{2+}-calmodulin, by post-translational modifications to MBP, and by membrane surface potential could be mechanisms to allow MBP to be involved in signaling in OLs and myelin (Boggs, 2006; Dyer et al., 1994).

Acknowledgments

The work from my laboratory on the interactions of MBP with actin was supported by the Canadian Institutes of Health Research and the Multiple Sclerosis Society of Canada. I thank Dr. George Harauz for a critical reading of the chapter, helpful discussions, and assistance with Fig. 4.

References

Arvanitis, D.N., Yang, Y., and Boggs, J.M. (2002). Myelin proteolipid protein, basic protein, the small isoform of myelin-associated glycoprotein, and p42 MAPK are associated in the triton X-100 extract of central nervous system myelin. *J. Neurosci. Res*, 70, 8-23.

Arvanitis, D.N., Min, W., Gong, Y., Meng, Y.M., and Boggs, J.M. (2005). Two types of detergent-insoluble, glycosphingolipid/cholesterol-rich membrane domains in isolated myelin. *J. Neurochem*, 94, 1696-1710.

Barylko B., and Dobrowolski Z. (1984). Ca^{2+}-calmodulin-dependent regulation of F-actin-myelin basic protein interaction. *Eur. J. Cell Biol*, 35, 327-335.

Bates, I.R., Libich, D.S., Wood, D.D., Moscarello, M.A., and Harauz, G. (2002). An Arg/Lys→Gln mutant of recombinant murine myelin basic protein as a mimic of the deiminated form implicated in multiple sclerosis. *Protein Expr. Purif*, 25, 330-341.

Behrisch, A., Dietrich, C., Noegel, A.A., Schleicher, M., and Sackmann, E. (1995). The actin-binding protein hisactophilin binds *in vitro* to partially charged membranes and mediates actin coupling to membranes. *Biochemistry*, 34, 15182-15190.

Benmerah, A., Scott, M., Poupon, V., and Marullo, S. (2003). Nuclear functions for plasma membrane-associated proteins? *Traffic*, 4, 503-511.

Boggs, J.M., and Wang, H. (2001). Effect of liposomes containing cerebroside and cerebroside sulfate on cytoskeleton of cultured oligodendrocytes. *J. Neurosci. Res*, 66, 242-253.

Boggs, J.M., and Wang, H. (2004). Co-clustering of galactosylceramide and membrane proteins in oligodendrocyte membranes on interaction with polyvalent carbohydrate and prevention by an intact cytoskeleton. *J. Neurosci. Res*, 76, 342-355.

Boggs, J.M., Rangaraj, G., and Koshy K.M. (1999). Analysis of the membrane-interacting domains of myelin basic protein by hydrophobic photolabeling. *Biochim. Biophys. Acta*, 1417, 254-266.

Boggs, J.M., and Rangaraj, G. (2000). Interaction of lipid-bound myelin basic protein with actin filaments and calmodulin. *Biochemistry*, 39, 7799-7806.

Boggs, J.M., Wang, H., Gao, W., Arvanitis, D.N., Gong, Y., and Min, W. (2004). A glycosynapse in myelin? *Glycoconj. J*, 21, 97-110.

Boggs, J.M., and Rangaraj, G. (2004). Effect of phosphorylation of myelin basic protein on its interaction with actin *in vitro*. *J. Neurochem*, 90 (Suppl. 1):73.

Boggs, J.M., Rangaraj, G., Hill, C.M.D., Bates, I.R., Heng, Y. M., and Harauz, G. (2005). Effect of arginine loss in myelin basic protein, as occurs in its deiminated charge isoform, on mediation of actin polymerization and actin binding to a lipid membrane *in vitro*. *Biochemistry*, 44, 3524-3534.

Boggs, J.M., Rangaraj, G., Gao, W., and Heng, Y.M. (2006). Effect of phosphorylation of myelin basic protein by MAPK on its interactions with actin and actin binding to a lipid membrane *in vitro*. *Biochemistry*, 45, 391-401.

Boggs, J.M., Gao, W., and Hirahara, Y. (2008a). Myelin glycosphingolipids, galactosylceramide and sulfatide, participate in carbohydrate-carbohydrate interactions between apposed membranes and may form glycosynapses between oligodendrocyte and/or myelin membranes. *Biochim. Biophys. Acta*, 1780, 445-455.

Boggs, J.M., Gao, W., and Hirahara, Y. (2008b). Signal transduction pathways involved in interaction of galactosylceramide/sulfatide-containing liposomes with cultured oligodendrocytes and requirement for myelin basic protein and glycosphingolipids. *J. Neurosci. Res*, 86, 1448-1458.

Bubb, M.R., Lenox, R.H., and Edison, A.S. (1999). Phosphorylation-dependent conformational changes induce a switch in the actin-binding function of MARCKS. *J. Biol. Chem*, 274, 36472-36478.

Chakraborty, G., Drivas, A. and Ledeen, R. (1999). The phosphoinositide signaling cycle in myelin requires cooperative interaction with the axon. *Neurochem. Res*, 24, 249-254.

Chan, K.F.J., Robb, N.D., and Chen, W.H. (1990). Myelin basic protein: interaction with calmodulin and gangliosides. *J. Neurosci. Res*, 25, 535-544.

Chou, F.C.H., Chou, C.H.J., Shapira, R., and Kibler, R.F. (1976). Basis of microheterogeneity of myelin basic protein. *J. Biol. Chem*, 251, 2671-2679.

Cook, W.J., and Bugg, C.E. (1975). Calcium-carbohydrate bridges composed of uncharged sugars. Structure of a hydrated calcium bromide complex of α-fucose. *Biochim. Biophys. Acta*, 389, 428-435.

DeBruin, L.S., Haines, J.D., Wellhauser, L.A., Radeva, G., Schonmann, V., Bienzle, D., and Harauz, G. (2005). Developmental partitioning of myelin basic protein into membrane microdomains. *J. Neurosci. Res*, 80, 211-225.

Deshmukh, D.S., Kuizon, S., Bear, W.D., and Brockerhoff, H. (1980). Distribution of phosphoinositides among subfractions of rat brain myelin. *Lipids,* 15, 14-18.

Dobrowolski, Z., Osinska, H., Mossakowska, M., and Barylko, B. (1986a). Ca^{2+}-calmodulin-dependent polymerization of actin by myelin basic protein. *Eur. J. Cell Biol*, 42, 17-26.

Dobrowolski, Z., Barylko, B., and Drabikowski, W. (1986b) Interaction of tropomyosin with myelin basic protein and its effect on the ATPase activity of actomyosin. *Eur. J. Cell Biol*, 41, 65–71.

Dominguez, R. (2004). Actin-binding proteins – a unifying hypothesis. *Trends Biochem. Sci*, 29, 572-5780.

Dyer, C.A. (1993). Novel oligodendrocyte transmembrane signaling systems. *Mol. Neurobiol*, 7, 1-22.

Dyer, C.A. (1997). *Myelin proteins as mediators of signal transduction*. In B.H.J. Juurlink, R.M. Devon, J.R. Doucette, A. J. Nazarali, D.J. Schreyer, V.M.K. Verge (Eds.), Cell Biology and Pathology of Myelin: Evolving Biological Concepts and Therapeutic Approaches (pp. 69-74). New York, NY, USA: Plenum Press.

Dyer, C.A. (2002). The structure and function of myelin: From inert membrane to perfusion pump. *Neurochem. Res*, 27, 1279-1292.

Dyer, C.A., and Benjamins, J.A. (1989). Organization of oligodendroglial membrane sheets. II. Galactocerebroside: antibody interactions signal changes in cytoskeleton and myelin basic protein. *J. Neurosci. Res*, 24, 212–221.

Dyer, C.A., Philibotte, T.M., Wolf, M.K., and Billings-Gagliardi, S. (1994). Myelin basic protein mediates extracellular signals that regulate microtubule stability in oligodendrocyte membrane sheets. *J. Neurosci. Res*, 39, 97-107.

Dyer, C.A., Philibotte, T.M., Billings-Gagliardi, S., and Wolf, M.K. (1995). Cytoskeleton in myelin-basic-protein-deficient shiverer oligodendrocytes. *Dev. Neurosci*, 17, 53-62.

Dyer, C.A., Phillbotte, T., Wolf, M.K., and Billings-Gagliardi, S. (1997). Regulation of cytoskeleton by myelin components: studies on shiverer oligodendrocytes carrying an MBP transgene. *Dev. Neurosci*, 19, 395-409.

Dyson, H.J., and Wright, P.E. (2002). Coupling of folding and binding for unstructured proteins. *Curr. Opin. Struct. Biol*, 12, 54-60.

Erickson, A.K., Payne, D.M., Martino, P.A., Rossomando, A.J., Shabanowitz, J., Weber, M.J., Hunt, D.F. and Sturgill, T.W. (1990). Identification by mass spectrometry of threonine 97 in bovine myelin basic protein as a specific phosphorylation site for mitogen-activated protein kinase. *J. Biol. Chem*, 265, 19728-19735.

Feijoo, C., Campbell, D.G., Jakes, R., Goedert, M., and Cuenda, A. (2005). Evidence that phosphorylation of the microtubule-associated protein Tau by SAPK4/p38delta at Thr50 promotes microtubule assembly. *J. Cell Sci*, 118, 397-408.

Fitzner, D., Schneider, A., Kippert, A., Mobius, W., Willig, K.I., Hell, S.W., Bunt, G., Gaus, K., and Simons, M. (2006). Myelin basic protein-dependent plasma membrane reorganization in the formation of myelin. *EMBO J*, 25, 5037-5048.

Fuxreiter, M., Simon, I., Friedrich, P., and Tompa, P. (2004). Preformed structural elements feature in partner recognition by intrinsically unstructured proteins. *J. Mol. Biol*, 338, 1015-1026.

Galiano, M.R., Andrieux, A., Deloulme, J.C., Bosc, C., Schweitzer, A., Job, D. and Hallak, M.E. (2006). Myelin basic protein functions as a microtubule stabilizing protein in differentiated oligodendrocytes. *J. Neurosci. Res*, 84, 534–541.

Gillespie, C.S., Wilson, R., Davidson, A., and Brophy, P.J. (1989). Characterization of a cytoskeletal matrix associated with myelin from rat brain. *Biochem. J*, 260, 689-696.

Gow, A., Southwood, C.M., Li, J.S., Pariali, M., Riordan, G.P., Brodie, S.E., Danias, J., Bronstein, J.M., Kachar, B., and Lazzarini, R.A. (1999). CNS myelin and sertoli cell tight junction strands are absent in Osp/claudin-11 null mice. *Cell,* 99, 649-659.

Grand, R.J.A. and Perry, S.V. (1980). The binding of calmodulin to myelin basic protein and histone H2B. *Biochem. J*, 189, 227-240.

Greengard, P., Benfenati, F., and Valtorta, F. (1994). Synapsin I, an actin-binding protein regulating synaptic vesicle traffic in the nerve terminal. *Adv. Second Messenger Phosphoprotein Res*, 29, 31-45.

Hakomori, S. (2002). The glycosynapse. *Proc. Natl. Acad. Sci, USA*, 99, 225-232.

Harauz, G., Ishiyama, N., and Bates, I.R. (2000). Analogous structural motifs in myelin basic protein and in MARCKS. *Mol. Cell. Biochem*, 209, 155-163.

Harauz, G., Ishiyama, N., Hill, C.M.D., Bates, I.R., Libich, D.S., and Farès, C. (2004). Myelin basic protein-diverse conformational states of an intrinsically unstructured protein and its roles in myelin assembly and multiple sclerosis. *Micron*, 35, 503-542.

Harauz, G., Musse, A., Boggs, J. M., and Gao, W. (2008). Myelin basic protein (MBP) as a "PIP_2-modulin": New biological function for an old CNS protein. *J. Neurochem*, 104 (Supplement 1), 44.

Haynes, C., and Iakoucheva, L.M. (2006). Serine/arginine-rich splicing factors belong to a class of intrinsically disordered proteins, *Nucl. Acids Res*, 34, 305-312.

Hill, C.M., Bates, I.R., White, G.F., Hallett, F.R., and Harauz, G. (2002). Effects of the osmolyte trimethylamine-N-oxide on conformation, self-association, and two-dimensional crystallization of myelin basic protein. *J. Struct. Biol*, 139, 3-26.

Hill, C.M.D., and Harauz, G. (2005). Charge effects modulate actin assembly by classic myelin basic protein isoforms. *Biochem. Biophys. Res. Commun*, 329, 362-369.

Hill, C.M.D., Libich, D.S., and Harauz, G. (2005). Assembly of tubulin by classic myelin basic protein isoforms and regulation by post-translational modification. *Biochemistry*, 44, 16672-16683.

Hirschberg, D., Radmark, O., Jornvall, H., and Bergman, T. (2003). Thr94 in bovine myelin basic protein is a second phosphorylation site for 42-kDa mitogen-activated protein kinase (ERK2). *J. Protein Chem*, 22, 177-181.

Isenberg, G., and Goldmann, W.H. (1992). Actin-membrane coupling: a role for talin. *J. Musc. Res. Cell Motility*, 13, 587-589.

Isenberg, G., and Goldmann, W.H. (1995). Actin-binding proteins - lipid interactions, in The Cytoskeleton, Structure and Assembly, Vol. 1, pp. 169-204. Greenwich, CT: JAI Press, Inc.

Janmey, P.A., Xian, W., and Flanagan, L.A. (1999). Controlling cytoskeleton structure by phosphoinositide-protein interactions: phosphoinositide binding protein domains and effects of lipid packing. *Chem. Phys. Lipids*, 101, 93-107.

Jo, E., and Boggs, J.M. (1995). Aggregation of acidic lipid vesicles by myelin basic protein: dependence on potassium concentration. *Biochemistry*, 34, 13705-13716.

Jones, L.S., Yazzie, B., and Middaugh, C.R. (2004). Polyanions and the proteome. *Mol. Cell. Proteomics*, 3, 746-769.

Kachar, B., Behar, T., and Dubois-Dalcq, M. (1986). Cell shape and motility of oligodendrocytes cultured without neurons. *Cell Tissue Res*, 244, 27-38.

Kahn, D.W., and Morell P. (1988). Phosphatidic acid and phosphoinositide turnover in myelin and its stimulation by acetylcholine. *J. Neurochem*, 50, 1542–1550.

Karthigasan, J., Kosaras, B., Nguyen, J., and Kirschner, D.A. (1994). Protein and lipid composition of radial component-enriched CNS myelin. *J. Neurochem*, 62, 1203-1213.

Laux, T., Fukami, K., Thelen, M., Golub, T., Frey, D., Caroni, P. (2000). GAP43, MARCKS, and CAP23 modulate PI(4,5)P(2) at plasmalemmal rafts, and regulate cell cortex actin dynamics through a common mechanism. *J. Cell Biol*, 149, 1455-1471.

Ledeen, R.W. (1992). *Enzymes and receptors of myelin.* In R. E. Martenson (Ed.), Myelin: Biology and Chemistry (pp. 531-570). Boca Raton, FL: CRC Press.

Li, X., Ionescu, A.V., Lynn, B.D., Lu, S., Kamasawa, N., Morita, M., Davidson, K.G., Yasumura, T., Rash, J.E., and Nagy, J.I. (2004). Connexin 47, connexin29 and connexin 32 co-expression in oligodendrocytes and Cx47 association with zonula occludens-1 (ZO-1) in mouse brain. *Neurosci*, 126, 611-630.

Libich, D.S., Hill, C.M.D., Bates, I.R., Hallett, F.R., Armstrong, S., Siemiarczuk, A., and Harauz G. (2003a). Interaction of the 18.5 kDa isoform of myelin basic protein with Ca^{2+}-calmodulin: effects of deimination assessed by intrinsic Trp fluorescence spectroscopy, dynamic light scattering, and circular dichroism. *Protein Sci*, 12, 1507-1521.

Libich, D.S., Hill, C.M., Haines, J.D., and Harauz, G. (2003b). Myelin basic protein has multiple calmodulin-binding sites. *Biochem. Biophys. Res. Commun*, 308, 313-319.

Lichtenberg, D., Goni, F.M., and Heerklotz, H. (2005). Detergent-resistant membranes should not be identified with membrane rafts. *Trends Biochem. Sci*, 30, 430-436.

LoPresti, P., Szuchet, S., Papasozomenos, S.C., Zinkowski, R.P., and Binder, L.I. (1995). Functional implications for the microtubule-associated protein tau: Localization in oligodendrocytes. *Proc. Natl. Acad. Sci, USA*, 92, 10369-10373.

Lunn, K.F., Baas, P.W., and Duncan, I.D. (1997). Microtubule organization and stability in the oligodendrocyte. *J. Neurosci*, 17, 4921-4932.

McMahon, H.T. and Gallop, J.L. (2005). Membrane curvature and mechanisms of dynamic cell membrane remodeling, *Nature*, 438, 590-596.

McLaughlin, S. and Murray, D. (2005). Plasma membrane phosphoinositide organization by protein electrostatics. *Nature*, 438, 605-611.

McLaughlin, S., Hangyas-Mihalyne, G., Zaitseva, I., and Golebiewska, U. (2005). Reversible – through calmodulin – electrostatic interactions between basic residues on proteins and acidic lipids in the plasma membrane. *Biochem. Soc. Symp*, 72, 189-198.

Martenson, R.E. (1980). *Myelin basic protein: What does it do?* In S. Kumar (Ed.), Biochemistry of Brain (pp. 49-79). Oxford: Pergamon Press.

Micu, I., Jiang, Q., Coderre, E., Ridsdale, A., Zhang, L., Woulfe, J., Yin, X., Trapp, B.D., McRory, J.E., Rehak, R., Zamponi, G.W., Wang, W., and Stys, P.K. (2005). NMDA receptors mediate calcium accumulation in myelin during chemical ischaemia. *Nature*, 439, 988-992.

Modesti, N.M. and Barra, H.S. (1986). The interaction of myelin basic protein with tubulin and the inhibition of tubulin carboxypeptidase activity. *Biochem. Biophys. Res. Commun*, 136, 482-489.

Moscarello, M.A. (1997). *Myelin basic protein, the 'executive' molecule of the myelin membrane.* In B.H.J. Juurlink, R.M. Devon, J.R. Doucette, A. J. Nazarali, D.J. Schreyer, V.M.K. Verge (Eds.), Cell Biology and Pathology of Myelin: Evolving Biological Concepts and Therapeutic Approaches (pp. 13-25). New York, NY, USA: Plenum Press.

Moscarello, M.A., Pang, H., Pace-Asciak, C.R., and Wood, D.D. (1992). The N terminus of human myelin basic protein consists of C2, C4, C6, and C8 alkyl carboxylic acids. *J. Biol. Chem*, 267, 9779-9782.

Muller, R., Heinrich, M., Heck, S., Blohm, D., and Richter-Landsberg, C. (1997). Expression of microtubule-associated proteins MAP2 and tau in cultured rat brain oligodendrocytes. *Cell Tissue Res*, 288, 239-249.

Murray, D., Ben-Tal, N., Honig, B., and McLaughlin, S. (1997). Electrostatic interaction of myristoylated proteins with membranes: simple physics, complicated biology. *Structure*, 5, 985-989.

Murray, D., Arbuzova, A., Hangyas-Mihalyne, G., Gambhir, A., Ben-Tal, N., Honig, B., and McLaughlin, S. (1999). Electrostatic properties of membranes containing acidic lipids and adsorbed basic peptides: theory and experiment. *Biophys. J*, 77, 3176-3188.

Niggli, V. (1995). Membrane Cytoskeleton, in The Cytoskeleton, Structure and Assembly, Vol.1, pp. 123-168, Greenwich, CT: JAI Press, Inc.

Pandey, N., Ganapath, M., Kumar, K., Dasgupta, D., Das Sutar, S.K., and Dash, D. (2004). Comparative analysis of protein unfoldedness in human housekeeping and non-housekeeping proteins. *Bioinformatics,* 20, 2904-2910.

Patil, A. and Nakamura, H. (2006). Disordered domains and high surface charge confer hubs with the ability to interact with multiple proteins in interactions networks. *FEBS Lett*, 580, 2041-2045.

Pedraza, L., Fidler, L., Staugaitis, S.M., and Colman, D.R. (1997). The active transport of myelin basic protein into the nucleus suggests a regulatory role in myelination. *Neuron*, 18, 579-589.

Pereyra, P.M., Horvath, E., and Braun, P.E. (1988). Triton X-100 extractions of central nervous system myelin indicate a possible role for the minor myelin proteins in the stability in lamellae. *Neurochem. Res*, 13, 583-595.

Pirollet, F., Derancourt, J., Haiech, J., Job, D., and Margolis, R.L. (1992). Ca^{2+}-calmodulin regulated effectors of microtubule stability in bovine brain. *Biochemistry*, 31, 8849-8855.

Polverini, E., Boggs, J.M., Bates, I.R., Harauz, G., and Cavatorta, P. (2004). Electron paramagnetic resonance spectroscopy and molecular modelling of the interaction of myelin basic protein (MBP) with calmodulin (CaM)-diversity and conformational adaptability of MBP CaM-targets. *J. Struct. Biol*, 148, 353-369.

Polverini, E., Rangaraj, G., Libich, D.S., Boggs, J.M., and Harauz, G. (2008). Binding of the proline-rich segment of myelin basic protein to SH3-domains - spectroscopic, microarray, and modelling studies of ligand conformation and effects of post-translational modifications. *Biochemistry*, 47, 267-282.

Prasad, K., Barouch, W., Martin, B.M., Greene, L.E., and Eisenberg, E. (1995). Purification of a new clathrin assembly protein from bovine brain coated vesicles and its identification as myelin basic protein. *J. Biol. Chem*, 270, 30551-30556

Richter-Landsberg, C. (2008). The Cytoskeleton in oligodendrocytes: Microtubule dynamics in health and disease. *J. Mol. Neurosci*, 35, 55-63.

Roth, G.A., Gonzalez, M.D., Monferran, C.G., De Santis, M.L., and Cumar, F.A. (1993). Myelin basic protein domains involved in the interaction with actin. *Neurochem. Int*, 23, 459-465.

Simons, M., and Trajkovic, K. (2006). Neuron-glia communication in the control of oligodendrocyte function and myelin biogenesis. *J. Cell Sci*, 119, 4381-4389.

Simons, M., and Trotter, J. (2007). Wrapping it up: the cell biology of myelination. *Curr. Opin. Neurobiol,* 17, 533-540.

Song, J., Goetz, B.D., Baas, P.W., and Duncan, I.D. (2001). Cytoskeletal reorganization during the formation of oligodendrocyte processes and branches. *Mol. Cell. Neurosci*, 17, 624-636.

Sonnenberg, A., and Liem, R.K. (2007). Plakins in development and disease. *Exp. Cell Res*, 313, 2189-203.

Stewart, R.J., and Boggs, J.M. (1993). The carbohydrate-carbohydrate interaction between galactosylceramide-containing liposomes and cerebroside sulfate-containing liposomes: dependence on the glycolipid ceramide composition. *Biochemistry*, 32, 10666-10674.

Taketomi, M., Kinoshita, N., Kimura, K., Kitada, M., Noda, T., Asou, H., Nakamura, T., and Ide, C. (2002). Nogo-A expression in mature oligodendrocytes of rat spinal cord in association with specific molecules. *Neurosci. Lett*, 332, 37-40.

Tompa, P., Szasz, C., and Buday, L. (2005). Structural disorder throws new light on moonlighting. *Trends Biochem. Sci*, 30, 484-489.

Tompa, P. (2005) The interplay between structure and function in intrinsically unstructured proteins. *FEBS Lett*, 579, 3346-3354.

Tompkins, T.A., and Moscarello, M.A. (1993). Stimulation of bovine brain phospholipase C activity by myelin basic protein requires arginyl residues in peptide linkage. *Arch. Biochem. Biophys*, 302, 476-483.

Uversky, V.N., Oldfield, C.J., and Dunker, A.K. (2005). Showing your ID: intrinsic disorder as an ID for recognition, regulation and cell signaling. *J. Mol. Recog.* 18, 343-384.

Van Itallie, C.M., and Anderson, J.M. (2004). The molecular physiology of tight junction pores. *Physiol*, 19, 331-338.

Wilson, R., and Brophy, P.J. (1989). Role for the oligodendrocyte cytoskeleton in myelination. *J. Neurosci. Res*, 22, 439-448.

Witt, A., and Brady, S.T. (2000). Unwrapping new layers of complexity in axon/glial relationships. *Glia,* 29, 112-117.

Wohnsland, F., Schmitz, A.A.P., Steinmetz, M.O., Aebi, U., and Vergères, G. (2000). Interaction between actin and the effector peptide of MARCKS-related protein. Identification of functional amino acid segments. *J. Biol. Chem,* 275, 20873-20879.

Wood, D.D., and Moscarello, M.A. (1989).The isolation, characterization, and lipid-aggregating properties of a citrulline containing myelin basic protein. *J. Biol. Chem*, 264, 5121-5127.

Worton, R. (1995). Muscular dystrophies: diseases of the dystrophin-glycoprotein complex. *Science*, 270, 755-756.

Yoshikawa, H. (2001). Myelin-associated oligodendrocytic basic protein modulates the arrangement of radial growth of the axon and the radial component of myelin. *Med. Electron Microsc*, 34, 160-164.

Zand, R., Li, M.X., Jin, X., and Lubman, D. (1998). Determination of the sites of posttranslational modifications in the charge isomers of bovine myelin basic protein by capillary electrophoresis-mass spectroscopy. *Biochemistry*, 37, 2441-2449.

Index

A

acceptor, 150
access, 76
acetylation, 3, 36, 50, 51, 60, 64, 65, 71, 148
acetylcholine, 18, 181
acid, 16, 25, 62, 74, 89, 168, 176, 181, 184
acidic, 7, 23, 44, 47, 71, 144, 147, 148, 160, 168, 172, 181, 182, 183
actin, x, 152, 159, 167, 168, 169, 170, 172, 173, 174, 175, 176, 177, 178, 179, 181, 182, 183, 184
action potential, 177
activation, 174
active compound, 73, 119
active transport, 183
acylated, 172
AD, 23, 59, 66, 90, 109
adaptability, 183
adenomatous polyposis coli, 10, 24, 25, 27, 29, 30, 32, 34, 35, 36, 37, 38, 39
adenosine, 89
adenosine triphosphate, 89
ADH, 115
adhesion, x, 10, 102, 106, 107, 108, 167, 168, 175
adhesions, 109
ADP, 43, 150, 151, 152, 163
adult(s), 152, 175
adverse effects, 122, 123
adverse event, 121
aerobic, 150
age, vii, 1, 21, 22, 90, 129
Age-Related Macular Degeneration (AMD), ix, 113
aggregation, viii, 8, 33, 41, 44, 45, 46, 48, 50, 55, 57, 63, 66, 92
alanine, 119, 168
aldolase, 149, 162
alkaloids, 56, 87, 115, 121
alpha, 158, 159, 160, 161
ALS, vii, 1, 22, 35
alternative, 168
alters, 37, 151
amide, 148
amino, 10, 74, 76, 89, 131, 162, 168, 184
amino acid(s), 10, 11, 74, 76, 162, 168, 184
anaerobic, 150
anchorage, 106, 108
anchoring, 19, 23, 152
androgen, 48, 59, 66
anemia, 122
angiogenesis, ix, 108, 109, 113, 114, 126, 129, 132
angiogenic process, 54
angiography, 126
animals, 149, 150, 151
anorexia, 121, 122, 123
anoxic, 150
anti-angiogenic agents, 124
anti-apoptotic role, 60
antibody, 52, 71, 75, 76, 77, 78, 79, 80, 126, 175, 177, 180
anti-cancer, 63, 133
anticancer drug, 57, 61, 109, 121, 138, 140
antigen, 75, 79, 83, 84
antitumor, 126, 131
antitumor agent, 131
anxiety, 122, 123
APC, vii, 1, 7, 10, 11, 13, 14, 16, 17, 18, 19, 20, 23, 24, 25, 26, 29, 30, 31, 33, 34, 35, 36, 38, 39
Apicomplexa, 163
apoptosis, 10, 14, 44, 56, 57, 58, 61, 108, 130, 131
Arabidopsis thaliana, 153, 155
archaea, 144, 147, 158
Argentina, 41
arginine, 179, 181
aromatics, 118
arrest(s), 57, 128, 132
arthritis, 129

artificial, 169
aspiration, 20
aspiration pneumonia, 20
assessment, 130
asthenia, 121, 122, 123
astrocytes, 16, 23, 32, 62
asymmetry, 43
ataxia, 13, 91, 121, 122
atherosclerosis, ix, 101
ATP, 43, 45, 46, 47, 48, 50, 51, 89, 149, 150, 151, 152, 162
ATPase, 180
atrophy, ix, 36, 46, 59, 85, 86, 91, 92, 98, 99, 100
attachment, 12, 32, 33, 71
autism, 39
autoantibodies, 84
autoimmunity, 84
autophagy, 152, 165
autosomal dominant, 22, 98
autosomal recessive, 91
axon, 164, 175, 177, 179, 184
axon terminals, 164
axonal, 165, 177
axons, 2, 3, 4, 14, 17, 18, 22, 26, 28, 38, 53, 56, 89, 90, 94, 96

B

bacteria, 44, 45, 144, 147, 151, 156, 157, 158
bacterial, 146, 147, 157, 159, 160, 163, 166
bacterial cells, 61
barrier, 151
base, 4
basement membrane, 117
batteries, 140
binding, x, 143, 144, 145, 148, 149, 150, 152, 157, 161, 162, 163, 165, 167, 168, 170, 172, 173, 174, 175, 177, 178, 179, 180, 181, 182
biochemistry, 82, 88, 150
bioenergetics, 150, 165
biogenesis, 161, 164, 183
biological, 168, 181
biological activities, 168
biological activity, 46, 47, 49, 54
biological processes, 54
biological responses, 102, 105
biology, 183, 184
bioluminescence, 141
biomolecules, 114
biosynthesis, 66, 72
biotechnology, 119
blindness, 121
blocks, 163
blood, ix, 34, 70, 71, 76, 77, 80, 106, 113, 114, 117, 120, 121, 122, 123, 124, 125, 128, 129, 141, 161
blood flow, 117, 122, 123, 124, 125, 128, 141
blood pressure, 34, 121, 122
blood supply, 114, 120
blood vessels, ix, 106, 113, 114, 118, 123, 129
bonding, 170
bonds, 47, 86
bone, 56
bovine, 170, 171, 176, 180, 181, 183, 184
bowel, 121
brain, 7, 8, 12, 13, 14, 16, 19, 23, 26, 30, 38, 39, 50, 52, 57, 64, 86, 92, 94, 100, 163, 168, 170, 174, 179, 180, 182, 183, 184
brain abnormalities, 13, 23
brain structure, 19
brain tumor, 52, 64
branching, 3, 17, 43
breakdown, 44, 49, 67
breast cancer, 56, 125, 141
bronchiectasis, 91
budding, 67, 165
buffer, 169
bundling, x, 162, 167, 172

C

Ca^{2+}, x, 167, 168, 170, 174, 175, 177, 178, 179, 182, 183
calcium, 49, 54, 179, 182
caliber, 22
calmodulin, x, 167, 168, 170, 171, 174, 178, 179, 181, 182, 183
cAMP, 176
Canada, 167, 178
cancer, viii, 42, 54, 56, 57, 60, 64, 65, 67, 114, 116, 117, 118, 120, 124, 125, 126, 127, 129, 135, 136, 138
cancer cells, viii, 42, 56, 57, 60, 118, 124
candidates, viii, 69, 120
CAP, 7, 8, 10, 22, 35, 37, 39
capillary, 103, 184
carbohydrate, 175, 179, 184
carboxyl, 89, 96, 144, 160
carboxylic, 183
carboxylic acid(s), 183
carcinogenicity, 128
carcinoma, 123, 129, 133
cardiomyocytes, 152
cardiovascular disease, 23, 102, 105, 108
cardiovascular physiology, 109
cargo, 149
cargoes, 43

carotid arteries, 108
cascades, 3, 4, 6, 102, 104, 105, 110
caspases, 44, 57
catastrophes, 7, 14, 37
Catharanthus roseus, 117
cation, 175
cattle, 70, 75
caveolin, x, 167, 176
CBS, 115, 116, 117, 118, 119, 120, 123, 127, 128
cDNA, 78
cell, x, 144, 148, 149, 150, 151, 158, 159, 165, 167, 168, 172, 173, 175, 181, 182, 184
cell biology, 26, 70, 184
cell body, x, 3, 17, 54, 89, 167, 172
cell culture, 93
cell cycle, viii, 34, 42, 50, 57, 62, 73, 81, 95, 106, 127, 128
cell death, vii, ix, 1, 22, 101, 130
cell differentiation, viii, 42, 54
cell division, 11, 12, 42, 44, 70, 75, 144
cell line(s), 7, 57, 65
cell membranes, 149
cell movement, viii, 27, 41, 43
cell signaling, 109, 184
central nervous system (CNS), ix, x, 25, 34, 35, 85, 86, 99, 167, 178, 183
centromere, 10
centrosome, 3, 4, 13, 14, 20, 21, 24, 37, 39, 42, 87, 88, 90, 96
ceramide, 184
cerebellar development, 13
cerebellum, 13, 19, 20
cerebral cortex, 14, 19, 21, 30, 39, 164
Chagas disease, 70, 76, 80, 84
challenges, 158
channels, 164
chaperones, viii, 41, 44, 45, 46, 47, 49, 50, 53, 54, 55, 57, 58, 59, 61, 62, 63, 64, 65
chemical, 81, 103, 120, 127, 182
chemotaxis, 165
chemotherapeutic agent, 64, 72, 73, 75, 79, 124, 125
chemotherapy, viii, ix, x, 56, 57, 58, 67, 69, 70, 113, 114, 122, 124, 125, 129, 133, 136
chicken, 145, 159
children, 135
chimera, 148
chloroplasts, 144
cholesterol, 176, 178
chromatography, 57
chromosome, 10, 12, 27, 29, 33, 38, 56, 90, 97, 98, 165
cilia, 51, 91
ciliate, 146
circular dichroism, 182
citrulline, x, 167, 184
CKD, 117, 120, 121, 123, 132
class switching, 79
classes, 56, 121
classical, 150
classification, 7, 9, 127, 140, 159
clathrin, 169, 183
cleavage, 20, 30, 50, 160
clinical application, 60
clinical trials, 56, 57, 114, 115, 117, 118, 119, 120, 121, 123, 124, 125, 126, 127, 129
CLIP-115 deficiency, vii, 1
cloning, 84
closure, 38, 150
clustering, 19, 36, 176, 179
clusters, 17, 18, 102
CNS, x, 35, 91, 92, 98, 122, 123, 167, 175, 181, 182
Co, 179
coding, 19
cognitive dysfunction, 55
cognitive function, 21
cognitive impairment, 20, 65
cognitive profile, 23
collagen, 110
combination therapy, 123
communication, vii, viii, 1, 41, 92, 99, 175, 177, 183
compaction, 176
comparative analysis, 105
competition, 48, 52, 59
compilation, vii
complexity, 92, 184
components, 151, 152, 170, 171, 180
composition, 25, 52, 175, 177, 182, 184
compounds, ix, 57, 62, 72, 73, 74, 75, 79, 81, 82, 113, 115, 118, 123, 127, 140
computation, 74
computer, 74
concentration, 147, 170, 181
conductance, 150
configuration, 6, 52, 103, 147
conformational, 175, 179, 181, 183
conformational states, 181
Congress, 80
connective tissue, 101, 109, 110, 111, 112
consensus, 55, 74, 168
conservation, viii, 69, 75, 78, 145, 146, 156, 159, 164
constant rate, 6
constipation, 122
constituents, 176, 177
constraints, 146
construction, 14, 39, 102

consumption, 152
control, 157, 176, 183
control group, 126
controversial, 7, 77
cooperation, 46
coordination, 23, 102, 175
corpus callosum, vii, 1, 26
correlation, 4, 31, 98, 128
correlations, 81
cortex, 2, 6, 11, 12, 13, 14, 15, 16, 19, 20, 21, 28, 32, 34, 36, 164, 182
cortical localization, 15
cortical neurons, 14, 20, 60
corticosteroids, 71
cost, 74, 83
coupling, 178, 181
covalent, 158
cranial nerve, 34
creatine, 163, 165
creatine kinase, 163, 165
crystal structure, 96, 126, 127, 159
crystallization, 127, 181
crystals, 145
CT, 181, 183
C-terminal, 145, 146, 147, 156, 161, 162
C-terminus, 161
cues, viii, 2, 4, 6, 14, 24, 30, 35, 41, 103
culture, 27, 54, 83, 92, 93, 94, 145, 164
culture conditions, 54
CV, 60, 64, 65, 67
cycles, 2, 3, 55, 125, 126
cyclooxygenase, 66, 110
cyclosporine, 47
CYT, 117, 121, 122, 123, 125
cytokinesis, 43
cytometry, 128
cytoplasm, 42, 53, 58, 60, 76, 90, 150, 162
Cytoskeletal, 184
cytoskeletal proteins, viii, 41, 49, 50, 58, 177
cytosol, 164, 173, 175
cytosolic, x, 164, 167, 170, 174, 175, 177
cytotoxic agents, 124

D

damages, 44, 49
damping, 55
database, 59, 127
deacetylation, 64
decay, 6
defects, vii, 1, 13, 14, 16, 19, 20, 21, 22, 23, 29, 44, 49, 91
deficiency, vii, 1, 23, 124
deficit, 20
deformation, 102, 103, 104, 105, 109
degradation, 12, 44, 45, 46, 50, 56, 59, 61, 65, 66, 73, 81
dehydration, 122
dehydrogenase, 149, 162
dementia, 90, 97, 98
denaturation, 50, 67
dendrites, 2, 3, 5, 14, 17, 18, 23, 26, 34, 89, 96
dendritic spines, 18, 30
density, x, 167, 176
Department of Labor, 167
dephosphorylation, 50, 52, 55
depolymerization, 43, 49, 87, 94, 95, 99, 106, 162, 168, 170, 171, 174, 175, 176
depolymerizing, 168
deposition, ix, 101
deposits, 23, 55, 90
depression, 18, 22, 39, 122
deprivation, 49
derivatives, 53, 57, 61, 73, 74, 75, 115, 120
destruction, 76, 114
detectable, 53, 76
detection, 76, 79
developing countries, 72
deviation, 146, 147, 156
diarrhea, 21, 122
dichotomy, 149
differentiation, 175
diffusion, 43, 163
dimer, 143, 144, 145, 147, 148, 149, 150, 151, 152, 157, 159, 160
dimerization, 7, 10, 12
directionality, 2, 13, 16, 36
disability, 20
discrimination, 159
discs, 33
disease model, 55
diseases, ix, 59, 63, 64, 66, 70, 72, 74, 85, 86, 90, 98, 120, 122, 125, 184
dislocations, 164
disorder, vii, 1, 20, 30, 39, 184
dissociation, 15, 54, 170, 174
distress, 20
distribution, viii, 20, 25, 41, 54, 71, 95, 98, 156, 165, 176
diversity, 7, 58, 103, 163, 183
division, 144, 159, 161, 165
DNA, ix, 44, 72, 73, 95, 101, 106, 163, 164
docetaxel, 121, 134
dogs, 79
domestication, 157
doublecortin (DCX), vii, 1, 20

down-regulation, 57
Drosophila, 15, 17, 33, 36, 60, 146, 153, 154
drug design, 166
drug resistance, 72, 126
drug targets, 72
drug treatment, 72
drugs, ix, 3, 47, 49, 56, 58, 63, 70, 72, 73, 81, 87, 103, 104, 106, 109, 113, 123, 126, 127, 128, 129
dynamin, 152
dysphoria, 122
dyspnea, 122
dystrophin, 184

E

E. coli, 147
E-cadherin, 109
ECM, ix, 101, 102, 103, 105
ectothermic, viii, 69, 79
elaboration, 33
electron, 86, 139, 145, 146, 150, 159
electrophoresis, 184
electrostatic, 168, 170, 172, 174, 182
electrostatic force, 174
electrostatic interactions, 170, 182
elongation, 3, 6, 15, 17, 21, 24, 25, 35, 37, 160
embryogenesis, 95
embryos, 159
encoding, vii, 1, 55, 91, 110
endocrine, 23
endogenous, 176
endometriosis, 129
endothelial cells, 102, 103, 105, 107, 108, 110, 111, 114, 117
endothelium, 111, 114
endothermic, 79
endurance, 22
energy, 149, 157
enlargement(s), vii, 1, 70, 150
environment, 47, 58, 66, 76, 107, 175
environmental stress(s), viii, 41
enzymatic activity, 47
enzyme(s), 8, 51, 55, 72, 149, 162
EPC, 117, 121, 122, 123, 125, 126
epilepsy, 20, 29
epithelial cell(s), 11, 39, 108, 109, 177
epitopes, 77, 78, 84
estrogen, 48
ethanol, 49
Euglena gracilis, 153, 155
eukaryote(s), x, 143, 144, 146, 148, 150, 151, 152, 157, 158, 160, 161, 164
eukaryotic, x, 8, 42, 44, 45, 46, 59, 61, 66, 70, 72, 78, 86, 87, 90, 143, 157, 158, 159, 162, 163
eukaryotic cell, x, 8, 42, 44, 45, 46, 59, 70, 72, 78, 86, 87, 90, 143, 157
evidence, 24, 49, 50, 51, 57, 58, 62, 74, 86, 90, 92, 94, 95, 98, 103, 146, 159, 166
evolution, vii, 7, 45, 46, 63, 80, 143, 144, 146, 148, 152, 157, 158, 159, 160, 163
excitability, 18, 27
exocytosis, 99
exogenous, 176
exons, 21
exposure, 44, 49, 103, 117, 147
external environment, viii, 41
extracellular, x, 167, 174, 175, 176, 177, 178, 180
extracellular matrix, ix, 14, 101, 106, 107, 109, 112, 175, 178
extraction, 176
extracts, 94, 97

F

FAA, 118
failure to thrive, 23
families, 90
family, 144, 147, 159, 161, 162
family members, vii, 1
FDA, 116, 127
fever, 121
fiber(s), 43, 49, 111, 152
fibrillation, 121
fibroblast growth factor, 114
fibroblasts, 24, 104, 106, 107, 110, 112
fibrosis, 101, 105
filament, ix, 26, 43, 44, 49, 63, 64, 85, 86, 88, 101
fish, viii, 69, 71, 77, 78, 80, 84
fission, 151, 152, 164, 165
FL, 182
flagellum, 71, 78
flexibility, 74, 146, 168
flow, 157
fluid, 76, 107
fluorescence, 103, 127, 182
folding, 144, 180
force, 6, 103
forebrain, 31, 33
formation, vii, ix, x, 1, 2, 3, 4, 11, 14, 15, 16, 17, 18, 20, 24, 25, 55, 56, 58, 60, 65, 85, 86, 87, 88, 92, 94, 100, 103, 107, 114, 115, 121, 122, 124, 128, 147, 162, 168, 170, 177, 180, 184
fragments, 84, 90
freedom, 55
freshwater, 71, 150, 151

fungal, 146, 152, 159
fungi, 72, 151, 159
fusion, 12, 93, 151, 152, 164, 165

G

ganglion, 16
gangliosides, 179
GDP, 49, 102, 145
GEF, 16, 102, 111
gelatinase A, 110
gene, 144, 146, 150, 157, 158, 159, 160, 166, 168, 172
gene arrays, 81, 146, 160
gene expression, ix, 12, 101, 102, 103, 104, 105, 106, 107, 108, 109, 110, 111, 112, 168
gene targeting, 55
gene transfer, 146, 150, 157, 166
genes, vii, 1, 7, 19, 22, 23, 29, 33, 37, 46, 47, 70, 71, 75, 81, 87, 91, 102, 103, 104, 105, 110, 112, 126, 144, 145, 156, 157, 159, 160, 161
genetic alteration, viii, 41
genetics, 19, 24, 26, 39, 80, 151
genome(s), 71, 75, 81, 149, 158, 163
genomic(s), 160, 161
genotype, 31, 98
genus, 146, 157, 160
geometry, ix, 85, 88, 162
glia, 14, 99, 183
glial, 184
glial cells, 14, 59
glioblastoma, 23, 117, 125, 126, 138
glioblastoma multiforme, 138
glioma, 122, 136
glucocorticoid, 48, 51, 54, 59, 62, 64, 66, 67
glucocorticoid receptor, 48, 51, 54, 62, 64, 66, 67
glucocorticoids, 47
glucose, 49
glutamate, 19, 51
glutamine, 66
glutathione, 47
glycerol, 171
glycine, 8, 18, 51
glycogen, 11, 56
glycoprotein, 178, 184
glycosphingolipid-enriched fraction of myelin, x, 167
gout, 118
gracilis, 153, 155
grants, 58
grasses, 159
groups, 146, 148, 150, 152, 172, 176
growth, viii, 2, 3, 4, 5, 6, 7, 14, 15, 16, 17, 23, 24, 28, 30, 31, 32, 33, 34, 35, 36, 37, 39, 41, 45, 49, 57, 62, 65, 70, 73, 76, 77, 81, 82, 83, 84, 87, 88, 95, 101, 102, 103, 104, 105, 106, 107, 109, 110, 111, 112, 114, 124, 132, 175, 177, 184
growth arrest, 49
growth factor(s), 62, 102, 103, 104, 105, 107, 109, 110, 111, 112, 114, 178
growth hormone, 175
growth rate, 14
GTPases, 13, 15, 102, 108, 159
guanine, 102, 111, 144
guidance, vii, 1, 6, 12, 15, 16, 30, 32, 33, 65

H

half-life, 4, 49
HAP, 22
haplotypes, 63
HCC, 132, 136, 137
headache, 121, 122, 123
healing, ix, 71, 101
health, 63, 183
heat shock protein, 44, 59, 60, 62, 65, 66, 67
heavy metals, 49
helix, 169, 171
hemodynamic stimuli, ix, 101
hepatocellular carcinoma, 132, 136, 137
herbicide, 82
heterodimer, 144
heterogeneity, 28, 77, 81, 98, 146
heterogeneous, 172
hippocampus, 20, 21, 49, 59
histone, 51, 181
history, 80, 98, 157, 164
HM, 62, 65, 66
homeostasis, viii, 41, 48, 54, 106, 107
homolog, 158
homology, 144, 160
horizontal gene transfer, 157, 166
hormone(s), 47, 60, 61, 175
host, viii, 69, 70, 71, 72, 73, 74, 75, 76, 78, 79, 80, 82, 83, 84, 150, 158, 159
House, v, 41
human(s), ix, 19, 20, 22, 23, 24, 27, 29, 33, 37, 43, 46, 50, 56, 59, 60, 63, 65, 66, 67, 80, 81, 85, 86, 91, 92, 94, 96, 97, 99, 108, 110, 112, 124, 132, 157, 159, 160, 165, 174, 175, 177, 183
human brain, 27, 50, 65, 94, 99, 174
Hunter, 2, 30, 107
hybrid, 7, 160
hydrocephalus, 20
hydrogen, 170

hydrolysis, 42, 45, 46, 90, 119, 149
hydrophobic, 170, 172, 173, 179
hydrophobic interactions, 173
hydrophobicity, 168
hyperglycemia, 122
hypersensitivity, 79
hypertension, 122, 123
hyperthermia, 59
hypertrophy, 55, 66, 91
hypotension, 98, 122
hypothesis, 165, 180
hypoxia, 105, 126, 138
hypoxic, 150

I

ID, 184
identification, viii, 61, 66, 69, 70, 72, 74, 75, 102, 103, 105, 127, 158, 161, 162, 165, 183
identity, 2, 4, 46, 48, 56, 74, 76, 77, 145
IgG, 173
images, 48, 51, 173
immune defense, 77
immune response, 70, 75, 77, 78, 79
immune system, 70, 71, 76
immunity, 71, 75, 80
immunization, viii, 69, 75, 76, 77, 79, 126
immunofluorescence, 71
immunogen, 70, 75, 76
immunogenicity, 77, 128
immunohistochemistry, 128, 170
immunophilins, viii, 41, 46, 47, 48, 52, 54, 55, 62, 63, 65
immunoreactivity, 32, 34, 98
impairments, 97
in situ, 163, 176
in vitro, v, x, 7, 12, 21, 25, 35, 38, 52, 58, 59, 63, 73, 75, 76, 77, 78, 79, 81, 84, 88, 94, 95, 96, 108, 127, 137, 140, 167, 170, 174, 178, 179
in vivo, x, 7, 12, 19, 21, 23, 31, 35, 36, 58, 61, 62, 63, 73, 76, 79, 81, 83, 95, 108, 119, 126, 127, 131, 132, 140, 159, 165, 167, 170, 172, 175, 177, 178
incidence, 20, 126
individuals, 71
inducer, 119
induction, viii, 66, 69, 78, 104, 105, 106, 110, 111, 112, 118
inert, 180
infancy, 23
infarction, 122
infection, 70, 71, 75, 76, 77, 78, 79, 80, 81, 83, 84, 121
infertility, 91
inflammation, 71, 105
inflammatory disease, 54
inhibition, 7, 11, 14, 15, 16, 17, 18, 21, 46, 50, 51, 56, 57, 62, 66, 67, 84, 92, 104, 110, 114, 123, 132, 140, 162, 182
inhibitor, 56, 101, 109, 111, 120, 127, 129, 131, 132, 133, 134, 136
inhibitory, 161
initiation, 3, 6, 14, 32
injury, 16, 53, 55, 111
innate immunity, 54
inositol, 104, 175
insertion, 19, 152, 172
integrin(s), ix, 101, 102, 103, 105, 106, 107, 126
integrity, 22, 49, 50, 52, 66, 112, 130, 176
interaction(s), x, 143, 148, 149, 151, 150, 152, 157, 162, 165, 166, 167, 168, 169, 170, 172, 173, 174, 175, 176, 177, 178, 179, 180, 181, 182, 183, 184
interface, 116, 127, 164
interference, 81
interferon, 79
interferon gamma, 79
intermolecular interactions, 45
internalization, 78
interneuron(w), 20, 34, 35
interphase, 43, 50, 72, 97
intramolecular, 168
intrinsic, 182, 184
intrinsically disordered proteins, 181
intron, 145
invertebrates, 150
ion channels, 18
ionic, 146, 169
ions, 152
irradiation, 49, 124
irreversible aggregation, 57
ischaemia, 182
ischemia, 55, 121, 122, 123
Islam, 80
isoforms, 144, 170, 175, 176, 181
isolation, 151, 184
isomers, 170, 172, 176, 177, 184
issues, 58, 72

J

Japan, 85, 94
joints, 23
Jordan, 109, 127, 132, 140

K

K$^+$, 28
keratin, 43
keratinocyte(s), 108, 110
kidney, 109, 111, 122
kinase, 149, 168, 176
kinase activity, 176
kinases, x, 167, 174, 176
kinetochore, 11, 12, 33, 37, 127
kinetoplastid group, viii, 69

L

labeling, 77, 172, 173
lactate dehydrogenase, 149
lactic acid, 122
lamella, 3, 32
lamellae, 183
laminar, 103, 105
lead, vii, x, 1, 4, 11, 16, 20, 21, 22, 23, 50, 57, 82, 102, 105, 113, 114, 115
leakage, 108
learning, 18
leishmaniasis, 71, 73, 78, 84
leukemia, 65
leukocytes, 152
life cycle, 71
life expectancy, 123
life-cycle, 148
lifetime, 17
ligand(s), 51, 53, 59, 105, 114, 168, 175, 183
light, 19, 89, 143, 147, 171, 182, 184
light scattering, 171, 182
linkage, 157, 162, 184
lipid(s), 150, 164, 169, 172, 173, 174, 175, 176, 177, 179, 181, 182, 183, 184
lipid rafts, 174
liposomes, 79, 174, 175, 176, 177, 178, 179, 184
liquid chromatography, 97
liver, 79, 163
livestock, viii, 69, 70
localization, viii, 12, 16, 17, 24, 26, 29, 32, 41, 44, 49, 59, 60, 81, 99, 109, 148, 151, 161, 162, 177
long period, 157
lower respiratory tract infection, 91
LTD, 18
lung cancer, 124, 125, 134, 138
Luo, 28, 33, 47, 50, 55, 63, 64, 141
lymphoma, 117
lysis, 76, 79, 121, 122

M

M.O., 184
machinery, 33, 45, 48, 56, 63, 65, 95, 151
macromolecules, 129
macrophages, 58, 82
macular degeneration, 129
magnetic resonance, 126, 128
magnetic resonance imaging, 128
magnitude, 79
majority, ix, x, 19, 113, 114, 143, 144, 150
malaise, 122
malaria, 163
malignant cells, 57
mammalian brain, 20, 55, 73
mammalian cell(s), 8, 30, 56, 59, 62, 152
mammals, 7, 14, 46, 148
man, 134
management, 45, 102
manipulation, 58
MAPK, 170, 174, 176, 178, 179
mapping, 74, 103
marrow, 122
Marx, 161
mass, 64, 95, 147, 180
mass spectrometry, 64, 95, 180
MAST, 32, 33
materials, 151
matrix, 43, 47, 96, 109, 152, 175, 178, 180
matrix metalloproteinase, 109
matter, 42
MB, 60, 64, 66, 80, 110
MBP, x, 167, 168, 169, 170, 171, 172, 173, 174, 175, 176, 177, 178, 180, 181, 183
measurement, 141
mechanical, 173
mechanical properties, 89
media, 93
mediation, 179
mediators, 180
meiosis, 43
MEK, 107, 109, 111, 176
melanoma, 123, 132
membranes, x, 60, 90, 150, 151, 157, 163, 167, 172, 175, 176, 178, 179, 182, 183
memory, 18, 23, 122, 123
memory processes, 18
mental retardation, vii, 2, 19, 20, 23, 29
mesangial cells, 108
messengers, 177
metabolism, 65, 72, 73, 83, 149, 150, 151, 157, 158, 162
metabolite, 151, 157, 164

metabolites, 151
metaphase, 36
metastasis, 108, 113, 114
mice, 13, 14, 17, 20, 21, 22, 23, 26, 30, 31, 35, 47, 53, 73, 75, 76, 79, 83, 84, 91, 94, 99, 112, 181
microarray, 183
microbes, 163
microcephaly, 20, 21
microcirculation, 128
microfilaments, x, 152, 167, 170, 172
microheterogeneity, 179
micropatterns, 109
microscope, 173
microscopy, 53, 141, 146
microtubule, 145, 146, 159, 160, 161, 162, 163, 168, 171, 172, 176, 180, 182, 183
microtubule interactions, vii, 1
Microtubule-associated proteins (MAPs), ix, 85, 88
Mieller-Dieker Syndrome (MDS), vii, 1
migration, vii, ix, 1, 2, 11, 12, 13, 14, 19, 20, 21, 24, 25, 26, 29, 30, 33, 34, 35, 37, 38, 39, 56, 60, 101, 102, 105, 114, 175
mimicry, 77, 84
mineralocorticoid, 48, 62
miniature, 163
mitochondria, vii, x, 45, 91, 97, 107, 143, 144, 148, 149, 150, 151, 152, 153, 154, 155, 156, 157, 163, 164, 165
mitochondria sensu stricto, x, 143
mitochondrial, 143, 144, 149, 150, 151, 152, 157, 158, 163, 164, 165
mitochondrial DNA, 163, 164
mitochondrial membrane, 163, 164
mitogen, 102, 107, 108, 110, 180, 181
mitogen-activated protein kinase, 180, 181
mitosis, 11, 22, 43, 44, 50, 57, 132
mitotic, 165
mitotic cell death, 63
MMP, 110
model system, 74, 75, 78
modeling, 160
modelling, 183
models, ix, 19, 21, 22, 24, 46, 50, 53, 65, 85, 86, 92, 94, 103, 127
modifications, viii, x, 3, 4, 27, 42, 44, 50, 51, 63, 67, 71, 74, 77, 95, 105, 148, 158, 160, 161, 167, 170, 178, 183, 184
modulation, 162
modules, 21
molar ratio(s), 171
mole, 169, 170, 171, 175
molecular biology, 28
molecular dynamics, 147
molecular weight, 47, 48, 79, 89, 96, 144
molecules, vii, viii, 45, 47, 50, 69, 70, 72, 73, 74, 77, 78, 80, 84, 88, 96, 102, 106, 114, 115, 120, 121, 123, 127, 128, 140, 151, 177, 184
MOM, 143, 150, 151, 152, 157
monolayer, 128
monomeric, 144
monomers, 8, 42
Moon, 131, 132, 139, 140
morphogenesis, 47, 67, 95, 165
morphology, 17, 18, 30, 37, 90, 103, 105, 140, 164
mortality, viii, 69, 70
mosaic, 20
Moses, 129
motif, 7, 8, 10, 11, 22, 29, 146, 147, 161, 171
motor neuron disease, vii, 1, 22, 33
motor neurons, vii, 1, 22, 26, 32
motors, 161
mouse, x, 145, 159, 167, 172, 176, 182
movement, 151, 152, 165
MR, 60, 63, 66, 111, 112
MRI, 128, 141
mRNA(s), 16, 81, 109, 110, 168
MS, 175
multicellular organisms, 148
multiple sclerosis, 178, 181
Multiple system atrophy (MSA), ix, 85
muscle, 152
muscle cells, 152
muscle contraction, 43
muscles, vii, 1, 12, 22
mutagenesis, 25
mutant, x, 13, 17, 20, 21, 22, 50, 59, 83, 103, 110, 167, 172, 178
mutation(s), vii, 1, 19, 20, 22, 24, 26, 29, 30, 32, 33, 35, 36, 37, 43, 55, 56, 61, 74, 83, 90, 91, 97, 98
myelin, x, 99, 167, 170, 173, 175, 176, 177, 178, 179, 180, 181, 182, 183, 184
myelin basic protein, 178, 179, 180, 181, 182, 183, 184
myelination, x, 168, 175, 178, 183, 184
myocardial infarction, 123
myosin, 13, 15, 43, 63, 89, 90

N

National Academy of Sciences, 131, 132, 139
National Institutes of Health, 143, 158
natural, 148
natural compound, 109
nausea, 121, 122
necrosis, ix, 113, 114, 118, 122, 123, 128
neocortex, 14, 24

nerve, 16, 52, 53, 89, 91, 93, 96, 181
nerve growth factor, 52, 54, 96
nervous system, vii, ix, x, 1, 2, 9, 13, 16, 19, 23, 24, 26, 27, 36, 85, 86, 87, 89, 167, 178, 183
nervous system disorders, vii, 1, 2, 19, 23
network, 165, 172, 177
neural development, 33
neuroblastoma, 14, 16, 53, 135
neuroblasts, 3, 20, 36
neurodegeneration, vii, ix, 1, 22, 64, 85, 86, 87, 91, 92, 93, 94
neurodegenerative diseases, 45, 46, 51, 55, 64, 86, 90, 99
neurodegenerative disorders, 19
neurofibrillary tangles, 59, 90
neurofilaments, 2, 44, 50, 54, 59, 86
neurogenesis, 13, 28
neuronal cells, 12, 14, 15, 16, 49, 51
neuronal differentiation, vii, 1, 5, 10, 16, 18, 26, 54, 65
neurons, vii, viii, ix, 1, 2, 3, 4, 13, 14, 15, 16, 17, 18, 19, 20, 21, 22, 24, 25, 27, 28, 29, 36, 37, 38, 42, 53, 54, 56, 85, 86, 87, 88, 89, 90, 91, 92, 93, 94, 99, 152, 176, 181
neuropathy, 121, 122, 123
neurotoxicity, 123, 124
neurotransmission, 165
neurotransmitter, 18, 93
neutral, 175
neutropenia, 121, 122, 123
New York, 163, 180, 182
NH_2, 111
nicotine, 98
Nielsen, 160, 161
nitrogen, 73
NMDA, 177, 182
NMDA receptors, 19, 177, 182
nonhereditary neurodegenerative disease, ix, 85, 91
normal, x, 154, 155, 156, 157, 167, 172
normal distribution, 52
NSA, 122
N-terminal, 145, 173
nucleation, 27, 37, 39, 42, 51, 88, 96, 147, 159, 171
nuclei, 47, 54, 98
nucleolus, 71
nucleotides, 144
nucleus, 13, 20, 21, 37, 43, 54, 60, 62, 65, 78, 87, 106, 150, 158, 183
null, 181
nutrients, ix, 113, 114, 118

O

observations, x, 168
occlusion, 130
ocular diseases, 129
oesophageal, 121
oligodendrocytes, ix, x, 85, 86, 91, 92, 98, 99, 167, 172, 173, 174, 178, 179, 180, 181, 182, 183, 184
oligomers, 50, 65, 88
optimization, 127
orbit, 30
organelle(s), viii, ix, 2, 4, 6, 18, 20, 23, 25, 33, 38, 41, 42, 43, 44, 49, 60, 66, 85, 88, 89, 90, 96, 144, 150, 157, 158, 163, 164
organism, 70, 75
organization, 177, 182
organs, 4, 70, 78
orientation, 147
ovarian cancer, 125, 138
overlap, 19
ox, 114
oxidative, 150, 151
oxidative stress, 105
oxide, 181
oxygen, ix, 113, 114, 150, 152
Oxygen, 163
oxygen consumption, 152

P

p53, 48, 62, 110
paclitaxel, 137, 138
pain, 121, 122, 129
parallel, 56, 59, 86
paralysis, vii, 1, 22
paramagnetic, 183
parasite(s), viii, 69, 70, 71, 72, 73, 74, 75, 76, 77, 78, 79, 80, 81, 82, 84, 146, 150, 159, 160, 163
parasitic infection, 70, 72, 75, 81, 83
parkinsonism, 22, 39, 90, 91
pathogenesis, ix, 85, 90, 105
pathology, 39, 54, 55, 60, 77
pathophysiological, viii, 41, 55, 86
pathophysiological roles, 55
pathophysiology, 21, 27, 86, 92, 94
pathways, 49, 50, 102, 104, 106, 108, 179
PDZ domains, 11
peptide, 45, 162, 172, 184
peptide(s), 45, 76, 84, 114, 143, 146, 162, 172, 183, 184
perfusion, 124, 128, 141, 180
perinatal, 47, 62

peripheral neuropathy, ix, 85, 91
permeability, 117, 128, 141, 157
permission, 171
peroxide, 107
personality, 23
personality characteristics, 23
pharmaceuticals, 120
pharmacokinetics, 137
pharmacological treatment, 43
pharmacology, 127
phenotype(s), 20, 22, 26, 29, 31, 32, 55, 59, 64, 72, 92, 98, 105
pheochromocytoma, 26, 54
phosphatases, 176
phosphate, 47, 118, 133, 137, 138, 140, 141, 149, 162, 176
phosphatidic acid, 176
phosphatidylcholine, 169, 175
phosphatidylserine, 174
phosphoinositides, 177, 179
phospholipase C, 176, 184
phosphoprotein, 162, 181
phosphorylates, 176
phosphorylation, x, 3, 10, 11, 15, 18, 24, 30, 32, 38, 50, 52, 54, 55, 60, 64, 67, 71, 102, 104, 109, 148, 150, 151, 167, 170, 172, 174, 175, 177, 179, 180, 181
phylogenetic, 159, 160
phylogeny, 159, 160
physical interaction, 49
physical mechanisms, 25
physical properties, 145
physics, 183
physiological, x, 146, 167, 168
physiology, 47, 184
PI3K, 17, 111, 174
PKC, 174
planar, 164
plants, 149, 151
plasma, 149, 175, 177, 178, 180, 182
plasma membrane, 19, 43, 105, 149, 175, 177, 178, 180, 182
plasminogen, 101, 106, 109, 110, 111
plasticity, 2, 18, 63, 110
platelet(s), 148, 161
platform, 6
platinum, 125, 138
play, 152, 175
playing, 47
pleiotropy, 31
PLP, 176
PM, 63, 67
point mutation, 19, 74, 91
polarity, 3, 4, 5, 14, 15, 17, 26, 27, 34, 36, 38, 39, 42, 43, 56, 67
polarization, 2, 4, 6, 15, 16, 24, 35, 38, 39, 43, 56, 67, 108
polyamine, 72
polymer, 6
polymerization, ix, x, 3, 4, 6, 7, 10, 12, 15, 16, 21, 30, 35, 42, 43, 49, 50, 51, 52, 57, 73, 75, 79, 85, 87, 92, 95, 109, 111, 118, 120, 127, 131, 132, 139, 140, 147, 160, 167, 169, 170, 172, 177, 179
polymers, 3, 21, 51, 133, 144, 145, 160
polymorphisms, 23, 39
polypeptide(s), 44, 45, 46, 49, 66
population, 73
pore(s), 162, 177, 184
positive correlation, 50
post-translational, x, 148, 167, 170, 172, 178, 181, 183
post-translational modifications, x, 148, 167, 170, 172, 178, 183
potassium, 18, 181
pregnancy, 70
preparation, 75, 78, 83
prevention, 46, 179
probe, 15, 103, 146
prodrugs, 131
production, 151, 175
progesterone, 48, 59
prognosis, 20
prokaryotes, 42, 157, 158
prokaryotic, 144, 158
prokaryotic cell, 44, 59
prolactin, 65
proliferation, ix, 20, 21, 26, 33, 56, 57, 75, 78, 92, 101, 102, 105, 109, 114, 127, 128
proline, 52, 169, 183
promoter, 92, 99, 109
proteases, 146
proteasome, 56, 65
protection, viii, 52, 55, 65, 69, 71, 75, 76, 77, 81, 83, 84
protective role, 46
protein, x, 143, 144, 145, 146, 147, 148, 149, 150, 151, 152, 156, 158, 159, 160, 162, 165, 166, 168, 169, 170, 172, 174, 176, 178, 179, 180, 181, 182, 183, 184
protein components, 59
protein family, 46
protein folding, 44, 45, 46, 58, 63, 66
protein function, 180
protein kinases, 107, 108
protein sequence, 158
protein structure, 74

protein synthesis, 45, 49, 58, 158
protein-protein interactions, 8
Proteins, 148, 149
proteobacteria, 150, 157
proteolipid protein, 178
proteolysis, 49, 56, 84, 162
proteome, 181
proximal, 177
pseudopodia, 43
psoriasis, 129
psychiatric disorders, 21
PTM, 148
purification, 57
pyrimidine, 117
pyruvate, 149

Q

quality control, viii, 42, 152, 165

R

race, 80
radiation, 124, 125, 126, 137, 141
radiation therapy, 124, 137
radiotherapy, 114, 129
range, 150
rat, 163, 168, 171, 173, 179, 180, 183, 184
RC, 182
RE, 82, 83
reactions, 58
reactive oxygen, 104
reactivity, 72, 126
reading, 178
real time, 128
reception, 6
receptors, viii, 18, 19, 24, 35, 36, 41, 46, 47, 48, 52, 61, 62, 102, 105, 114, 177, 182
recognition, 15, 46, 76, 180, 184
recovery, 52, 122, 124
recruiting, 24
red wine, 57
redistribution, 19, 53, 165, 175, 176
reduction, 170, 171
regeneration, 16
regression, 124
regrowth, 124
regulation, 148, 151, 157, 160, 164, 168, 174, 178, 181, 184
relationship(s), 146, 158, 159, 160, 162, 165, 184
relatives, 158
relevance, 91
remodeling, 182
repair, 110, 111
replication, 42, 74, 81
reproduction, 128
repulsion, 15, 168, 175
requirements, 49
researchers, 73
residues, 7, 8, 10, 11, 12, 44, 51, 52, 74, 77, 144, 145, 146, 147, 153, 156, 157, 168, 170, 173, 174, 182, 184
resistance, 52, 63, 71, 72, 73, 74, 75, 77, 83, 129, 138
resolution, 96, 159
respiration, 163, 165
response, 2, 3, 4, 17, 18, 24, 43, 44, 49, 51, 54, 56, 58, 61, 65, 70, 71, 76, 77, 78, 79, 93, 102, 103, 104, 105, 107, 108, 109, 110, 111, 122, 125, 126, 141, 175
responsiveness, 14
restrictions, 163
restructuring, 81, 103, 146, 160
resveratrol, 57, 61
reticulum, 47, 90, 151
ribosome, 58
rings, 45, 53
risk, 66
RNA(s), 81
RNAi, 72
root(s), 16, 158
Roses, 98

S

Saccharomyces cerevisiae, 152, 153, 155, 164
safety, 127, 134, 138
salt, 170
salt concentration, 170
scattering, 171, 182
schizophrenia, 21, 30, 33, 36
scleroderma, 112
sclerosis, 178, 181
sediments, 150
segregation, 10, 42, 56, 164
selectivity, 57, 60, 118, 123
sensitivity, 44, 67, 98, 146
sensitization, 58
sensors, 6, 63
septic shock, 121, 122
series, 170
serine, 10, 11, 12, 50
serum, 54, 71, 76, 78, 103, 108, 110

shape, ix, 2, 6, 42, 43, 44, 46, 56, 59, 70, 72, 75, 86, 87, 101, 103, 104, 105, 106, 108, 109, 110, 111, 114, 117, 151, 164, 165, 181
shear, ix, 101, 102, 103, 105, 106, 107, 110, 111
shock, v, viii, 41, 42, 44, 45, 46, 49, 50, 52, 55, 59, 60, 62, 64, 65, 121
showing, 9, 10, 14, 54, 115, 116, 127, 129, 157
side chain, 171, 172, 173
signal transduction, 46, 65, 102, 103, 106, 107, 109, 111, 168, 175, 176, 180
signaling, x, 167, 173, 174, 175, 176, 177, 179, 180
signaling pathway, ix, 14, 16, 17, 101, 103, 104, 105, 109, 110
signalling, 65, 108
signals, 2, 3, 6, 15, 17, 19, 23, 43, 87, 93, 102, 103, 105, 172, 173, 174, 175, 177, 178, 180
signs, 22, 123
similarity, 144, 145
simulation, 74, 147
sinusitis, 91
siRNA, 176
sites, 145, 157, 162, 170, 172, 174, 182, 184
situs inversus, 91
skeletal muscle, 12, 152
skin, 47, 112, 124
sleeping sickness, 81
smooth muscle, 102, 105, 107, 108, 109, 110, 112
smooth muscle cells, 102, 105, 107, 108, 110, 112
sodium, 46, 62, 120, 136, 137
solid tumors, 125, 129, 134, 135, 136, 137
solubility, 44, 50, 60, 123
solution, 55, 169
South Africa, 118
SP, 107, 110
Spain, 1
spastic, 20, 162
specialization, 150, 161
species, viii, x, 15, 17, 46, 50, 55, 69, 74, 75, 76, 77, 78, 104, 143, 144, 146, 148, 156, 157, 176
specificity, 165
spectroscopy, 182, 183, 184
spectrum, x, 143
sperm, 149, 160
spin, 172
spinal cord, 184
spindle, 11, 12, 20, 21, 22, 25, 28, 29, 30, 32, 33, 34, 36, 43, 50, 51, 57, 67, 70, 75, 165
spine, vii, 1, 2, 18, 24, 28, 30
spleen, 70, 79
sprouting, 3
stability, ix, 4, 12, 25, 38, 44, 49, 55, 56, 63, 74, 83, 85, 88, 89, 149, 176, 180, 182, 183
stabilization, 2, 3, 4, 10, 11, 12, 15, 16, 19, 21, 27, 34, 39, 52, 56, 57, 67, 124
stabilize, 168
stabilizers, 133
stages, 146
standard deviation, 147, 156
stasis, 141, 160
state(s), 45, 47, 49, 50, 54, 55, 57, 102, 105, 111, 164, 181
statistics, 156
steel, 147
stimulation, 25, 53, 104, 181
stimulus, 104
stoichiometry, 50, 127
strain, 173
strategies, 161
strength, 146, 169
stress, 44, 45, 46, 48, 49, 50, 52, 55, 57, 59, 61, 62, 63, 64, 66, 75, 102, 103, 105, 106, 107, 108, 110, 111
stress response, 105
stroke, 90, 97
structural changes, 89
structural protein, 102, 178
structural transitions, 15
structure, vii, viii, ix, 11, 13, 29, 31, 37, 41, 42, 43, 45, 47, 53, 59, 60, 67, 70, 72, 75, 84, 86, 87, 89, 90, 94, 97, 101, 108, 120, 122, 126, 127, 143, 144, 145, 148, 151, 157, 159, 161, 162, 163, 165, 168, 180, 181, 184
subdomains, 171
substitution, 22, 73
substrate(s), 3, 18, 45, 46, 49, 50, 51, 57, 58, 59, 60, 61, 103, 108, 109, 151, 162, 168
subtilisin, 146, 162
sugar(s), 175, 179
sulfate, 178, 184
sulfatide, 174, 175, 176, 177, 179
sulfonamide, 73, 120
sulfonamides, 82, 83
Sun, 64, 133, 140
supernatant, 176
suppression, 57, 58, 60, 63, 103, 122, 129
surfactant, 47
survival, viii, 22, 28, 46, 47, 62, 66, 70, 72, 114, 118, 126
survival rate, 126
survivors, 77
susceptibility, 21, 72, 92, 111
swelling, 13
symbiont, 150, 151
symbiotic, 160
symptoms, vii, 1, 38, 91, 123

synapse, 37, 93, 165
synaptic plasticity, 18, 30, 86
synaptogenesis, 21
syndrome, 23, 28, 29, 30, 33, 34, 35, 39, 91, 98, 121
synthesis, ix, 44, 49, 61, 95, 101, 106, 158
systems, 180

T

T cell, 84
tandem repeats, 21, 71
tangles, 55, 90, 100
target(s), viii, 2, 6, 16, 57, 64, 69, 70, 71, 72, 74, 75, 78, 79, 82, 94, 102, 103, 108, 109, 114, 115, 118, 123, 124, 129, 163, 168, 183
tau, 3, 50, 52, 54, 55, 59, 61, 63, 64, 65, 66, 67, 89, 90, 96, 97, 149, 162, 168, 172, 182, 183
taxane, 115
techniques, 44
telomere, 165
temperature, 21, 48
tensile strength, 43
tension, 13
terminals, 89, 93, 94, 99, 162, 164
testis, 146, 148, 160
textbooks, 150
TGF, 104, 105, 108, 111, 112
theoretical, 172
theory, 183
therapeutic approaches, 91, 94
therapeutic targets, viii, ix, 42, 81, 86, 105
therapy, 114, 121, 124, 125, 126, 129
thermodynamics, 162
theta, 154, 155
thinking, 164
threonine, 50, 180
thrombocytopenia, 122, 123
thrombosis, 121, 123
throws, 184
thyroid, 122, 125, 133
thyroid cancer, 122
TID, 172, 173
tight junction, 170, 177, 181, 184
time, x, 143, 152
tissue, ix, 38, 43, 47, 49, 50, 61, 76, 92, 100, 101, 105, 106, 107, 109, 124, 148, 164, 170
TNF, 61, 111
TNF-alpha, 111
toxicity, viii, 63, 69, 72, 73, 74, 118, 121, 122, 123, 126, 128, 130
toxicology, 127
tracks, 2, 6, 21, 24, 25, 43, 177
traffic, 181
trafficking, 21, 22, 37, 48, 54, 62, 65, 66, 67
trans, 175
transcription, 57, 103, 104, 105, 110
transcription factors, 57, 104
transcripts, 31, 104
transduction, 43, 46, 65, 168, 175, 176, 177, 179, 180
transfer, 146, 150, 160, 173
transformation, 108, 150
transforming growth factor, 109, 110, 112
transgene, 180
transient ischemic attack, 121
transition, 157
translational, 161, 170
translocation, 6, 13, 51
transmembrane, 180
transmission, 164, 174, 175, 177, 178
transport, vii, ix, 1, 2, 3, 4, 6, 8, 12, 20, 21, 22, 24, 28, 29, 31, 32, 33, 35, 36, 37, 38, 39, 42, 43, 44, 49, 51, 62, 65, 70, 85, 86, 88, 89, 90, 96, 97, 161, 165, 177, 183
transport processes, 44, 49
transportation, viii, 41, 91
treatment, viii, 52, 53, 57, 69, 70, 72, 73, 81, 82, 86, 92, 102, 108, 114, 115, 116, 118, 120, 121, 123, 124, 125, 126, 129, 138
trial, 120, 125, 133, 135, 136, 137, 138
triggers, 63, 70
trimethylamine, 181
Trp, 182
trypanosomiasis, 70, 73, 80, 81, 83
tubulin inhibitors (TIs), ix, 113, 115
tumor, 10, 11, 24, 25, 31, 33, 37, 56, 57, 58, 104, 121, 122, 124, 125, 130, 132, 134, 137, 138, 141
tumor cells, 56, 57, 58, 124
tumor growth, 132
tumorigenesis, 16
tumors, viii, 42, 64, 66, 131, 137
tumour growth, 113, 118
tumours, ix, 57, 113, 114, 117, 118, 120, 123, 125, 128, 129, 134, 138
turnover, 15, 33, 49, 176, 181
two-dimensional, 181
tyrosine, 8, 18, 27, 32, 42, 51, 102, 103, 104, 107, 109

U

ubiquitin, 46, 56
ubiquitous, 177
UK, 63, 64
unfolded, 146, 147
uniform, 3, 17, 34, 77, 105, 111

unique features, 114
urine, 76, 83
urokinase, 109, 110
USA, 95, 96, 97, 110, 131, 139, 158, 180, 181, 182

V

vaccine, vii, viii, 69, 70, 75, 78, 79, 84
Valencia, 59
variation, 143, 151, 157, 161
variations, 159
Vascular Disrupting Agents (VDAs), ix, 113
vascular endothelial growth factor (VEGF), 114
vascular occlusion, 114
vascular system, ix, 101, 102, 105
vascular wall, 102
vascularization, 132
vasculature, 105, 114, 115, 120, 124, 128, 130, 137, 141
vasoconstriction, 123
vasovagal syncope, 121, 122
vector, 71
vein, 121, 123
velocity, 15, 128
ventricle, vii, 1
vertebrates, viii, 20, 69, 77, 78, 79
vesicle, viii, 18, 20, 41, 70, 91, 93, 181
vessels, 114
video microscopy, 128
voltage dependent anion channel, 143
vomiting, 121, 122
vulnerability, 92

W

Washington, 135
water, 131, 177
weakness, 22, 91, 122
weight loss, 22, 39
western blot, 77
white matter, 98
wild type, 176
Williams Syndrome (WS), vii, 1, 23
Wiskott-Aldrich syndrome, 51
Wnt signaling, 7, 10
worms, 45
wound healing, 54

X

X chromosome, 20

Y

yeast, 34, 59, 67, 82, 145, 146, 150, 152, 159, 164, 165
yield, 80, 106

Z

Zea mays, 153, 155
zinc, 8
ZO-1, 170, 182